VOLATILES IN THE EARTH AND SOLAR SYSTEM

AIP
CONFERENCE
PROCEEDINGS 341

VOLATILES IN THE EARTH
AND SOLAR SYSTEM

PASADENA, CA SEPTEMBER 1994

EDITOR: KENNETH A. FARLEY
CALIFORNIA INSTITUTE OF
TECHNOLOGY

American Institute of Physics   **New York**

L.C. Catalog Card No. 95-77911
ISBN 1-56396-409-0
DOE CONF-9409315

Printed in the United States of America.

CONTENTS

3. EXPERIMENTAL STUDIES

PREFACE

Volatiles in the Earth and Solar System is the outgrowth of an international conference held at Caltech in September of 1994. The objective of the *Deep Earth and Planetary Volatiles* conference and the present volume was to bring together a broad audience of scientists interested in the behavior and significance of volatile elements in the earth and solar system. The conference was sponsored by the Lunar and Planetary Institute, the National Science Foundation, and the U.S. Coordinating Committee for Cooperative Studies of the Deep Earth Interior and was organized by a committee consisting of T. J. Ahrens (chair), K. A. Farley, K. O'Nions, U. Ott, and K. Zahnle. The conference benefitted from the efforts of David Black and Rebecca Simmons at LPI, and Susan Yamada and David Stevenson at Caltech. The editor gratefully acknowledges the numerous reviewers who improved the quality of this volume.

1. TERRESTRIAL VOLATILES

STORAGE AND RECYCLING OF H_2O AND CO_2 IN THE EARTH

BERNARD J. WOOD, Department of Geology, University of Bristol, Bristol BS8 1RJ U.K.

ABSTRACT

The concentrations of H_2O and C in the upper mantle can be estimated as approximately 200 ppm and 50 ppm respectively from their concentrations in Mid Ocean Ridge Basalts. Estimates for the bulk silicate earth are less precise, but, from geochemical and cosmochemical arguments values of 550-1900 ppm for H_2O and 900-3700 ppm for C are plausible. The implication, on the 2-reservoir model, is that the (undegassed) lower mantle is enriched relative to the upper mantle in these volatile components, but that concentrations there are still only of the order of 2000 ppm.

An analysis of available phase equilibrium data for hydrated peridotite shows that no known hydrate is stable in the asthenosphere and that recycling of water, by subduction, into the deeper parts of the mantle is only likely in the colder parts of rapidly subducting oceanic lithosphere. In the absence of stable hydrates, water must reside in nominally anhydrous phases such as olivine and β-$(Mg,Fe)_2SiO_4$ both of which have recently been shown to dissolve some H_2O. The important question is whether or not storage of large amounts of H_2O in the earth by β-$(Mg,Fe)_2SiO_4$ which has been shown to be both theoretically and experimentally feasible can be tested.

It has been found that the water contents of olivine and β-phase in the mantle can be constrained by the width of the seismic discontinuity at 410 Km, provided that the latter corresponds to the olivine to β-phase transformation. Given reasonable models for the solution of H_2O in the two phases the seismically observed width of the discontinuity constrains the water content of upper mantle olivine to be 0-500 ppm. A similar type of constraint can probably be applied to the γ-spinel to perovskite plus magnesiowüstite transformation. Thus, arbitrary amounts of water cannot be assigned to the deeper parts of the mantle without considering the seismological implications.

In contrast to hydrates, carbonates in peridotite are extremely refractory; they are likely to survive subduction under most conditions and are stable over a wide range of mantle P-T conditions. Storage of carbon depends on redox relationships with Fe, however and recent experimental results indicate that the deeper parts of the upper mantle and transition zone are relatively reduced. Subducted carbonate should therefore be reduced by Fe to diamond and stored in the mantle as diamond rather than carbonate. Re-oxidation to CO_2 or carbonate occurs in the shallower parts of the asthenosphere or in the lithosphere. The reduction of subducted carbonate to C is a hypothesis consistent with the 'eclogitic suite' of diamond inclusions, minerals included within diamond which could reasonably be the remnants of subducted basaltic crust.

INTRODUCTION

The abundances of water and carbon dioxide in the mantle have potentially important effects on its viscosity, the nature of convection and on related geologic processes such as partial melting and volcanic degassing. The concentrations of these species in the uppermost mantle can be estimated from corresponding abundances in Mid-Ocean Ridge Basalt (MORB) glasses, while those in the whole mantle can be derived from geochemical arguments based on, for example, assumptions about the manner in which the earth accreted [1,2], or the K/H ratio of the exosphere [3] or the $C/^4He$ ratios of MORB's, ultramafic xenoliths and hotspot basalts [4].

If we assume that MORB are the products of approximately 10% partial melting of depleted upper mantle then their mean H_2O contents of about 0.2% lead to a depleted upper mantle concentration of approximately 200 ppm H_2O by weight. In terms of CO_2 content MORB concentrations are less reliable because of pre-eruptive degassing, but $C/^3He$ ratios combined with total ^3He fluxes lead to estimates of 0.1-0.3% CO_2 in undegassed MORB [5,6]. The figure for CO_2 translates to 100-300 ppm CO_2 or 27-80 ppm carbon in the depleted mantle. Jambon and Zimmermann [3] argue that the bulk silicate earth water content (including crust and exosphere) is in the range 550-1900ppm with the lower bound corresponding to the oceans plus the water concentration in the MORB source and the upper bound coming from the assumption that the K_2O/H_2O ratio of the exosphere (0.14) applies to the bulk silicate earth. In the former case current water concentrations in the whole mantle would be similar to those for the MORB source region (200ppm) and would correspond reasonably well to those derived from cosmochemical arguments by Wänke and Dreibus [1,2]. In the latter case the mantle would currently contain about 1500 ppm water and if this were distributed between a degassed upper mantle containing 200ppm and a lower undegassed mantle, the latter would currently contain about 2000 ppm H_2O.

Although volatiles are depleted to varying degrees in meteorites and planetary bodies, there is a good correlation between C and [36]Ar concentrations indicating similar volatility during planetary formation processes. Assuming that the atmosphere now contains between 50 and 99% of the terrestrial inventory of [36]Ar, one can estimate that the earth contains between 3 and 12 x 10^{23} moles of C [7]. If we subtract the 8x 10^{21} moles in the external reservoirs and crust, then the average mantle concentration should be 900-3700 ppm if the core is C-free. The MORB source concentration of 50 ppm, if applicable to whole mantle would yield a lower mantle C concentration of 1200-5000 ppm, similar to the range derived by Trull et al. [4] from their two-reservoir model of C/[4]He and C/[3]He systematics. In practise some of the earth's carbon must be present in the core [8] but this simple analysis doesn't provide any constraint on the concentration.

These geochemical arguments indicate that the concentrations of H_2O and C in both upper and lower mantles are <<1%, so that they are unlikely to affect the elastic properties or density of the mantle except where conditions are close to the solidus. Direct constraints on concentration are difficult to derive, therefore, although, as I show below, the widths of seismic discontinuities provide a possible means of placing upper limits on water content. I will proceed with a discussion of the ways in which water is stored in and recycled to the mantle and continue with an attempt to use seismological observations to constrain mantle water content.

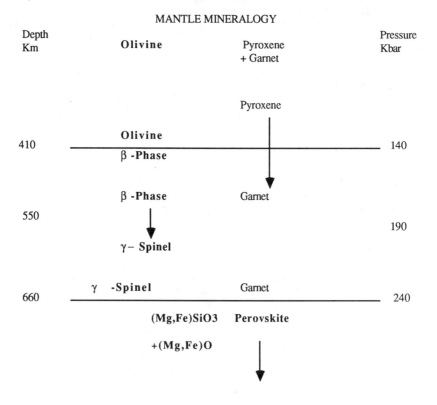

MANTLE MINERALOGY

Fig. 1.   Phase transformations in anhydrous peridotite as a function of depth and pressure. The major transformations of olivine → β-phase and γ-spinel to perovskite + (Mg,Fe) O are correlated with the seismic discontinuities at 410 and 660 Km depth.

Given that the upper mantle is dominantly peridotitic in composition, containing approximately 70% olivine, then the stable anhydrous mineralogy as a function of depth through the Transition Zone may be taken to be that shown schematically in Fig. 1.   The most important mineralogic transformations between 100 and 700 Km depth are the reactions of $(Mg,Fe)_2SiO_4$ olivine to $(Mg,Fe)_2SiO_4$ β-phase at about 140 Kbar and the breakdown of $γ-(Mg,Fe)_2SiO_4$ spinel to $(Mg,Fe)SiO_3$ perovskite plus $(Mg,Fe)O$ magnesiowüstite at about 240 Kbar. These transformations correlate in pressure with the seismic discontinuities at 410 and 660 Km depth and, as will be discussed below, the sharpness of the discontinuities may place important constraints on the water content of the Transition Zone.

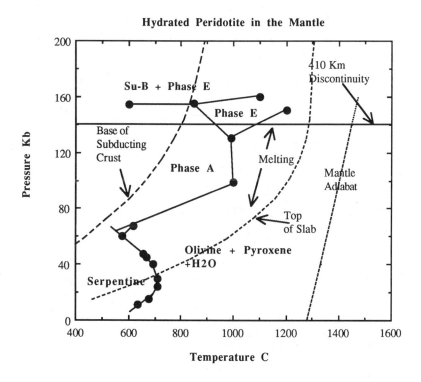

Fig 2.   Stabilities of hydrates in peridotitic composition compared to a typical mantle adiabat and to temperatures in a rapidly subducting (10cm/yr) oceanic lithosphere (range of values after Peacock [55]). (N.B. Phase diagram is simplified by the exclusion of all but the most important reactions).

Addition of water to anhydrous peridotite produces stability fields of hydrated magnesium silicates as shown in Fig. 2, redrawn and simplified from the experimental data of Yamamoto and Akimoto [9] and Gasparik [10]. It may readily be seen that no known hydrous phase is stable under 'normal' mantle pressure-temperature conditions (mantle adiabat of Fig. 2) and that, given the form of the water-release curve, hydrated magnesium silicates only remain stable under conditions such as those found in cold subduction zones. Thus, some of the water subducted in hydrated lithosphere (estimated as 9 x 10^14 gm/yr [11] may be recycled deep into the mantle as hydrated magnesium silicates, but most should be released to cause melting and/or metasomatism in the shallow mantle wedge overlying the subduction zone (Fig. 2). A more precise estimate of the stability fields of hydrated minerals in subducting lithosphere would be given by the phase diagram for the hydrated basalt component of the uppermost oceanic crust [11]. Apart from the fact that amphibole dehydrates above 30Kbar [12] and that epidote and lawsonite are stable to much higher pressures, however only qualitative deductions may be made for lack of data. The most important observation [13] is that lawsonite is stable at least to 120 Kbar under the thermal regime of subducting oceanic crust, providing a mechanism for deep recycling of some of the water tied-up in the oceanic lithosphere. If we assume that the water associated with arc magmatism comes dominantly from subducting hydrated lithosphere then at least 10% of water is returned to the shallow mantle (< 150 Km depth) [11] while at least some of the remainder undergoes long-term recycling. Since the amount of water released through Mid-Ocean Ridge volcanism is about the same as that released by arc magmatism [11] it is possible that the bulk water content of the mantle below 150Km depth is stable or even slowly increasing.

The observation that depleted mantle must contain some water and the apparent lack of any major stable hydrous phase means that most mantle water must reside in one of the nominally anhydrous phases shown in Fig. 1. It is now well-established that all of the major minerals of the upper mantle, olivine, pyroxene and garnet can dissolve several hundred ppm of water associated with point defects [14]. Recent experiments [15] indicate that at pressures close to the 410 Km discontinuity olivine may contain up to 1000 ppm. Furthermore, it is now evident that the β-phase contains substantially more water than coexisting olivine [16,17] and could provide a host for large amounts of recycled water in the Transition Zone [18].

WATER STORAGE IN ANHYDROUS MINERALS

From a comparison with oxygen and hydroxyl sites in a wide range of minerals, Smyth [18] showed that the electrostatic potential of the O1 site in anhydrous β-Mg_2SiO_4 is intermediate between those of 'normal' oxygen sites in silicates and those of hydroxyl sites. Furthermore, the O1 site is, like hydroxyl sites in silicates, not bonded to an Si atom. Smyth [18] therefore argued that the O1 site was a potential host for hydrogen as an OH defect, probably coupled to Mg vacancies to provide charge balance. The hypothetical fully hydrated end-member would have the formula $Mg_7Si_4O_{14}(OH)_2$ [19] and would contain 3.33 wt% H_2O. Measurements of the water contents of β-$(Mg,Fe)_2SiO_4$ produced experimentally from olivine confirm Smyth's suggestion that the β-phase can dissolve substantially more water than comparable anhydrous silicates. Young et al. [16] found up to 0.4 wt % H_2O while Inoue [17] in a study of the system Mg_2SiO_4-$MgSiO_3$-H_2O found, by electron microprobe, β-phase compositions close to the hypothetical hydrated end-member in terms of apparent deficiency in MgO-i.e an inferred H_2O content of 3%. In both these studies other silicates coexisting with β-phase have low water contents. Inoue [17] inferred H_2O contents of olivine of less than 0.1% while Young et al [16], used infra-red spectroscopy to determine that the ratio of water in β-phase to that in nearby olivine is about 40:1. This lends further support to the suggestion that the Transition Zone may be the host for substantially more water than the upper, degassed part of the mantle.

WATER AND THE 410 KM DISCONTINUITY

Potentially, one of the most important constraints on the water contents of the Transition Zone and Lower Mantle are the observed depth intervals over which low pressure assemblages transform to their high pressure analogs. Recent seismological studies using high frequency reflected waves indicate that both 410 and 660 km discontinuities are very sharp and that the change in physical properties associated with them occur over very small depth intervals. For the 660 discontinuity, for example, a width of 4 to 5 Km is consistent with the seismic data [20,21,22]. Most authors seem to favour a slightly broader 410 km discontinuity, on the order of 10 Km wide [20], but a recent study of precursors to P'P' [21] implies that this too is locally extremely sharp, there being of the order of 4 Km between upper and lower boundaries in some areas. The implication for the phase change hypothesis is considerable. The chemical system of the mantle is multicomponent and simple phase relations show that the transformation from olivine to beta-phase must occur in a divariant loop (Fig. 3), giving a somewhat smeared-out transition interval. In order to explain the reflections observed by Benz and Vidale [21] a maximum interval of around 8 Km (3 kbar) is required [23]. In a recent article Helffrich and Wood [23] argued that the experimental data on the olivine to beta transformation is consistent with a minimum transition interval of 4 Km, although values close to 8 Km provide better fits to the experimental data. A similar analysis for the γ-spinel to perovskite plus magnesiowüstite transformation [24] show that, within experimental uncertainty on the Fe-Mg partitioning relationships between the phases, the transformation could appear univariant or seismically discontinuous (Fig.1). These analyses are predicated, however, on the assumption that the chemical system of the earth is MgO-FeO-SiO$_2$ and that other components such as CaO, Al$_2$O$_3$, Fe$_2$O$_3$ etc. are either insoluble in the major phases or partition equally between them so that they have no effect on the positions and widths of the phase transformations.

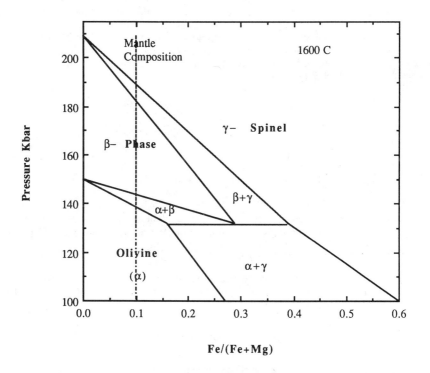

Fig 3.   Phase diagram [29] for Mg$_2$SiO$_4$ - Fe$_2$SiO$_4$ at high pressures. Note that the transformations of olivine → β-phase, β→γ and so on are divariant and occur over a range of pressures at any given temperature. Peridotitic mantle has Fe/Fe+Mg of about 0.1

As discussed above, water partitions strongly into the β-phase relative to other silicates, and no known hydrous phases are stable under the P-T conditions of the 410 Km discontinuity (140 Kbar and about1500°C respectively). These observations do not require, however that β-phase in the mantle is a host for large quantities of water. The effect of water on phase transformations involving the beta-phase should be dramatic and I will show that the observed seismic discontinuities, if correctly correlated with phase transitions, constrain the concentration of water stored in the mantle to be 0-500 ppm.

In order to calculate the effect of water on the transition of olivine to beta phase it is necessary to develop models for the free energy of solution of H_2O in the two phases. Smyth [18] has proposed, based on calculated electrostatic potentials of the different oxygen sites in β-Mg_2SiO_4, that replacement of O1 by an OH group is energetically favoured. Charge compensation would come via creation of cation vacancies in the Mg sublattices. Since the prediction of high solubility has proven correct, I adopted a solution model based on Smyth's [18] calculations which would give a hydrous end-member containing 3.3 wt % H_2O with a formula $Mg_7Si_4O_{14}(OH)_2$. In a mantle saturated with $(Mg,Fe)_2SiO_4$ and the $(Mg,Fe)SiO_3$ component of pyroxene, the formation of the end-member may be represented by the equilibrium :

$$3Mg_2SiO_4 + MgSiO_3 + H_2O \quad = \quad Mg_7Si_4O_{14}(OH)_2 \qquad (1)$$
$$\beta\text{-phase} \qquad\qquad \text{pyroxene} \qquad\qquad\qquad \beta\text{-phase}$$

This creates a β-phase solid solution in which 1/8 of the available oxygen sites (O1) may become hydroxyls and in which charge-compensating vacancies on the Mg sites may either be disordered or ordered close to the OH defects [19]. In order to calculate the effect of water on the 2-phase loop, it is necessary to obtain the partial molar free energies of Mg_2SiO_4 and Fe_2SiO_4 components in the hydrated β-phase solid solution. At low concentrations of OH the solution of this component will be in the Henry's Law region where its only effect on the major components is entropic. In that case the chemical potentials of Mg_2SiO_4 and Fe_2SiO_4 components are given by :-

$$\mu^{\beta}_{Mg_2SiO_4} = \mu^{o}_{Mg_2SiO_4} + RT\ln\{X^2_{Mg}\gamma_{Mg}.(1-X_{OH})^{0.5}\}$$

$$\qquad (2)$$

$$\mu^{\beta}_{Fe_2SiO_4} = \mu^{o}_{Fe_2SiO_4} + RT\ln\{X^2_{Fe}\gamma_{Fe}.(1-X_{OH})^{0.5}\}$$

In equation (2), X_{Mg} and X_{Fe} refer to atomic fractions of magnesium and iron respectively on the large cation positions while X_{OH} is the fraction of O1 sites which are hydroxyl positions. The standard state chemical potentials $\mu^{o}_{Mg_2SiO_4}$ and $\mu^{o}_{Fe_2SiO_4}$ refer to the free energies of the pure end-member β-phases at the pressure and temperature of interest. The activity coefficients for the Mg-Fe sites, γ_{Mg} and γ_{Fe} refer to nonideal Mg-Fe mixing, combined with (nominally) ideal mixing of vacancies, which have a concentration :-

$$V^{ll}_{Mg} \quad = \quad 0.125X_{OH} \qquad\qquad (3)$$

Equations (2) and (3) enable calculation of the effect of water on the two major components, given the assumptions of random mixing of OH groups with O1 and of vacancies with Mg and Fe. Other assumptions will be discussed later. The nonideal part of Mg-Fe mixing was obtained by assuming a symmetric interaction parameter W_{MgFe} of 4.0 Kj per gram atom [25].

For olivine, Bai and Kohlstedt [26] have shown that H_2O solubility is dependent on fH_2, fO_2 and the activity of $MgSiO_3$ in the system. Based on their low pressure data they derived a solution model for H_2O in which H complexes with interstitial oxygen atoms, O^{ll}_i the concentrations of the latter being dependent on the thermodynamic state of the system. In the stability field of β-$(Mg,Fe)_2SiO_4$, Young et al [16] find a partitioning of H_2O between β-phase and olivine of greater than 10:1, while

Kohlstedt et al. [15] report approximately 0.1 wt% of H_2O in olivine at 8GPa and 1100 to 1300 ^oC. A simple representation of the Bai and Kohlstedt model is, in Kröger-Vink notation :-

$$O_i^{ll} \quad + \quad H_2O \quad = \quad 2OH_i^{l} \tag{4}$$

In this case calculation of the chemical potentials of Mg_2SiO_4 and Fe_2SiO_4 components in the partially hydrated olivine, relative to water-free olivine under the same conditions of intensive variables are given by:-

$$\mu_{Mg}^{\alpha} = \mu_{Mg_2SiO_4}^{o} + RTln\{X_{Mg}^2 \cdot \gamma_{Mg} \cdot (1-X_{OH})^4/(1-0.5X_{OH})^4\} \tag{5}$$

With a similar form for Fe_2SiO_4 component. In equation (5), I have taken explicit account of the fact that creation of OH defects destroys half as many pre-existing oxygen interstitials (denominator on the right hand side) and the standard state values μ^o refer to hypothetical olivine free of oxygen interstitials. Activity coefficients, γ_{Mg}, γ_{Fe} were calculated from the symmetric solution model using an experimentally measured interaction parameter of [27] adjusted to 140 Kbar from the excess volumes on the forsterite-fayalite join [28], giving a value of + 5.0 Kj per gram atom.

Effect of 500 ppm Water on Olivine-Beta loop

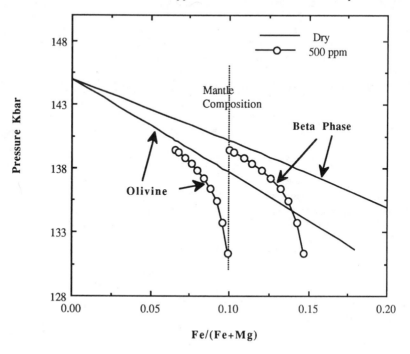

Fig 4.   The effect of 500ppm water on the olivine-β transition loop (note that this is strictly

a projection from Mg_2SiO_4 - Fe_2SiO_4 - H_2O space back onto the Mg_2SiO_4 - Fe_2SiO_4 face). The first β-phase to appear at 130 Kbar has 5000ppm water and coexists with olivine with 500ppm. Olivine disappears when it has 50ppm water and β-phase has 500ppm, together with Fe/Fe+Mg of 0.1

The combined effect of water and iron on the olivine- β transition may be calculated by fixing the bulk Fe/(Fe+Mg) and water content of mantle olivine at a point just above the depth where β-phase appears and then calculating the shift in equilibrium pressure relative to the value for pure anhydrous Mg_2SiO_4. The water contents and Fe/(Fe+Mg) ratios of the first β-phase to appear should then be calculated by equating the chemical potentials of H_2O and Fe_2SiO_4 in the two coexisting phases. For simplicity, however, I assumed that the ratio of water contents (by weight) in β-phase to olivine is 10:1, a value consistent with the low end of measured values [16]. Higher partition coefficients slightly increase the effect of water on the transformation interval, so this is a conservative approximation. Iron-magnesium exchange equilibrium was fixed from the standard state free energy change,ΔG^O of the equilibrium :-

$$\text{Mg(olivine)} + \text{Fe(β-phase)} = \text{Fe(olivine)} + \text{Mg(β-phase)} \tag{6}$$

The experimental results of Katsura and Ito [29] on this equilibrium indicate minimum values of ΔG^O of about 5.5 KJ yielding a 2-phase loop in the anhydrous system which is about 2.5 kbar (7 Km) wide (Figs 4 and 5). For comparison I repeated the calculations with ΔG^O of 8.0 KJ which gives an anhydrous loop 4.8 Kbar (13 Km) in width (Fig. 6).

Proportions of phases through the 2-phase region were calculated as follows :-The Fe/(Fe+Mg) ratio of the olivine was set at $\leq$ the value above the transition zone. The Fe/(Fe+Mg) ratio of the coexisting β-phase was then obtained from ΔG^O of reaction (6), taking into account Fe-Mg nonideality. Then, from the Fe/(Fe+Mg) of the initial mantle olivine, phase proportions were simply solved from the lever rule. Given proportions of olivine and beta, the concentrations of water in the two phases were calculated from the partition coefficient of 10:1 and the fixed water content of olivine above the transition zone. The equilibrium pressure was then calculated from [24]

$$(P-P^O)\Delta V^O = -RT\ln\{X_{Mg}^2 \gamma_{Mg} \cdot (1-X_{OH})^{0.5}\}_{beta} + RT\ln\{X_{Mg}^2 \gamma_{Mg} \cdot (1-X_{OH})^4/(10.5X_{OH})^4\}_{ol} \tag{7}$$

where P^O refers to the equilibrium pressure for the end-member reaction :-

$$\begin{array}{ccc} Mg_2SiO_4 & = & Mg_2SiO_4 \\ \text{olivine} & & \text{β-phase} \end{array} \tag{8}$$

and ΔV^O refers to the volume change of this reaction at the P,T conditions of interest. I performed the calculations at 1773K and took P^O to be 145 Kbar (Fig. 4). The volumes of the two phases at P and T were calculated from the Murnaghan equation of state using the thermodynamic properties of Table 1, as discussed by Wood [24].

TABLE 1

Thermophysical Properties of Olivine and β-Phase

	Olivine	β-Phase
Volume (cc/mol)	43.670	40.520
K_S (298K) GPa	128.0	173.0
K'	5.2	4.8
a $(\times 10^5)$ K^{-1}	4.00	3.57
δ_S	4.0	4.3
γ_{th}	1.20	1.30

Effect of Water Concentration on the Olivine-Beta Phase Transformation

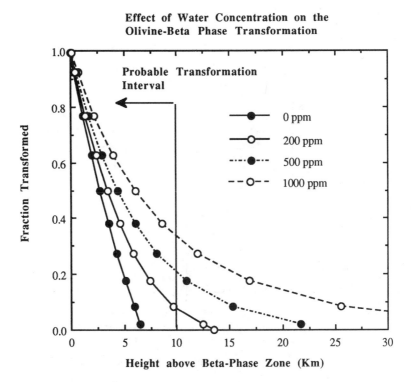

Fig 5.   The olivine - β-phase transition interval from Fig. 4 showing the effects of varying water content. Note that 500ppm water produces a transition interval substantially greater than 10Km, the maximum likely from seismic observations.

Fig. 4 shows a comparison of the olivine-beta phase transition with 0 ppm and with 500 ppm by weight of H_2O in mantle olivine above the transition zone. The effect of as little as 500 ppm water is dramatic, β-phase appearing approximately 6 Kbar lower in the water-bearing than in the anhydrous system. The first β-phase to appear has, of course 5000 ppm water, while olivine disappears at high pressure as its water content reaches 50ppm and that of β-phase 500 ppm. The width of the two-phase region is enlarged from 2·5 Kbar in the dry system to 8 Kbar when there are 500 ppm water (Fig. 5), so that the transition interval would be approximately 22 Km wide if upper mantle olivine contained 500 ppm water. Fig. 5 shows the effects of varying water content between 0 and 1000 ppm on the transformation interval from olivine to β-phase. In this case the anhydrous interval is taken to be 7 Km (2·5 Kbar) as discussed above. It can clearly be seen that, given this 'intrinsic' width to the phase transformation, water contents in upper mantle olivine of 500 or 1000 ppm are inconsistent with seismological observations at high frequency which imply that the 410 Km discontinuity is less than 10Km wide [20,21,22]. Thus, the seismic discontinuity is strongly affected by the water content of the mantle and should not be regarded as a passive bystander as arbitrary amounts of water are assigned to β-phase of the Transition Zone. If we give a greater intrinsic width to the phase transformation (Fig. 6), then it is difficult to see how the mantle at the top of the Transition Zone could contain even 200 ppm water.

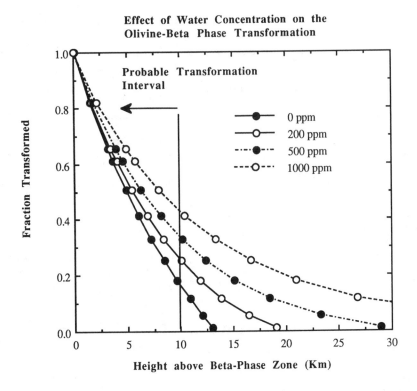

Effect of Water Concentration on the Olivine-Beta Phase Transformation

Fig 6. Similar to Fig. 5 except for a broader 2 phase loop in the water - absent system. Note that the water content of the mantle must be close to zero to match the seismic observations.

The major uncertainties which arise in these calculations are in the assumed partitioning of water between the two phases and in the solution model for water in β-phase and olivine. A sensitivity analysis shows that errors in the solution model for β-phase would have by far the largest effect. I have assumed that both cation vacancies V_{Mg} and OH species on O1 sites mix randomly in the structure. This gives the effect shown. The other end-member assumption would be that vacancies cluster with OH to maintain local charge balance and that the entropy of solution is lower than I have assumed. If we model clustering by completely excluding the entropy of vacancy-Mg-Fe mixing then we still obtain a calculated effect which is 70% of that shown in Figs 4-6. i.e to a good approximation we can replace the curves labelled 1000ppm by 1400 ppm, 500 ppm by 700 and 200 ppm by 300 ppm. The effects are still dramatic and cannot be ignored. In practise the completely disordered assumption is reasonable at the low water contents compatible with the observed sharpness of the discontinuity, while substantial ordering is likely at compositions close to the hydrated end-member [19].

Results of seismic tomography in regions of subduction zones [30] suggest that H$_2$O may be released from subducting lithosphere in two depth intervals, one shallower than 100 Km and the other about 300 Km deep. If the latter low-velocity zone is indeed a zone where fluids or hydrous melts are present then, from Figs 4-6, one would expect the high water content to affect the olivine β-phase transition. In particular, one would expect the 410 Km discontinuity to be elevated and smeared-out relative to more normal regions of the mantle where the water contents are lower. This suggests the need for further seismic experimentation on the nature of the 410 and perhaps 660 Km discontinuities in the region of subduction zones. A final point about the calculated effect of water is that it may explain the apparent breadth of the phase transition measured experimentally, which is about 5 Kbar at 1600°C [29]. Calculations based solely on Fe-Mg partitioning and assuming completely anhydrous conditions [31,23] suggest transition intervals in the range 2-3 Kbar. Since it is virtually impossible to completely exclude water from these high pressure experiments, it is quite likely that some of the experimentally observed width of the 2-phase loop is due to the presence of small amounts of H_2O in the sample cell, so that the discrepancy between calculation and experiment may be an experimental artifact rather than a problem with the theory.

WATER AND THE 660 KM DISCONTINUITY

Since water has such a large effect on the width of the olivine-β phase transformation, it is reasonable to consider whether its influence on the reaction :-

$$\begin{array}{lll} (Mg,Fe)_2SiO_4 & = & (Mg,Fe)SiO_3 + \quad (Mg,Fe)O \\ \gamma\text{-spinel} & & \text{perovskite} \qquad \text{magnesiowüstite} \end{array} \qquad (9)$$

is comparable. In this case the phase transformation is correlated with the 660 Km discontinuity [32], but it is possible that there is also a change in bulk composition associated with the discontinuity [33]. If there is a change in bulk composition, then nothing can be deduced about the water content at this depth in the mantle since the transformation would be essentially discontinuous and controlled by the bulk composition of the major components. If , on the other hand the discontinuity at 660 Km is entirely due to reaction (9), then we may derive some idea of how it would be affected by water. Meade et al. [34] have observed small amounts of OH in synthetic $MgSiO_3$ perovskite and, by analogy with olivine, it is likely that addition of Fe^{3+} and Al will greatly enhance water solubility in this lower mantle phase.

As an illustration of the possible effect of water on the γ-spinel-perovskite reaction (9) I used a model in which H_2O combines with oxygen vacancies, a common defect in perovskites [35] via a reaction such as :-

$$O_O^x + V_O^{oo} = 2OH_O^o \qquad (10)$$

The partial molar free energies of the major $MgSiO_3$ and $FeSiO_3$ components were than calculated using the methodology described above for β-phase [24]. I used a similar model for the γ-spinel, and assumed that the partition coefficient of H_2O between perovskite and spinel is 14:1 and that the magnesiowüstite phase can dissolve no water. The results, which are purely illustrative, are shown in Fig. 7. It can be seen that this reaction is less sensitive to water than is the olivine-β-phase transformation but that, given a likely large partition coefficient for water, 1000-2000 ppm of water would broaden the transformation beyond the width required by observed high frequency reflections [21,22]. This calculation is for an intrinsic 0 ppm width of 1Km, which is within the range required by the Fe-Mg partitioning data of Ito and Takahashi [32,24] and greater transformation widths would place tighter constraints on the water content of the lower mantle.

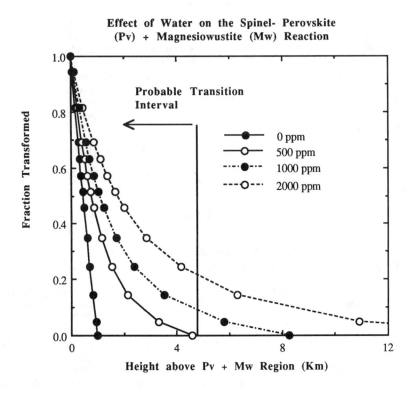

Fig 7.   Possible effect of water on the γ-spinel → perovskite + magnesiowüstite reaction. Note that constraints on water content would be looser than for the olivine → β-phase reaction, but that it is difficult to reconcile more than 1000ppm water with the seismic observations.

In conclusion, although water is very soluble in the β-phase and conceivably quite soluble in perovskite, suggestions that there are high concentrations of H_2O in the nominally anhydrous phases of the lower mantle and Transition Zone must take into account the effects of these components on the well-established phase transformations. If the 410 and 660 Km seismic discontinuities are due solely to phase transformations then the water contents of β-phase in the Transition Zone and perovskite in the lower mantle are limited to <500ppm and about 1000 ppm respectively. If there is release of H_2O from subducting lithosphere at depths greater than 300Km, as suggested by seismic tomography [30] then these should be regions where the olivine and β-phase have relatively high water contents. In that case the 410 Km discontinuity should, based on my analysis, be elevated and smeared-out in these regions relative to what is observed in 'normal' mantle. Clearly the hypothesis of the large effect of water on the phase transition is testable experimentally and seismic data may help constrain the water content of the mantle in different regions.

CARBON AND CARBONATE

As discussed earlier, analysis of the CO_2 contents of Mid Ocean Ridge Basalt glasses leads to an estimated carbon content of the degassed upper mantle of about 50ppm. Estimates of carbon content of the bulk silicate earth vary from 900 to 3700 ppm based on the correlation between C and [36]Ar in chondritic meteorites [7], assuming that the C content of the core is negligible. A recent analysis of $C/^4He$ and $C/^3He$ in peridotite xenoliths and in basalts by Trull et al. [4] led to the conclusion that, using a two-reservoir model, the upper mantle contains 35-400 ppm C and the lower, undegassed mantle about 2500 ppm C. From the ratio of C to ^3He and the ^3He flux at ridge crests, one can estimate a CO_2 flux at ridges of approximately 4×10^{13} gm/yr [7,36]. Measurements of carbonates in altered oceanic crust by Staudigel et al. [37] lead to a bulk CO_2 content of about 0.3 wt%, mostly as calcite and a subduction rate of CO_2 of about 10^{15} gm/yr. This suggests that the amount of CO_2 being evolved at ridges and hotspots is less than that being recycled to the mantle at subduction zones. The important questions center on how much CO_2 is returned to the exosphere in arc magmas (very uncertain) whether any of the recycled carbonate remains stable to greater depths than do hydrates in the earth (Fig. 2) and on how primordial and recycled carbon is stored in the mantle.

The presence of CO_2-rich fluids and carbonate magmas demonstrate that oxidised carbon occurs at depth. Similarly, native carbon as both graphite and diamond occurs in kimberlites and in deep-seated peridotite and eclogite xenoliths associated with kimberlites. This suggests that the form in which carbon is stored may be related to tectonic environment or depth and it is appropriate to consider the interactions between P, T and oxidation state.

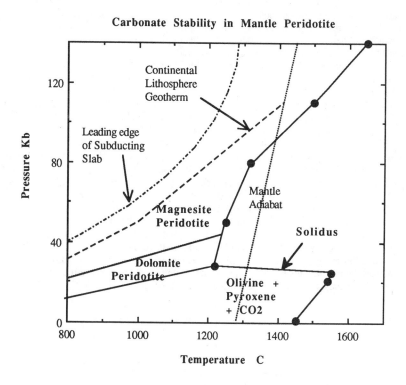

Carbonate Stability in Mantle Peridotite

Fig 8.   The stability of carbonate (magnesite and dolomite) in the mantle. Note that carbonates are much more refractory than hydrates and are thermally stable in subduction zones and throughout most of the mantle. Line labelled 'Leading edge of subducting slab' from Fig. 2.

In terms of stability within the mantle, carbonates are much more refractory than hydrates, as may be seen from Fig. 8, which is re-drawn from data in White and Wyllie [38] and Brey et al. [39]. At moderate depths (< 100 Km) within the mantle dolomite $Ca(Mg,Fe)(CO_3)_2$ is stable coexisting with the anhydrous peridotite assemblage shown in Fig 1. At greater depths dolomite transforms to magnesite $(Mg,Fe)CO_3$ through a reaction which may be approximated as :-

$$Ca(Mg,Fe)(CO_3)_2 + 2(Mg,Fe)SiO_3 = Ca(Mg,Fe)Si_2O_6 + (Mg,Fe)CO_3$$

| Dolomite | Orthopyroxene | Clinopyroxene | Magnesite |

Magnesite then appears to be stable in the peridotitic assemblage from 100 Km depth down to lower mantle pressures, as shown in Fig. 8 and demonstrated experimentally at lower mantle pressures [40,41].

When compared to mantle temperatures, carbonate is stable anywhere in the continental lithosphere and could occur on the mantle adiabat at any depth below 200 Km (Fig. 8, [40]). When subduction zone P-T paths are plotted on Fig. 8, it appears that magnesite would be stable in any peridotitic assemblage virtually anywhere within subducting lithosphere. The normal form of subducted carbonate in the basaltic part of the oceanic plate is calcite rather than magnesite [37], but high pressure experiments indicate that this is converted to stable magnesite or dolomite in basaltic compositions, not disappearing before the beginning of melting at 15-35Kbar [42]. Therefore we can assume that a reasonable proportion of subducted carbonate is returned to the mantle in subduction zones.

Given that reasonable quantities of carbonate are returned to the mantle, it is of interest to consider whether it remains stored as carbonate or is converted into diamond or graphite. Diamonds commonly contain inclusions of high pressure silicate minerals such as garnet, sodic pyroxene, and mantle olivine and orthopyroxene. There are two major suites of inclusions [43], the peridotitic suite which is correlated with the major component of the mantle and the eclogitic suite which may have its origin as subducted basaltic crust [44]. Carbon isotope analysis is generally equivocal, but some diamonds of the eclogitic suite have very light $^{12}C/^{13}C$ ratios, indicatinng a crustal component of carbon. It is quite plausible that diamonds of these suite formed from a mixture of subducted carbonate and organic carbon [44]. This would require a reduction reaction to have occurred.

The two most important elements involved in redox reactions in the earth are C and Fe. In the upper mantle Fe is dominantly present as reduced 'FeO'(about 8 wt%), but the concentration of Fe^{3+} is also considerable (about 0.2% as Fe_2O_3, [45]). Given these concentrations, which are robust and based on analyses of peridotites by Mössbauer spectroscopy, then the oxidation state of Fe should dominate that of carbon, given that the latter is present at a level of 50 ppm. The oxidation state of Fe is reflected by the oxygen fugacity which has been measured for peridotites from a wide range of localities by oxygen thermobarometry [45,47] for example. The results demonstrate that oxygen fugacity depends on tectonic environment [45,48]. At oceanic ridges and in continental rift zones the mantle is relatively reduced, with fO_2 values extending from 2 log units below the Fayalite-Magnetite-Quartz (FMQ) reference buffer up to about FMQ. Mantle peridotite from above subduction zones and in non-rift continental environments is more oxidised, extending to about 1 log unit above FMQ. This leads to the spread of asthenosphere (rift) and lithosphere values shown in Fig. 9.

Oxygen Fugacity as a function of Depth

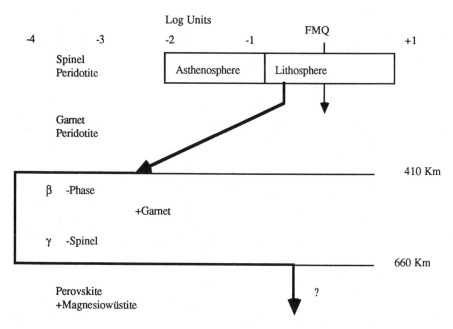

Fig 9.   Oxygen fugacity in the mantle relative to the FMQ buffer as a function of depth. Note that the deeper parts of the upper mantle (Garnet peridotite) and the transition zone are relatively reduced. Vertical line separates reduced from oxidised carbon. Carbon is oxidised in the lithosphere (CO_2 or carbonate) and stored as reduced diamond in the deeper parts of the upper mantle and Transition Zone.

The results shown in Fig. 9 all apply to the (aluminous) spinel peridotite facies of the upper mantle, which extends from the surface to depths of about 90 Km. Deeper samples of the lithosphere are provided by Garnet peridotite xenoliths in Kimberlites and oxygen barometry for limited suites of such samples suggest that the deeper parts of the lithosphere may have slightly more reduced fO_2 conditions than the spinel lherzolite facies of the asthenosphere and continental lithosphere [49]. If we take the bulk composition of the upper mantle to be approximately constant with 0.2% Fe_2O_3 distributed at low pressure between aluminous spinel (main host for Fe^{3+}) and pyroxene and at higher pressures between garnet (main host for Fe^{3+} and pyroxene) then there is a simple pressure effect on the relative fO_2 of the mantle.

At low pressure the dependence of fO_2 on pressure may be represented by the equilibrium :

$$6Fe_2SiO_4 + O_2 = \quad 2Fe_3O_4 \quad + \quad 6FeSiO_3 \tag{11}$$
$$\text{olivine} \qquad\qquad \text{spinel} \qquad\qquad \text{orthopyroxene}$$

The volume change of the solids in this reaction ($+8.6$ cm^3) is about half that for the FMQ buffer ($+17.95$ cm^3) with the result that, at fixed composition fO_2 declines by about 0.25 log units relative to FMQ for every 10 Kbar increase in pressure. In the garnet stability field, the pressure effect on fO_2 may be estimated from the equilibrium :-

$$4Fe_2SiO_4 + 2FeSiO_3 + O_2 = 2Fe_3^{2+}Fe_2^{3+}Si_3O_{12}$$

olivine   opx   garnet

In this case the volume change of the solids is - 8.6 cm^3 [49], meaning that ferric iron has a relatively low partial molar volume in garnet and that fO_2 in the garnet field, at fixed composition, declines by about 0.9 log units relative to FMQ for every 10 kbar increase in pressure. This effect is purely crystallochemical in origin and must occur even at fixed composition. It is compounded by the breakdown of pyroxene to majorite garnet which dilutes the Fe^{3+} in garnet and hence lowers fO_2 still further. The overall effect, a gradual decline in fO_2 relative to FMQ with increasing depth is shown schematically in Fig. 9.

To determine whether native C or CO_2(or carbonate) are stable under mantle conditions, we may plot the position of the equilibrium :-

$$C + O_2 =   CO_2   (CCO)$$

on Fig. 9 and compare it with the curve for mantle fO_2. Currently available data suggest that (CCO) lies 1-2 log units below FMQ under typical upper mantle conditions with a weak pressure dependence [45]. This means that, if the C content of the upper mantle is on the order of 50 ppm and Fe dominates the redox properties then, in much of the asthenosphere and lithosphere CO_2 and carbonates should be stable, while deeper in the mantle CO_2 (or carbonate) will be reduced by the FeO- rich mantle minerals to produce diamond/graphite and higher concentrations of 'Fe_2O_3' in major mantle minerals :-

$$4'FeO' +   CO_2   =   2'Fe_2O_3'   +   C$$

The implication is that, although carbonate is thermally stable in subduction zones (Fig. 8), it will generally be reduced to diamond within the upper 200-300 km of the mantle and will be stored and recycled as diamond rather than carbonate, as suggested by Kesson and Ringwood [44].

Recent studies of Fe^{3+} in β-phase and in γ-spinel by Hazen et al. [50] and O'Neill et al. [46] indicate that both these phases may have $Fe^{3+}/(Fe^{2+}+Fe^{3+})$ ratios of close to 0.1 in contrast to the very low values (<0.02) found in their low pressure analogue, olivine [51]. This means that, in the Transition Zone, ferric iron should be present at moderate concentrations in the major phases rather than being forced at high concentrations into minor phases such as aluminous spinel. The implication [46] is that fO_2 in the Transition Zone is much lower than in the upper mantle, since the ease with which ferric iron is accomodated in the major phases means, even at fixed composition, a very low chemical potential for Fe^{3+}-components. In the lower mantle, in contrast, most Fe^{3+} should be concentrated in the minor magnesiowüstite phase [46] with the implication that lower mantle fO_2 should be higher than Transition Zone fO_2 and, relatively, more like the values pertaining in the upper mantle. This is shown schematically in Fig. 9. From the point of view of carbon, storage in the Transition Zone should be as diamond but ought, in the lower mantle, to be oxidised as magnesite $MgCO_3$. Magnesite is thermally stable under lower mantle conditions [40,41] and if these conclusions about fO_2 are correct, should store carbon in the oxidised form in the deep mantle.

A potential test of the fO_2-depth relations in the deep mantle is provided by electrical conductivity data [52,53,54]. Although there are still discrepancies between the different laboratories, magnesiowüstite and Fe-bearing perovskite have electrical conductivity behaviour typical of an electron-hopping mechanism. In such cases the Fe^{2+}/Fe^{3+} ratio governs conductivity and is, in turn controlled by fO_2. Given improved experimental and geomagnetic data it should be possible to place some constraints on the oxidation state of the deep mantle from its electrical conductivity.

SUMMARY

The important point that I wish to re-iterate is that although H_2O and C are present in only trace amounts in the upper mantle and likely only at concentrations of 1000ppm in the deep mantle, the properties of the mantle should not be regarded as independent of the concentrations and chemical form of these two components. I have shown that the width of the $\alpha-\beta$ transition in $(Mg,Fe)_2SiO_4$ is extremely sensitive to the water contents of the anhydrous minerals in the mantle and hence that the position and width of the 410 Km seismic discontinuity constrains the water content of olivine to be 0-500 ppm. A similar constraint probably applies at 660 Km, but quantification awaits data on partitioning of H_2O between the different phases.

Carbon is much more refractory than H_2O and an analysis of the stabilities of carbonates suggests that most carbon in subducting oceanic lithosphere is returned to the deep mantle. Because, however, of the redox relationship between carbon and iron, carbonate is reduced to diamond or graphite within the upper mantle and storage in the Transition Zone is as diamond rather than carbonate. In the lower mantle preliminary results suggest that carbonates are thermally stable and that fO_2 conditions are high enough to stabilise magnesite rather than diamond. Potential clues to the oxidation state of the deep mantle are provided by electrical conductivity data, which appear to corroborate the hypothesis that the lower mantle is relatively oxidised [54].

ACKNOWLEDGEMENTS

NSF supported my attendance at the Pasadena Conference on 'Deep Earth and Planetary Volatiles'. Many of the ideas presented here were bounced-off my colleague George Helffrich and I acknowledge his comments with thanks.

REFERENCES

1.   H. Wänke, G. Dreibus, Phil. Trans. Roy. Soc. Lond. A325, 545-557 (1988)

2.   H. Wänke, G. Dreibus, Abstract. Conference on Deep Earth and Planetary Volatiles. LPI Contrib. 845, p.46 (1994)

3.   A. Jambon, J. L. Zimmermann, Earth Planet. Sci. Letts. 101, 323-331 (1990)

4.   T. Trull, S. Nadeau, F. Pineau, M. Polvé, M. Javoy, Earth Planet. Sci. Letts. 118, 43-64 (1993)

5.   P. Sarda, D. Graham, Earth Planet. Sci. Letts. 97, 268-289 (1990)

6.   Y. Zhang, A. Zindler, Earth Planet. Sci. Letts. 117, 331-345 (1993)

7.   A. Jambon, Reviews in Mineralogy 30, 477-517 (1994)

8.   B.J. Wood, Earth Planetary Sci. Letts. 117, 593-607 (1993)

9.   K. Yamamoto, S. Akimoto, Amer. J. Sci. 277, 288-312 (1977)

10.   T. Gasparik, Jour. Geophys. Res. 98, 4287-4299 (1993).

11.   E. Ito, D.M. Harris, A.T. Anderson, Geochim. Cosmochim. Acta 47, 1613-1624 (1983)

12.   R.E.T. Hill, A.L. Boettcher, Science 167, 980-982 (1970).

13.   A.R. Pawley, Contrib. Mineral. Petrol. (in press) (1994)

14.   D.R. Bell, G.R. Rossman, Science 255, 1391-1397 (1992).

15.   D.L. Kohlstedt, D.C. Rubie, H. Keppler, Terra Abstracts 6, p.29 (1994)

16.   T.E. Young, H.W. Green II, A.M. Hofmeister, D. Walker, Phys. Chem. Mineral. 19, 409-422 (1993)

17.   T. Inoue, Phys. Earth Planet. Ints. 85, 237-263 (1994)

18.   J. R. Smyth, Am. Mineral. 72, 1051-1055 (1987)

19.   J. R. Smyth, Abstract. Conference on Deep Earth and Planetary Volatiles. LPI Contrib. 845, p.43 (1994)

20.   H. Paulssen, Jour. Geophys. Res. 93, 10489-10500 (1988)

21.   H. M. Benz, J. E. Vidale, Nature 365, 147-150 (1993).

22.   A. Yamazaki, K. Hirahara, Geophys. Res. Letts 21, 1811-1814 (1994)

23.   G. R. Helffrich, B. J. Wood, Nature (1994) (submitted).

24.   B. J. Wood, Jour. Geophys. Res. 95, 12681-12685 (1990)

25.   M. Akaogi, E. Ito, A. Navrotsky, Jour. Geophys. Res. 94, 15671-5685 (1989).

26.   Q. Bai, D. L. Kohlstedt, Phys. Chem. Mineral. 19, 460-471 (1993).

27.   N. M. Wiser, B. J. Wood, Contrib. Mineral. Petrol. 108, 146-153 (1991)

28.   B. J. Wood, Reviews in Mineralogy 17, 71-95 (1987)

29.   T. Katsura, E. Ito, Jour. Geophys. Res. 94, 15663-15670 (1989)

30.   G. Nolet, EOS Trans. Amer. Geophys. Union 75, 232 (1994)

31.   C. R. Bina, B. J. Wood, Jour. Geophys. Res. 92, 4853-4866 (1987).

32.   E. Ito, E. Takahashi, Jour. Geophys. Res. 94, 10637-10646 (1989)

33.   Y. Zhao, D. L. Anderson, Phys. Earth Planet. Ints. 85, 273-292 (1994)

34.   C. Meade, J.A. Reffner, E. Ito, Science 264, 1558-1559 (1994)

35.   D. M. Smyth, Amer Geophys. Union Geophys. Monogr. 45, 99-103 (1989)

36.   P. Sarda, D. Graham, Earth Planet. Sci. Letts. 97, 268-289 (1990)

37.   H. Staudigel, S. R. Hart, H.-U. Schmincke, B. M. Smith, Geochim. Cosmochim. Acta 53, 3091-3094 (1989)

38.   B. S. White, P. J. Wyllie, J. Volcanol. Geotherm. Res. 50, 117-130 (1992)

39.   G. Brey, W. F. Brice, D. J. Ellis, D. H. Green, K. L. Harris, I. D Ryabchikov, Earth Planet. Sci. Letts 62, 63-74 (1983).

40.   S. A. T. Redfern, B. J. Wood, C. M. B. Henderson, Geophys. Res. Letts. 20, 2099-2102 (1993)

41. C. Biellmann, P. Gillet, F. Guyot, J. Peyronneau, B Reynard, Earth Planet. Sci. Letts 118, 31-41 (1993).

42. G. M. Yaxley, D. H. Green, H. Klápová, Mineral. Mag. 58A 996-997 (1994)

43. H. O. A.Meyer, Ch. 34 in P. H. Nixon, ed 'Mantle Xenoliths' Wiley & Sons New York (1987)

44. S. E. Kesson, A. E. Ringwood, Chem. Geol. 78, 97-118 (1989)

45. B. J. Wood, L. T. Bryndzia, K. E. Johnson, Science, 248, 337-345 (1990)

46. H. St.C. O'Neill, D. C. Rubie, D. Canil, C. A. Geiger, C. R. Ross II, F. Seifert, A. B. Woodland, Amer. Geophys. Union Geophys. Monogr. 74, 73-88 (1993)

47. L.T. Bryndzia, B. J. Wood, Amer. Jour. Sci. 290, 1093-1116 (1990).

48. D. A. Ionov, B. J. Wood, Contrib. Mineral. Petrol. 111, 179-193 (1992)

49. G. Gudmundsson, B. J. Wood, Contrib. Mineral. Petrol. (1994) (in press).

50. R. M. Hazen, J. Zhang, J. Ko, Phys. Chem. Min., 17, 416-419 (1990).

51. M. D. Dyar, A. V. McGuire, R. D. Ziegler, Amer. Mineral. 74, 969-980 (1989).

52. J. Peyronneau, J-P. Poirier, Nature 342, 537-539 (1989)

53. X. Li, R. Jeanloz, Jour. Geophys. Res. 95, 5067-5078 (1990)

54. B. J. Wood, J. Nell, Nature. 351, 309-311 (1991)

55. S. M. Peacock, Phil. Trans. Roy.Soc. Lond. A 335, 341-353 (1991)

SEISMIC EVIDENCE FOR THE OCCURRENCE OF VOLATILES BELOW 200 KM DEPTH IN THE EARTH

G. Nolet

Department of Geological and Geophysical Sciences, Princeton University, Princeton, NJ 08544

ABSTRACT

Strong low velocity anomalies in regions of active seismic subduction are indicative of either fluids, partial melts or a temperature close to the melting point. Since the solidus for dry peridotite is far above the probable geotherm, low velocities imply the occurrence of volatiles to lower the solidus. Seismic evidence indicates that subducting slabs in the northwest Pacific have a regime of devolatilization that extends to 400 km depth. Observations in central Europe indicate that deeply penetrating fluids have eroded the deep roots at the edge of the Russian platform.

INTRODUCTION

The convective machinery of Earth creates about 20 km^3/yr of oceanic crust, while it consumes a comparable amount through subduction of the cooled oceanic lithosphere. However, the process is ill-balanced for the volatile components of the crust: whereas mid ocean ridge basalts (MORB) are dry[1], the crust and sediments returning to the mantle are estimated to contain as much as 10^{12} kg H_2O and 2×10^{11} kg CO_2, of which only a small fraction is known to be released in arc and back arc magmatism, such that some 85% of the water ends up 'missing'[2].

The stability fields of common hydrous minerals such as serpentine, amphibole and talc do not extend beyond pressures of 7 GPa, even at very low temperatures. The volcanism that occurs in the back arcs is located approximately where the slab reaches depths corresponding to this pressure range, and there is no doubt that a major dehydration event takes place around 120 km depth[3]. It has long been assumed that virtually all volatiles would leave the slab at this point, and it is probably possible to account for the missing water by assuming that it exsolves and escapes during volcanism, escapes along the subduction shear zone, and is incorporated in the wedge without reaching the surface in the back arc.

This view has recently been challenged. The presence of water stored in the mantle is needed to explain the origin of the Earth's atmosphere and oceans[4,5]. Staudigel and King[6] show how slab temperatures in fast subducting slabs are low enough for hydrous magnesian silicates to remain stable to large depth, and may reach the stability field of hydrous A, the first of the dense hydrous magnesium silicates to become stable at higher pressures, about 7 GPa[7,8,9,10]. Another possible vehicle to transport volatiles into the deeper regions of the upper mantle is formed by the nominally anhydrous phases. Pyroxenes, in particular, may contain up to 1000 ppm H_2O[11]. Although less is known about the behaviour of carbonates, it has been suggested that these would follow a pattern very similar to the magnesium silicates as long as the water content of the rock remains high[6].

The presence of volatiles at depth in the upper mantle would be of great signif-icance. Meade and Jeanloz[12] developed a model for the occurrence of deep earthquakes involving martensitic phase transformations during dehydration. Pockets of volatile-enriched rock, floating around in the upper mantle, could be responsible for a number of observations, such as the inclusion of majorite in diamonds found in the Monastery Kimberlite[13,14], the trace element and isotope ratios found in Mesozoic continental flood basalts[15], and the variable strength of the Monteregian hotspot[16,17].

Until recently the presence of volatiles in the deep upper mantle has only been a convenient explanation for such processes. Now, however, a number of tomographic studies are showing strong indications for the occurrence of deep volatile-rich regions. An added surprise is that these are not only found in regions of active subduction[18,19], but also in areas of ancient subduction[17].

VOLATILES AND LOW SEISMIC VELOCITIES

In the technique of seismic tomography, slight deviations in the arrival time or phase of seismic waves are inverted to produce a three-dimensional image of fast and slow regions in the Earth[20]. In regions of active subduction, a cool ($< 700°C$) slab enters the upper mantle. In tomographic images the slab shows up as a region of distinctly higher velocity. Since the slab has a thickness of about 100 km, which is usually less than the tomographic resolution allows for, the image of the slab is somewhat smeared and the velocity amplitudes consequently damped, but nevertheless compressional velocity (V_P) anomalies of +3% with respect to 'normal' mantle values are commonly observed. Direct measurements using very high frequency waves seem to indicate an average V_P contrast of +5% for a thin crustal layer[21]. This is interpreted as an effect of composition: the crustal basalt transforms to high velocity eclogite. Toward the interior of the slab, however, the composition is peridotitic like the rest of the mantle, and the anomaly must be the effect of temperature. Estimates for the temperature derivative $\partial V_P/\partial T$ from the effects of anharmonicity are in the range of 0.5 m/s/K, so that the lower bound of +3% implies a velocity contrast of some 500 degrees at least.

The cold slab acts as a heat sink for the ambient mantle, and if heat transfer were purely conductive, low temperatures would be expected to occur next to the slab after sufficient time has elapsed. Actually, there is flow of material in the wedge which is driven by the downgoing plate and any cooling effects are minor[22]. On might therefore expect tomographic images to consist of the high velocity anomaly of the slab, embed-ded in normal mantle velocities. This turns out not to be the case: negative velocity anomalies are observed with anomalies in excess of several percent. Since it is unlikely that the mantle temperature is abnormally high - certainly not by hundreds of degrees - this cannot be a simple temperature effect. Although the shear in the wedge is likely to align olivine crystals, the low velocities observed cannot be the effect of a resulting velocity anisotropy since the seismic rays traverse the wedge with many different angles and this will average out any effects of anisotropy. Likewise, compositional changes can have only minor effect on the seismic velocity.

The remaining possible cause for inducing low seismic velocities is also the most likely one: the shear modulus is reduced by the presence of melts or volatiles. The

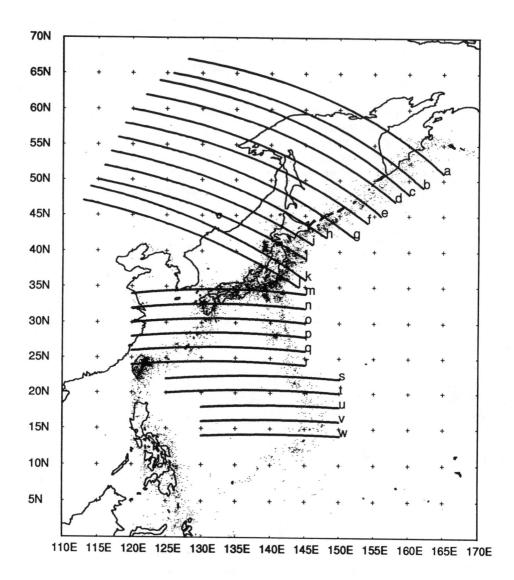

Figure 1: Locations of cross sections in Figure 2

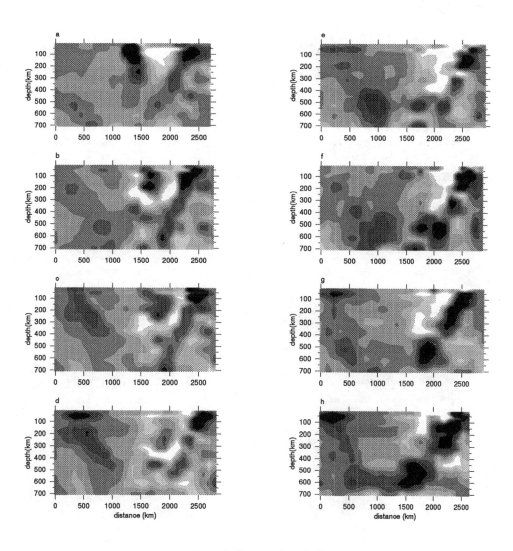

Figure 2: Cross sections 2a-2h

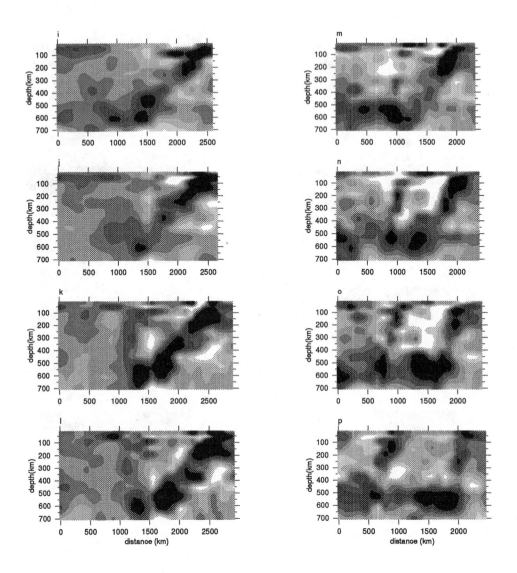

Figure 2: Cross section 2i-2p

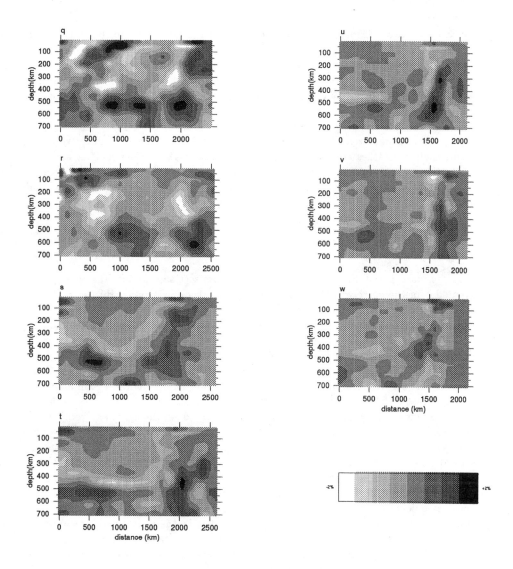

Figure 2: Cross sections 2q-2w

arc and back arc volcanism is a clear indication of the presence of melts in the upper mantle. The melt by itself implies the presence of volatiles, since the solidus for dry peridotite is several hundreds of degrees above any likely mantle temperature, but is strongly lowered in the presence of H_2O or CO_2[23,8]. Since a deterioration of the shear modulus may even occur as the temperature approaches the solidus (depending on the seismic frequency), the presence of melt is not always strictly required[24,25] but we would still need the volatiles to reduce the melting temperature of the rock.

Low seismic velocities in areas of subduction are therefore indicative of the occurrence of volatiles. In the next section we shall take a close look at an active subduction system in the Northwest Pacific and and an ancient continental collision zone in central Europe, and investigate how deep such volatiles penetrate into the upper mantle.

DEEP LOW VELOCITY ZONES

Figure 1 and 2 show a sequence of 21 cross sections through the tomographic V_P model for the northwest Pacific resulting from a study by Van der Hilst et al.[18]. The cross sections cut the subducting slab almost perpendicularly. The velocity scale is from -2% to +2%. Resolution tests[26] show that the deeper (> 200 km) part of the model is well resolved near all subducting slabs. At more shallow levels, the features near the slab are equally well resolved for sections 2a-2p, although a slight loss of resolving power occurs for sections 2q-2w in the Marianas. In particular, the resolution in the depth interval between 200 and 400 km is particularly good in the mantle wedge above the slab as a result of the inclusion of pP phases in the inversion.

Figure 2a shows the slab located on the right side of the figure and recognizable by the high compressional velocity (dark). Directly above the slab is a large area of low velocity (white) extending to some 200 km depth. Both high and low velocity anomalies saturate the velocity scale and we interpret the white area as the source region for the volcanic activity on Kamchatka. The high velocities left of that indicate the continental mantle of eastern Siberia. This structure probably extends further west, but resolving power is quickly lost away from the area of high seismicity. In this figure there is no indication of low velocities deeper than 250 km, but in the next several images there are some characteristic changes in the velocity pattern on the leading edge of the slab.

Figures 2b-2d show a diminishing of the shallow low velocities, but a 300 km deep anomaly now extends to below the Sea of Okhotsk. Note that the image of the slab seems to weaken appreciably. In figures 2e-2f the low velocities are located closer to the slab. At the transition between the Kurile and Japan island arcs (2h-2j) the deep velocities dissapear or at least become strongly diminished in amplitude, but they reappear in the middle of the Japan subduction zone and persist along the Izu-Bonin subduction zone as far as 25°N (2k-2r). The deep low velocity zones seem to extend to 400 km, but not much deeper, and in view of the limited resolving power of tomographic images it is very well possible that the 400 km discontinuity acts as a definite boundary for low velocities. A similar pattern of low velocities was observed along the New Hebrides subduction zone[19].

A striking parallel to these low velocities in active subduction zones was observed by Nolet and Zielhuis[17] in a study of the Tornquist-Teisseyre zone, the old suture zone

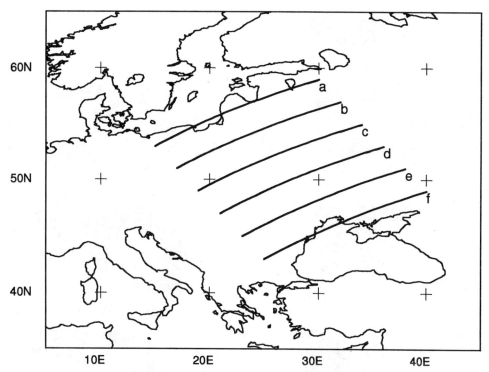

Figure 3: Locations of cross sections in Figure 4

in central Europe, resulting from the collision of eastern Avalonia and Baltica in the Ordovician[27], followed by the accretion of younger terranes and the closure of the Tethys ocean in more recent times[28]. Figures 3 and 4 show the cross sections through their V_S model. Since the area itself is aseismic, surface waves have been used to delineate the upper mantle structure, and care should be taken in comparing Figures 2 (based on high frequency P waves) and these images. The low frequency surface waves are dominated by shear energy. Subsolidus effects may be much more drastic for these waves[24], so that low velocities merely imply a lowering of the solidus, and partial melt is not necessary even to explain strong anomalies. The resolving power of these images is also less - typically a few hundred km horizontally and 100-200 km in depth.

Figure 4a shows a low velocity minimum centered at 400 km depth below the high velocities of the Russian Platform lithosphere. The deep root under the Baltic shield is visible in the right of the figure, and is a persistent feature in all cross sections. The velocity minimum below the Russian Platform attains its strongest anomalous value in cross section 4b. Nolet and Zielhuis[1] note a correlation of this minimum and the location of the Etna hotspot track, active in the area between 150 and 200 my ago. the minimum diminishes further south, but picks up again in the last cross section (2f), which may correlate with a much more recent passage of the hotspot at 50 my, or the closure of the Tethys ocean. The low velocities at shallow depth (100 km) to the left of figures 4c-4f mark the young asthenosphere under the Carpathians/Pannonian basin and are probably the remnants of the much more recent westward subduction of young,

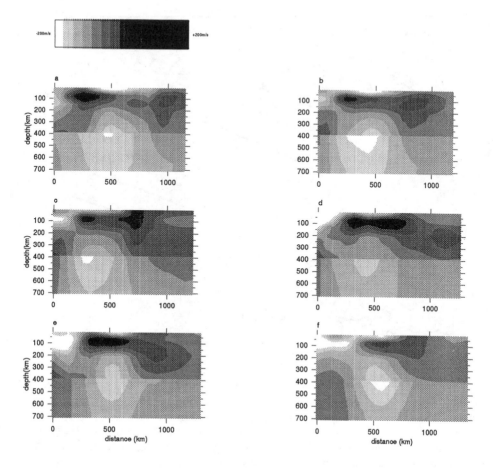

Figure 4: Cross sections 4a-4f

accreted terranes in the Tertiary[28,29].

The resemblance between cross section 4b with some of the northwest Pacific sections like 2d-2g or 2l-2p is suggestive of a similar mechanism. Nolet and Zielhuis[17] discard a number of alternative mechanisms to explain the low velocities in Figure 4, and conclude that the low S velocities can only be caused by the presence of volatiles. This points to a common cause for deep low velocity zones in active and old subduction zones, but does not explain how low velocities can persist for such a long geologic time. Nolet and Van der Hilst[19] suggest that volatiles may have been incorporated in a cool cratonic root under the Russian Platform in the form of hydrous-A[8], only to be released much later as the root was subjected to a thermal event such as the passage of the Etna hotspot.

DISCUSSION

Recent results from seismic tomography show clearly that at least some volatiles are able to survive the major dehydration event in the slab near 100 km depth, and penetrate as deep as 400 km. The minerals that are responsible for the transport of volatiles are not yet uniquely identified, although several candidates exist and hydrous-A is likely to play a dominant role for pressures above 7 GPa, while hydrous-β is a most probable storage site for water below 400 km.

As for the $P - T$ path, no hydrous minerals will remain stable unless the slab is very cold[9], which probably limits the deep transport to regions of mature subduction zones. However, following a suggestion by Fyfe[30], the process of deep volatile transport must become more efficient as the Earth cools, and this could eventually lead to the disappearance of the oceans on the Earth's surface.

Acknowledgement: This research was supported under NSF grant EAR 9204386.

REFERENCES

1. J.E. Dixon, E. Stolper and J.R. Delaney, Earth Plan. Sc. Lett. 90, 87, 1988.

2. S.M. Peacock Science 248, 329, 1990.

3. P.J. Wyllie, Geol. Soc. Am. Bull. 93, 468, 1982.

4. L.W. Finger, J. Ko, R.M. Hazen, T. Gasparik, R.J. Helmley, C.T. Prewitt and D.J. Weidner, Nature 341, 140, 1989.

5. T.J. Ahrens, Nature 342, 122, 1989.

6. H. Staudigel and S.C. King, Earth Plan. Sc. Lett., 109, 517, 1992.

7. Yamamoto, K. and S. Akimoto, Solid State Chem. J., 9, 187, 1974.

8. A.B. Thompson, Nature 358, 295, 1992.

9. T. Gasparik, J. Geophys. Res., 98, 4287, 1993.

10. K. Bose, P. Burnley and A. Navrotsky, in: Conference on Deep Earth and Plane-
    tary Volatiles, LPI Contr. 845, 5, 1994.

11. D.R. Bell and G.R. Rossman, Science 255, 1391, 1992

12. C. Meade and R. Jeanloz, Science 252, 68, 1991.

13. A.E. Ringwood, S.E. Kesson, W. Hibberson and N. Ware, Earth Plan. Sc. Lett.
    113, 521, 1992.

14. Moore and Gurney, Nature 318, 553 , 1985.

15. K. Gallagher and C. Hawkesworth, Nature 358, 57, 1992.

16. N.H. Sleep, J. Geophys. Res. 95, 6715, 1990.

17. G. Nolet and A. Zielhuis, J. Geophys. Res. 99, 15813, 1994.

18. R. Van der Hilst, R. Engdahl, W. Spakman and G. Nolet, Nature 353, 37, 1991.

19. G. Nolet and R. van der Hilst, subm. for publ., 1994.

20. G. Nolet, J. Geophys. Res. 95, 8513, 1990.

21. D. Gubbins and R. Snieder, J. Geophys. Res. 96, 6321, 1991.

22. J.H. Davies and D.J. Stevenson, J. Geophys. Res. 97, 2037, 1992.

23. M. Olafsson and D.H. Eggler, Earth Plan. Sc. Lett. 64, 305, 1983.

24. S.-i. Karato, Geophys. Res. Lett. 20, 1623, 1993.

25. H. Sato, I.S. Sacks, T. Murase and C.M. Scarfe, Geophys. Res. Lett. 11, 1227,
    1988.

26. R.D. van der Hilst, E.R. Engdahl and W. Spakman, Geophys. J. Int. 115, 264,
    1993.

27. T.H. Torsvik, A. Trench, I. Svensson and H.J. Walderhaug, Geophys. J. Int. 113,
    651, 1993.

28. P.A. Ziegler, A.A.P.G. Mem. 43, 198pp., 1988.

29. A. Zielhuis and G. Nolet, Science 265, 79, 1994.

30. W.S. Fyfe, Geosc. Can. 3, 82, 1976.

Subduction and Volatile Recycling in Earth's Mantle

Scott D. King[*] and Joel J. Ita[†]

Department of Earth and Atmospheric Sciences, 1397 CIVIL
Purdue University, West Lafayette, IN 47907-1397
[†]*Seismological Laboratory, MS 252-21, California Institute of Technology[1]*
Pasadena, CA 91125

The subduction of water and other volatiles into the mantle from oceanic sediments and altered oceanic crust is the major source of volatile recycling in the mantle. Until now, the geotherms that have been used to estimate the amount of volatiles that are recycled at subduction zones have been produced using the hypothesis that the slab is rigid and undergoes no internal deformation after subduction. We consider the effects of the strength of the slab using two-dimensional calculations of a slab-like thermal downwelling with an endothermic phase change. Because the rheology and composition of subducting slabs are uncertain, we consider a range of Clapeyron slopes which bound current laboratory estimates of the spinel to perovskite plus magnesiowüstite phase transition and simple temperature-dependent rheologies based on an Arrhenius law diffusion mechanism. Phase transitions can have two pronounced effects on subducting slab deformation and the resulting geotherms. First, an endothermic phase transformation can inhibit the vertical descent of the slab. If the slab is weak, this can lead to large deformation, even in the upper 200 km of the slab. Second, the phase transformation can slow the subduction velocity (and plate velocity) of the entire slab, a more pronounced effect than the slab deformation. Because the initial thermal structure of the descending lithosphere, the volatile content, and subduction velocity all affect the viscosity of the slab, it is likely that subduction zones may behave differently—some with more pronounced pile-up and avalanche periods and some where the subduction velocity is more uniform with time. These mechanisms create a highly uneven distribution of recycled components in the mantle within relatively short periods of time in Earth's history.

INTRODUCTION

The thermal state of descending oceanic lithosphere plays a major role in the recycling of volatiles in the Earth's mantle. Early thermal models of subduction zones sought to explain arc volcanism by producing enough heat to

---

[1]Presently at: Department of Earth and Atmospheric Sciences, Purdue University

melt the crust at the top of the slab (e.g., (1)). It is now accepted that the temperatures at 100-150 km depth at the top of the slab are not sufficient to melt oceanic crust. However, the volatiles, mostly water driven out of hydrated minerals in the oceanic crust, interact with the hot mantle wedge to form the source melts for arc volcanism. These models use analytic or finite-difference solutions to model the thermal structure of the descending oceanic lithosphere at a uniform velocity. They generally make the assumption that heat is transferred down-dip only by advection and perpendicular to slab dip by pure conduction. Simple advection-conduction models, like those mentioned above, are still used in conjunction with dehydration reactions to understand the fate of water and other volatile components in the downgoing slab (e.g., (2,3)). While these models provide some guidance in our understanding of the thermal state in and around subduction zones, there are two important limitations that must be considered when applying them.

First, the slab is forced to descend along a specific geometry by an imposed, external, uniform, velocity field and is not dynamically consistent with the density contrast between the cold slab and surrounding mantle. Because the velocity fields used in these models are often steady-state fields, these models may miss significant effects associated with the initiation of subduction or the time-dependence of plate motions. One dynamical mechanism that may produce strongly time-dependent plate velocities is the interaction of the subducting material and the phase transformations at 410 and 660 kilometers depth on the downgoing slab. The endothermic phase transformation of spinel to perovskite and magnesiowüstite inhibits the downward descent of the downgoing limb can can lead to a style of convection with periods of slow, uniform plate velocities as the subduction material collects above the phase transformation boundary. This quiet phase is interrupted by short periods with high plate velocities when the pool of cold, dense material becomes unstable and flushes into the lower mantle (e.g., (4,5)). As we will show, this has important consequences for the P-T path in a subduction zone.

The second limitation of the advection-conduction models is that they do not account for deformation within the downgoing slab. In previous models, the geometry of the flow is simple and the slab is assumed to behave rigidly. These models assume either a uniform velocity distribution or a corner flow solution, and these are assumed constant both in space and time. Folding, bending, thickening and thinning of the slab, all of which can alter its thermal structure, are not considered in the advection-conduction models. The rigid slab assumption is the opposite end-member of most constant viscosity mantle convection calculations (e.g., (4,5)). In constant viscosity convection calculations the slab has no greater, or lesser, strength than the surrounding mantle and, thus, as it subducts the slab deforms substantially. The Earth lies somewhere in between these two end-member views.

In this paper, we will explore the effect of temperature-dependent viscosity on the dynamics of convection in the presence of an endothermic phase change. While this topic has been studied in detail in several recent papers (6), we will focus our attention on the thermal structure of the downgoing limb in an attempt to understand the slab geotherm (i.e., the pressure-temperature-time paths of subducting material). To understand the first-order effects of temperature-dependent viscosity on the dynamics of convection in the presence of an endothermic phase transition, we consider a model with one phase transition (to represent the 660 km discontinuity) at one-half the depth of the box. The effect of the exothermic reaction of olivine to spinel at 410 km depth appears to weaken the effect of the 660 km phase transformation (e.g., (5)). Thus, including only one phase transformation we can capture the first-order effects of the phase transformations on subduction. Because we focus on the geotherm in the upper 300 km of the slab, the complexities of the transition zone probably have little influence, to first-order, on our results. While ignoring many of the complexities in a subduction zone environment, such as thrust faults, viscous heating, and dehydration reactions, our model allows us to concentrate on the effect of slab rheology in a model that is driven by the density contrast between the slab and the mantle in a self-consistent manner. We will calculate geotherms through the subduction zone from these calculations to address the impact of mantle layering on subduction geotherms, something that the advection-conduction models with an imposed flow have not addressed.

EQUATIONS AND METHOD

We use a modified version of the two-dimensional, Cartesian, finite element convection code ConMan, described in Ref. (7). We incorporate spinel-perovskite phase transformation using a method similar to that in Ref. (8) which includes the effect of latent heat released from the major endothermic phase transformation. We do not include the effect of the latent heat of the metamorphic dehydration reactions. Ref. (2) demonstrates that these have an almost negligible effect on the temperature of the slab. Further details on this formulation can be found in Ref. (9).

The phase transformation occurs, on average, at a normalized depth of 0.5. If our results are scaled such that the phase transition is half completed at 660 km depth, the phase transition would occur over a 70-km depth range. This was done to minimize the computational expense and to allow the highest possible resolution, without a highly irregular grid, around the phase transformation. The Clapeyron slope and density change associated with the phase transition are incorporated through the nondimensional phase buoyancy parameter, P, such that

$$P = \frac{\gamma \Delta \rho}{\rho_o^2 \alpha g h},$$
(1)

where γ is the Clapeyron slope, $\Delta \rho$ is the density change across the transition, ρ_o is the reference density, α is the thermal expansivity, g is the force of gravity, and h is the depth of the box. In the following calculations, $\Delta \rho$ is fixed at 0.35 Mg/m^3. This value is consistent with seismically inferred mantle density models (10). Thus, P will measure the change in the Clapeyron slope.

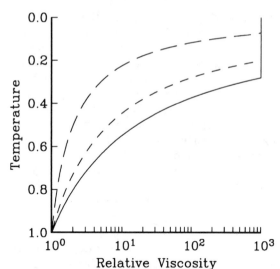

FIG. 1. Viscosity laws used in this study from Equation 2; solid line - uniform viscosity, long dash - weakly temperature-dependent ($E^* = 1.0$), medium dash - moderately temperature-dependent ($E^* = 2.0$), short dash - strongly temperature-dependent ($E^* = 3.0$)

We use an Arrhenius form for the temperature-dependent part of our viscosity law:

$$\eta(\theta) = \eta^* \exp \left[\frac{E^*}{\theta + T_o} - \frac{E^*}{1.0 + T_o} \right] \tag{2}$$

where η^* is the normalized pre-exponential viscosity, $\eta(\theta)$ is the effective viscosity, θ is the dimensionless temperature, E^* is the activation energy divided by $R\Delta T$, where ΔT is the temperature scaling factor (for reference $E^* \approx 6$ for olivine if $\Delta T \approx 3000°K$). T_o is the temperature offset; a temperature offset is needed because when $\theta = 0$, the first term becomes singular. The second term in the exponential scales $\eta(\theta)$ such that $\eta(1.0) = \eta^*$. The viscosity profiles used in this investigation resulting from this functional form are shown in Figure 1.

The calculations are performed in a 4 by 1 box with free-slip bottom and reflecting boundary conditions on the sides. On the top, the majority of the surface is free-slip, but several nodes are pinned at the far right-hand side of the box to keep the boundary layer on the right side of the trench immobile. This helps to create an asymmetry in the downwelling at the trench. For all the problems we consider here, the grid size is 384 by 96 uniformly spaced elements. Temperature boundary conditions along the top and bottom are $\theta = 0.0$ and $\theta = 1.0$, respectively. Internal heating is incorporated in the model with approximately 50% of the heat flux at the surface coming from the bottom of the box. All calculations used the same initial condition, a 4 by 1 box with a downwelling in the center and no phase transition. Because

TABLE 1. Physical Properties

Symbol[†]	Name	Value
ρ_o	density	3.5×10^3 kg/m^3
h	depth	1340 km
g_o	magnitude of gravity	10 m/s^2
ΔT	temp. drop across h	1600 °K
κ_o	thermal diffusivity	1×10^{-6} m^2/s
α_o	thermal expansivity	3×10^{-5}°K^{-1}
η_o	viscosity	10^{22} Pa-s

[†]Subscript o refers to the value of the variable at the upper surface.

we use several temperature-dependent viscosity laws, it is difficult to specify an exact Rayleigh number. We attempted to keep the logarithmic volume averaged Rayleigh number as close as possible in all of our calculations. The volume averaged Rayleigh number for each calculation is within the range $2 \times 10^5 - 4 \times 10^5$ based on the reference values shown in Table 1.

To create plate-like surface velocities, we impose weak zones at the trench and ridge (11). This is accomplished by reducing the pre-exponential term, η^* in equation (2) in a specific geometric zone of the computational domain to a value denoted by the variable η_{wz}. In these calculations, the weak zones are located at the upper left hand corner of the box, the model spreading center, and at the top center of the box, the model trench. In the upper left hand corner, the weak zone is 3 by 3 elements. The weak zone at the top center of the box is 9 by 20 elements. In the constant viscosity models, we used a two element thick high viscosity layer across the top between the weak zones to enforce the piecewise uniform velocity.

RESULTS

In uniform viscosity convection models, subducted material collects above the phase transformation boundary until the pool of cold, dense material becomes gravitationally unstable and sinks into the lower mantle. This style of convective motion has been observed in both 2-D and 3-D calculations in spherical and Cartesian geometries (e.g., (4)). Figure 2 illustrates the effect on the downgoing limb of increasing the Clapeyron slope of the phase change (increasing the phase buoyancy parameter). From the contours of temperature, it is clear that for the weakest phase change ($P = -0.19$ or $\gamma = -3.0$ MPa/°K), the downgoing limb penetrates the phase change with little or no deformation, while for the stronger phase changes, ($P = -0.29$ or $\gamma = -4.5$ MPa/°K and $P = -0.39$ or $\gamma = -6.0$ MPa/°K), the downgoing limb is strongly deflected along the phase transformation boundary. The surface velocities also decrease systematically with the increasing magnitude of the Clapeyron slope. As the slope varies from -3.0 MPa/°K to -4.5 MPa/°K to -6.0 MPa/°K, the surface velocity decreases from approximately 450 mm/yr to 325 mm/yr to 150 mm/yr respectively (see Figure 2).

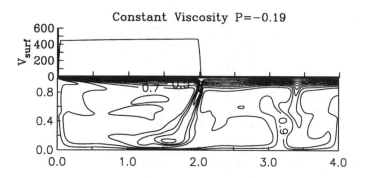

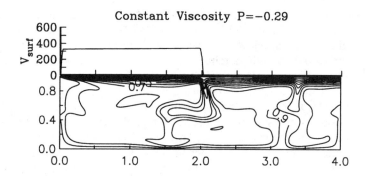

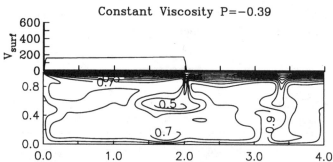

FIG. 2. Temperature contours and surface velocities from three constant viscosity plate models with varying phase buoyancy parameters (Clapeyron slope); a) $P = -0.19$ or $\gamma = 3.0$ MPa/°K, b) $P = -0.29$ or $\gamma = -4.5$ MPa/°K, c) $P = -0.39$ or $\gamma = -6.0$ MPa/°K. Plate velocities are in mm/yr.

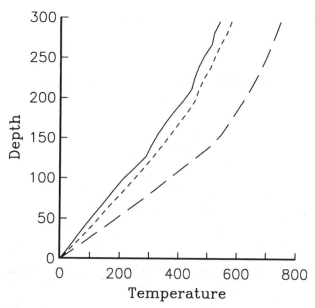

FIG. 3. Slab geotherms from the from the temperature fields shown in Figure 2; solid line - $P = -0.19$ or $\gamma = 3.0$ MPa/°K, short dash - $P = -0.29$ or $\gamma = -4.5$ MPa/°K, long dash - $P = -0.39$ or $\gamma = -6.0$ MPa/°K.

From the calculations in Figure 2, we can construct 'slab' geotherms. Because these are incompressible calculations, there is no adiabatic temperature gradient. We compensate for this by adding a 0.3 °K/km gradient to the temperature fields from these calculations. Because the subduction is one-sided (from the velocities in Figures 2a-2c the right hand side of the box has a zero surface velocity), to a first approximation the top of the slab will co-incide with the coldest temperature. This is not exactly correct because, as the cold top of the slab comes in contact with warm mantle, the top of the slab will heat up by diffusion and the coldest part of the slab will migrate inwards as a function of depth (time). Because the time interval of these calculations is short compared to the diffusion time, the coldest part of the slab will be within 10 km of top of the slab. This is smaller than the grid resolution of our calculations. Figure 3 illustrates the "slab" geotherm for the three temperature fields shown in Figure 2. From Figure 3 it is clear that in addition to its effect on the deep structure of the slab (i.e., whether or not the slab penetrates the phase transformation), the phase boundary influences the shallow temperature structure of the slab. For this particular choice of parameters, the temperature at 150 km depth increases by 250 °K when the Clapeyron slope varies from −3.0 MPa/°K to −6.0 MPa/°K. We also note the negative correlation between surface or "plate" velocity and slab geotherm; as the velocity increases, the slab geotherm decreases. This agrees with previous work (2,3).

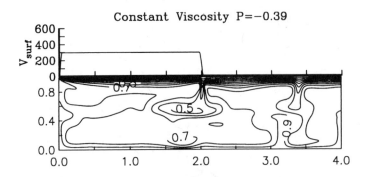

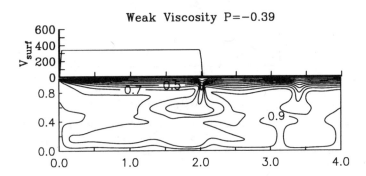

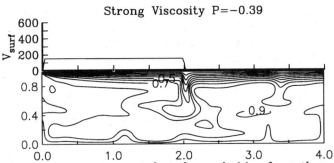

FIG. 4. Temperature contours and surface velocities from three temperature-dependent plate models with $P = -0.39$ or $\gamma = -6.0$ MPa/°K; a) constant viscosity b) weakly temperature-dependent viscosity c) strongly temperature-dependent viscosity. Plate velocities are in mm/yr.

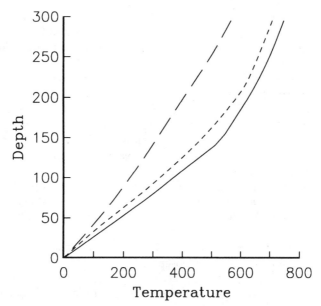

FIG. 5. Slab geotherms from the from the temperature fields shown in Figure 4; solid line - weak slab, short dash - intermediate slab, long dash - strong slab.

Choosing one phase buoyancy parameter ($P = -0.39$ or $\gamma = -6.0$ MPa/°K), we now investigate the influence of viscosity on the structure of the downgoing limb in the presense of a phase transformation. These calculations are shown in Figures 4a-4c. It is important to note that the thickness of the downgoing limb in the center of the box near the surface is noticeably different for each calculation (Figure 4). In Figure 4a, the downgoing limb is thinner than the thermal boundary layer on the upper left-hand side prior to subduction. In contrast, the downwelling in Figure 4c is about the same thickness as the plate. This is because as the strength of the temperature-dependent viscosity increases, the maximum depth of the stiff portion of the slab increases. For the constant rheology case shown in Figure 4a, the viscosity of the slab and the surrounding fluid are almost identical. In the weakly temperature-dependent case (Figure 4b), again, the deeper part of the slab is still approximately the same viscosity as the surrounding fluid. In the strong viscosity case, (Figure 4c), the viscosity of the slab is ten to thirty times the background viscosity. In this case, a noticeable contrast between the slab and the background fluid persists over the entire depth extent of the slab. Thus, while the viscosity law acts over the entire fluid, the major effect of changing the viscosity law is to change the strength of the deep slab.

We can again construct slab geotherms from these temperature fields. We follow the same procedure we used in constructing the geotherms in Figure 3; adding a 0.3 °K/km gradient to the temperature field and using the coldest temperature of the slab as the approximation for the top of the slab. The slab geotherms for the fixed phase buoyancy parameter and varying viscosity structure are shown in Figure 5. The effect of rheology on the slab geotherms is as strong as the effect of the Clapeyron slope illustrated in Figure 3. There is approximately a 250 °K temperature difference between these three models at 150 km depth. The strong slab is thicker and the interior of that slab is colder than the weak and constant viscosity slabs which appear to thin significantly while subducting.

When we consider the surface velocity, we find that it decreases by a factor of two when the viscosity model varies from constant or weakly temperature-dependent to strongly temperature-dependent. This means that the slab geotherm decreases with decreasing plate velocity. Thus, for this value of the Clapeyron slope, variations in rheology lead to a positive correlation between plate velocity and slab geotherm; as the plate velocity increases, so does the slab geotherm. Note that this is the inverse of the relationship we found in models in which the rheology was held constant and the Clapeyron slope varied. This arises from the fact that increases in velocity brought on by variations in rheology are accompanied by increases in internal deformation of the slab. Not surprisingly, this is not observed in the simple advection-conduction models because they proscribe the style of slab deformation a priori.

The difference in slab geotherm between the different rheologies is most pronounced in the cases where the magnitude of the Clapeyron slope is large enough to effectively stop the cold, downgoing slab from penetrating the phase transformation. In cases where the Clapeyron slope is smaller, the results from the four rheologies look much more similar, so the case in Figure 4 should be considered as an end-member case. In Figure 4a and to a lesser extent in Figure 4b, the layering of the phase change sets up a small scale convection cell near the slab (notice the thinning of the isotherms just to the left of the slab). This small scale circulation enhances the larger plate scale circulation and is responsible for thinning of the slab.

If we had created a subduction zone using a narrow fault, rather than a broad weak zone, it is possible that the deformation would have been less dramatic in the weak and constant viscosity cases; however, a recent study suggests that the effect of viscous dissipation in temperature-dependent fluids may be significant enough to increase the temperature of the slab by as much as several hundred degrees (12). Because the region of greatest deformation coincides with the weak zone in our models, the weak zone provides the effect of viscous dissipation on the rheology (but not the temperatures). The importance of the slab rheology and parameterization of the plate on mass transport across the phase transformation are discussed in greater detail in (6).

CONCLUSIONS

While simple analytical models of can provide limited insight into the thermal structure of subduction zones, these simple models do not account for the effect of the 410 and 660 km phase transformations on subducting slabs. If the 660 km discontinuity does act as a barrier to convection, and subducted slabs are slowed and deformed in the upper mantle, then these simple models could underestimate the temperatures of the shallow slab by up to several hundred degrees. However, it is important to note that the range of slab geotherms we have presented are all within the range of geotherms presented from the advection-conduction models with various plate velocities. This is not surprising in light of the fact that in a fully dynamic calculation with temperature-dependent rheology and an endothermic phase transformation, the slab geotherm is highly correlated with the plate (and thus slab) velocity. However, the nature of this correlation is dependent upon the choice of parameters selected as we demonstrate above. Thus it is not clear that the variation in geotherm with plate velocity seen in the kinematic models is representative of that in the Earth.

The time-dependent nature of plate motions also has an important effect on slab geotherms. While we don't have the space to discuss it in detail, we can make the following observation: during the pile-up period, the slab geotherm is characteristic of geotherms produced by steady-state slab models with slow plate velocities; however, during the avalanche period, when the material that has been stored above the phase change begins to flush into the lower mantle, the subducting slab geotherm approaches the geotherms from steady-state models with fast velocities. The difference between the subducting slab velocities of the two periods (the pile-up period and the avalanche period) decreases with increasing viscosity of the slab.

We also draw attention to the fact that the temperature-pressure curves presented in this paper should be treated with great caution. Because these calculations are incompressible and are performed at a grid resolution of 10's of kilometers, they should not be used to evaluate subduction zone chemistry in detail. What they do illustrate is that the effects of slab rheology and the 660 km phase transformation can significantly alter the P-T path of the shallow slab, but this seems to be primarily in the case where the Clapeyron slope is large enough to strongly layer the flow. Neither rheology or phase transformations are accounted for in the simple models currently used to make inferences on dehydration reactions in slabs; however, the isotherms from the strong temperature-dependent viscosity slab look much more like the advection-conduction models than the weak viscosity models do. Thus, our results are in agreement with the general characterists of the advection-conduction models except for weak viscosity slabs with strong phase changes. If slabs are truly weak and the phase change strongly layers the flow, then the advection-conduction models may underestimate the deformation in the slab and underestimate the temperature in the slab.

This directly impacts the amount of volatiles being recycled back into the mantle. Weak slabs and/or strongly layered flows will lead to high slab geotherms and a shallow devolatilization of the subducting lithosphere. Whereas, strong slabs and/or weakly layered flows will lead to lower slab geotherms and and a deeper emplacement of volatile materials. It is very probable that the characteristics of subduction zones are spatially and temporally dependent and the resulting concentration of volatiles within the Earth will not lend itself to a simple analysis.

ACKNOWLEDGMENTS

We gratefully acknowledge support from NSF grants EAR-9218621 and EAR-9206036. JI also acknowledges support from NSF grant EAR-9218390. Reviews by S. Peacock and an anonymous reviewer noticeably improved the manuscript.

REFERENCES

1. E. R. Oxburgh and D. L. Turcotte, Geol. Soc. Am. Bull. **81**, 1665 (1970); J. W. Minear and N. M. Toksoz, J. Geophys. Res. **75**, 1397 (1970); M. N. Toksoz, J. W. Minear, and B. R. Julian, J. Geophys. Res. **76**, 1113 (1971); D. L. Turcotte and G. Schubert, J. Geophys. Res. **78**, 5876 (1973).
2. S. Peacock, Tectonics **9**, 1197 (1990).
3. H. Staudigel and S. D. King, Earth Planet. Sci. Lett. **109**, 517 (1992).
4. S. Honda, S. Balachandar, D. A. Yuen, and D. Reuteler, Science **259**, 1308 (1993); P. J. Tackley, D. J. Stevenson, G. A. Glatzmaier, and G. Schubert, Nature **361**, 699 (1993).
5. P. J. Tackley, D. J. Stevenson, G. A. Glatzmaier, and G. Schubert, J. Geophys. Res. **99**, 15877 (1994).
6. S. Zhong, and M. Gurnis, J. Geophys. Res. **99**, 15903 (1994); S. D. King and J. J. Ita, J. Geophys. Res. **Submitted**, (1995).
7. S. D. King, A. Raefsky, and B. H. Hager, Phys. Earth Planet. Int. **59**, 196 (1990).
8. F. M. Richter, Rev. Geophys. Space Phys. **11**, 223 (1973).
9. J. J. Ita and S. D. King, J. Geophys. Res. **99**, 15919 (1994).
10. A. M. Dziewonski and D. L. Anderson, Phys. Earth Planet. Int. **25**, 297 (1981).
11. M. Gurnis and B. H. Hager, Nature **335**, 317 (1988); G. F. Davies, Geophys. J. **98**, 461 (1989).
12. V. Steinbach and D. A. Yuen, Phys. Earth Planet. Int. **86**, 165 (1994).

MANTLE DEVOLATILIZATION AND RHEOLOGY IN THE FRAMEWORK OF PLANETARY EVOLUTION

S. Franck and Ch. Bounama
Potsdam University, PF 601632, 14416 Potsdam, F. R. Germany

ABSTRACT

We investigate the thermal history of an Earth-like planet with the help of a parameterized mantle convection model including the volatile exchange. The weakening of mantle silicates by dissolved water is described by a functional relationship between creep rate and water fugacity. We use flow law parameters of diffusion creep in olivine under dry and wet conditions. The mantle degassing rate is considered as directly proportional to the seafloor spreading rate which as well is dependent on the mantle heat flow. To calculate the spreading rate, we assume that the heat flow under the mid-ocean ridges is double the average mantle heat flow. The rate of regassing also depends on the seafloor spreading rate as well as on other factors like the efficiency of volatile recycling through island arc volcanism. Both mechanisms (de- and regassing) are coupled self-consistently with the help of the parameterized convection model under implementation of a temperature and volatile-content dependent mantle viscosity.

We calculate time series for the Earth's and Venusian evolution over 4.6 Gyr. In the case of Venus, there is the possibility that the Venusian mantle convection might have changed from oscillatory to quasi-steady circulation, i.e., Venus changed from an Earth-like planet to a Mars-like planet at around 500 Myr ago as far as its tectonic style is concerned. Based on this view we also discuss the importance of our model for the investigation of the degassing history of Venus.

INTRODUCTION

The thermal history of a planet depends mainly on the initial mantle temperature, the distribution of the radiogenic heat sources, and the mechanism of the heat transport within the mantle. It is well known that subsolidus convection is the dominant mechanism of the heat transport within the mantle. With the help of parameterized convection models [1-6] temporal variations of the average mantle temperature and the surface heat flow in terms of the Rayleigh number can be calculated for certain initial conditions.

The basic equation of the parameterized convection model is the equation of the conservation of the energy in terms of the time rate of change of the average mantle temperature $\dot{T}$:

$$\frac{4}{3}\pi\varrho c(R_m^3 - R_c^3)\dot{T} = -4\pi R_m^2 q + \frac{4}{3}\pi Q(R_m^3 - R_c^3) \tag{1}$$

where ϱ is the density, c is the specific heat at constant pressure, q is the heat flow from the mantle, and R_m and R_c are the outer and the inner radii of the mantle, respectively. For simplicity our model does not include core heat flow which at present is about 10% of the surface heat flow. The core heat flow has no qualitative influence on the mantle devolatilization. The quantity Q describes the energy production rate by decay of radiogenic heat sources in the mantle:

$$Q = Q_0 \cdot e^{-\lambda t} \tag{2}$$

where Q_0 and λ are constants and t is the time. The Rayleigh number Ra for a convecting mantle is :

$$Ra = \frac{g\alpha(T-T_s)(R_m - R_c)^3}{\kappa \nu} \tag{3}$$

where g is the acceleration due to gravity, α is the coefficient of thermal expansion, T_s is the surface temperature, κ is the thermal diffusivity, and ν is the viscosity. The mantle heat flow q is parameterized in terms of Ra:

$$q = \frac{k(T-T_s)}{(R_m - R_c)}\left(\frac{Ra}{Ra_{cr}}\right)^\beta \tag{4}$$

where k is the thermal conductivity, Ra_{cr} is the critical value of Ra for the onset of convection, and β is an empirical constant.

In the next chapter we show how the parameterized convection model can be coupled with the process of the volatile exchange between mantle and surface reservoirs. In this case it is very important to take into account the negative feedback between kinematic viscosity and volatile content of the mantle (chapter 3). The results for both the Earth and the Venus are presented in the 4th part. In the last chapters of this paper we test the stability of our results and infer the main points following from our investigations.

DEGASSING AND REGASSING OF VOLATILES

The first self-consistent model that couples the thermal and degassing history of the Earth was proposed by McGovern and Schubert [6]. In this model volatiles from the mantle degas at mid-ocean ridges from a certain volume (degassing volume) that depends on the areal spreading rate S and the melt generation depth d_m (Figure 1).

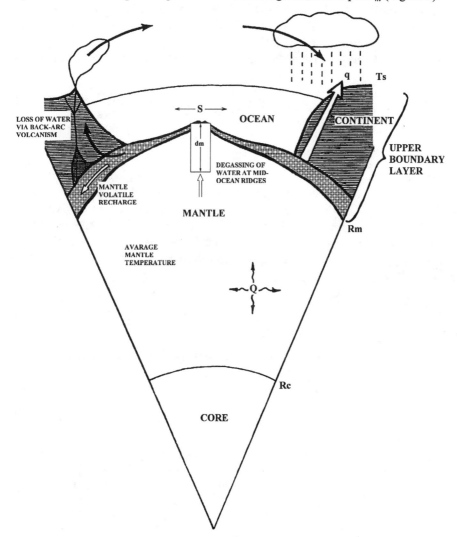

Fig. 1. Global volatile cycle coupled with a parameterized mantle convection model. S is the seafloor spreading rate and d_m the depth of partial melting from which volatiles are released. Regassing occurs at subduction zones after the loss of some volatiles via arc and back-arc volcanism.

The degassing rate of volatiles at mid-ocean spreading centers $[\dot{M}_{mv}]_d$ is given by:

$$[\dot{M}_{mv}]_d = \rho_{mv}\, d_m\, S \tag{5}$$

where ρ_{mv} is the density of volatiles in the mantle (mass of mantle volatiles per mantle volume), d_m is the melt generation depth, and S is the areal plate spreading rate. The melt generation depth d_m is defined by the depth where ascending mantle material intersects the basalt eutectic and extensive melting and melt segregation occur.

The rate of regassing at subduction zones $[\dot{M}_{mv}]_r$ is:

$$[\dot{M}_{mv}]_r = f_{bas}\, \rho_{bas}\, d_{bas}\, S\, \chi_r \tag{6}$$

where f_{bas}, ρ_{bas}, and d_{bas} are the mass fraction volatile content, the average density, and the thickness of the seafloor crust and sediments before subduction, respectively. χ_r is an efficiency factor representing the fraction of volatiles that actually enter the deep mantle instead of returning to the surface through arc and back-arc volcanism or offscraping (Figure 1). The balance equation for the mass M_{mv} of mantle volatiles is given by:

$$\dot{M}_{mv} = [\dot{M}_{mv}]_r - [\dot{M}_{mv}]_d \tag{7}$$

The initial value of M_{mv} is the number n of ocean masses M_{ocean} originally in the mantle:

$$M_{mv}(t=0) = n \cdot M_{ocean} \tag{8}$$

The equations for the parameterized convection model (1)...(4) can be coupled to the degassing / regassing model (5)...(8) via the relation between heat flow q and the average age of the subducting oceanic crust which as well is a function of the spreading rate S:

$$q = \frac{\sqrt{S}\, 2k(T-T_s)}{\sqrt{\pi \kappa A_0(t)}} \tag{9}$$

where $A_0(t)$ is the area of the ocean basins at the time t. In our model [7] we assume that the heat flow under the mid-ocean ridges is twice the average heat flow of the parameterized convection model.

MANTLE RHEOLOGY AND VOLATILE CONTENT

It is well known that the kinematic viscosity of the mantle depends on the temperature, the pressure, and the volatile content. The parameterization of activation temperature T_A for solid-state creep as a function of water weight fraction x was expressed in the model [6] as follows:

$$v = \bar{v} \, \exp(T_A/T) \tag{10}$$

$$T_A = \alpha_1 + \alpha_2 x \tag{11}$$

where $\bar{v}$, α_1, α_2 are constants. The physical background of the formulas (10) and (11) is the idea that dissolved volatiles like water weaken the minerals by reducing their activation temperature T_A for solid-state creep.

On the other hand, there are some experiments on the effect of water fugacity on the deformation of olivine. They indicate the higher fugacity of water and hydrogen under wet conditions in difference to that under dry conditions [8, 9, 10]. A power law dependence between creep rate $\dot{\epsilon}$ and water fugacity f_{H2O} was found [11]:

$$\frac{\dot{\epsilon}_{wet}}{\dot{\epsilon}_{dry}} - 1 \sim f_{H_2O}^r \tag{12}$$

where the power law exponent r is in the range 1/3..1/5. Such a power law dependence can be appropriate for a model of water weakening where the creep rate is proportional to the concentration of water related point defects. Therefore, we assume that the water fugacity is proportional to the water weight fraction x used in (11). We can write down equation (12) in the following way:

$$\dot{\epsilon}_{wet} = \dot{\epsilon}_{dry} + K \cdot (x)^r \cdot \dot{\epsilon}_{dry} \tag{13}$$

The unknown constant K in (13) can be determined from the flow law parameters of dry and water saturated olivine given by Karato and Wu [12]. We write the steady state strain rate $\dot{\varepsilon}$ under both dry and water saturated conditions as a function of a reference temperature T_{ref}, a reference pressure P_{ref}, and a reference shear stress σ

$$\dot{\varepsilon}_{sat/dry} = A_{sat/dry} \left(\frac{\sigma}{\mu}\right)^n \cdot \left(\frac{b}{d}\right)^m \cdot \exp\left(-\frac{E^*_{sat/dry} + P_{ref} V^*_{sat/dry}}{R \cdot T_{ref}}\right) \quad (14)$$

where A is the prexponential factor, μ is the shear modulus (~ 80 GPa), b is the length of the Burgers vector (~ 0.5 nm), n is the stress exponent, m is the grain-size exponent, E^* is the activation energy, V^* is the activation volume, and R is the gas constant. We use flow law parameters for diffusion creep in olivine of Karato and Wu [12]. The main parts of the upper mantle are believed to flow by this mechanism. The reference pressure P_{ref} is determined with the help of (11) and the assumption that water saturated conditions correspond to ~ 0.03 % water by weight [12], i.e. $x = 0.0003$.

$$P_{ref} = \frac{A_{sat} \cdot R - E^*_{sat}}{V^*_{sat}} \quad (15)$$

$$A_{sat} = \alpha_1 + \alpha_2 \cdot 0.0003 \quad (16)$$

The corresponding parameters and the result for P_{ref} are given in [7]. With the help of the eqns. (14), (15), and (16) we can determine the value of K from (13) under water saturated conditions, i.e. $x = 0.0003$.

$$K = 2.115 \quad (17)$$

Now it is possible to use (13) for the describtion of the dependence of creep rate $\dot{\varepsilon}$ on the volatile weight fraction x.

In the case of a linear rheology ($n = 1$) the effective viscosity η is defined as

$$\eta = \frac{\sigma}{2 \cdot \dot{\varepsilon}_{wet}} \quad (18)$$

with $\dot{\varepsilon}_{wet}$ taken from (13). The kinematic viscosity v is given as

$$v = \frac{\eta}{\bar{\varrho}_{mantle}} \tag{19}$$

where $\bar{\varrho}_{mantle}$ is the mean density of the mantle.

THERMAL AND DEGASSING HISTORY OF EARTH AND VENUS

The coupled system of differential equations for the average mantle temperature, the mantle viscosity, the mantle heat flow, the Rayleigh number, the Urey ratio, and the volatile loss are solved numerically with the help of a fourth order Runge-Kutta scheme [13]. The Urey ratio defines the ratio of the amount of heat generated by radioactive decay to the heat flux at the surface. Using different initial average mantle temperatures between 3000 K and 3600 K we find that all the physical quantities of the thermal history begin to converge rapidly during the early stage of the planetary evolution [7]. This is the effect of the so-called readjustment time [5]. After this time (about 800 Myr) the Earth-like planet has "forgotten" its initial thermal state of the mantle.

Figure 2 shows the thermal and degassing history model for the Earth with an initial average mantle temperature of 3300 K as the so-called standard case. All the other necessary parameters can be found in [7]. After readjustment, the average mantle temperature, the mantle heat flow, and the Rayleigh number decrease monotonically, while the mantle viscosity increases as a result from the combined effects of the decreasing mantle temperature and the volatile loss. The present value of the Urey ratio is about 0.75 and is in the range of 0.65 to 0.85 given by Schubert et al. [2]. The degassing history of the Earth is described by a rapid outgassing event at the beginning of the planetary evolution. The timescale for the outgassing of one ocean mass is about 100 Myr. At present time degassing and regassing are in dynamical equilibrium.

In the case of Venus, there is the possibility that the Venusian mantle convection might have changed from oscillatory to quasi-steady circulation, i. e. Venus changed from an Earth-like planet to a Mars-like planet around 500 Myr ago as far as its tectonic style is concerned [14]. Therefore, in the evolution model of Venus 4 Gyr after the start we set the spreading rate S equal to zero. This takes into account the transition from an Earth-like scenario with crustal recycling to a Mars-like scenario of a one-plate planet without crustal recycling.

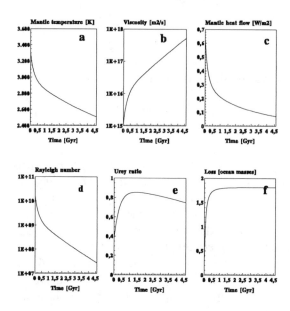

Fig. 2. Thermal and volatile history of the Earth in our standard model [7]. a: average mantle temperature, b: viscosity, c: mantle heat flow, d: Rayleigh number, e: Urey ratio, f: volatile loss

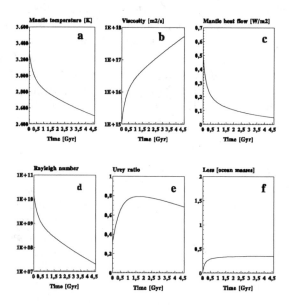

Fig. 3. Thermal and volatile history of Venus with Earth-like plate tectonics ending 4 Gyr after starting the planetary evolution.

In Figure 3 we show the thermal and degassing history of Venus. In our calculation [15] we have used the same values for material constants, rheological and tectonic parameters as for the Earth. Acceleration of gravity, mantle outer and inner radius, mantle mass, and recent mantle heat flow have the well-known values accepted in planetological literature. The main difference in the two models derives from the respective surface temperatures (273 K for Earth, 730 K for Venus) and average depths from which volatiles are released from the mantle (10^5 m for Earth, $5.3 \cdot 10^4$ m for Venus).

We found that the volatile loss is very sensitive against variations of the average depth d_m from which volatiles are released from the mantle. The upper 100 km of Venus are "dried out" [16] with the result of a raised solidus. As well known the surface temperature of Venus is much higher. We conclude that d_m is smaller than in the Earth's case. The value of d_m is adjusted in such a way that only 0.35 ocean masses have been outgassed from the mantle in accordance with the amount of radiogenic noble gases in the atmosphere (see below). Therefore, the recent Venusian mantle is more volatile-rich than the Earth's mantle.

SENSITIVITY TEST OF OUR RHEOLOGICAL MODEL

In chapter 3 we introduced our rheological model for the calculation of the influence of the volatile content on the mantle rheology based on laboratory results. Our main equation is (13) describing the creep rate of wet mantle material as a function of the volatile content x. Testing the sensitivity of our results we varied the constant K (assessed value 2.115) in the range of 0.1 to 100.

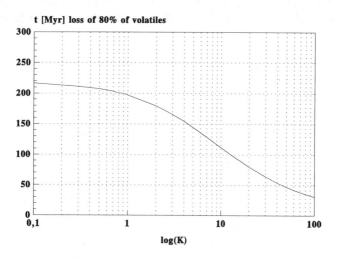

Fig. 4. Dependence of the time t of 80% volatile loss on the value of K.

Figure 4 shows that 80% of the recent surface volatiles can be outgassed in less than 220 Myr if K > 0.1. A variation of ± one order of magnitude $(0.2 \leq K \leq 20)$ leads to a corresponding time t for an early outgassing event between 215 Myr and 65 Myr, respectively. The assessed value of $K = 2.1$ gives a time $t \approx 175$ Myr. Figure 5 presents the corresponding variation of the recent average mantle temperature T_m that is in a range of about 220 K.

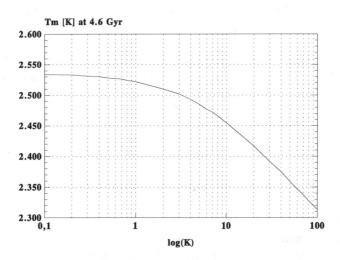

Fig. 5. Dependence of the recent average mantle temperature T_m on the value of K.

The calculated time series' show that with increasing K the influence of volatiles is stronger, the final average mantle temperature lower, and the degassing process more rapid. But nevertheless the results are relatively insensitive to variations of the constant K.

CONCLUSIONS

The description of the influence of dissolved volatiles on the mantle rheology by a nonlinear functional relationship between creep rate and water fugacity results in evolution models for the average mantle temperature, the mantle heat flow, the mantle viscosity, the Urey ratio, the Rayleigh number, and the volatile loss which are in the generally accepted range.

In the case of the Earth we find an extremly rapid outgassing event in the early history which is in agreement with $^{129}Xe/^{130}Xe$ systems which leads to a strong volatile loss in less than about 170 Myr [17].

Even if the tectonic style of the Venusian thermal evolution changed from Earth-like to Mars-like about 500 Myr ago Venus should have much less efficient outgassed. The relative volatile loss between Earth and Venus (i.e. 1.8 ocean masses to 0.35 ocean masses) corresponds to the observed radiogenic noble gases in the atmosphere [18].

$$\frac{volatile\ loss\ \oplus}{volatile\ loss\ ♀} = \frac{^{40}Ar\oplus}{^{40}Ar♀} \approx 5$$

REFERENCES

1. H. N. Sharpe and W. R. Peltier, Geophys. J. R. Astron. Soc. 59, 171 (1979).
2. G. Schubert, P. Cassen, and D. J. Stevenson, J. Geophys. Res. 85, 2531 (1980).
3. D. J. Stevenson, T. Spohn, and G. Schubert, Icarus 54, 466 (1983).
4. M. J. Jackson and H. N. Pollack, J. Geophys. Res. 89, 10103 (1984).
5. U. R. Christensen, J. Geophys. Res. 90, 2995 (1985).
6. P. J. McGovern and G. Schubert, Earth Planet. Sci. Lett. 96, 27 (1989).
7. S. Franck and Ch. Bounama, submitted to Phys. Earth. Planet. Inter. (1994).
8. B. Poumellec and O. Jaoul, Mat. Sci. Res. 18, 281 (1984).
9. S. J. Mackwell, D. L. Kohlstedt, and M. S. Paterson, J. Geophys. Res. 90, 11319 (1985).
10. S.-I- Karato, M. S. Paterson, and J. D. Fitz Gerald, J. Geophys. Res. 91, 8151 (1986).
11. S.-I. Karato and M. Toriumi, Rheology of Solids and of the Earth (Oxford University Press, Oxford, N.Y., Tokyo, 1989), p. 190.
12. S.-I. Karato and P. Wu, Science 260, 771 (1993).
13. W. H. Press, B. P. Flannery, S. A. Teukolsky, and W. T. Vetterling, Numerical Recipes in Pascal (Cambridge University Press, N.Y., 1992), p. 759.
14. J. Arkani-Hamed, J. Geophys. Res. 99 (E1), 2019 (1994).
15. S.Franck and Ch. Bounama, accepted for publication in Adv. Space Res. (1994).
16. D. R. Williams and V. Pan, Geophys. Res. Lett. 17, 1397 (1990).
17. T. Staudacher and C. J. Allègre, Earth Planet. Sci. Lett. 60, 389 (1982).
18. J. B. Pollack and D. C. Black, Icarus 51, 169 (1982).

A UNIFIED MODEL FOR TERRESTRIAL RARE GASES

D. Porcelli and G.J. Wasserburg
The Lunatic Asylum of the Charles Arms Laboratory
Division of Geological and Planetary Sciences
California Institute of Technology, Pasadena, CA 91125

ABSTRACT

A steady state upper mantle model for the rare gases has been constructed which explains the available observational data of mantle He, Ne, Ar, and Xe isotope compositions and provides specific predictions regarding the rare gas isotopic compositions of the lower mantle, subduction of rare gases, and mantle rare gas concentrations. The model incorporates two mantle reservoirs; an undegassed lower mantle (P) and a highly degassed upper mantle (D). Chemical species are transferred into D within mass flows from P at plumes and from the atmosphere by subduction. Rare gases in D are derived from mixing of these inflows with radiogenic nuclides produced *in situ*. The upper mantle is degassed at mid-ocean ridges and hotspots. Flows of each isotope into D are balanced by flows out of D, so that upper mantle concentrations are in steady state. Rare gases with distinct $^3He/^4He$, $^{20}Ne/^{22}Ne$, $^{129}Xe/^{130}Xe$, and $^{136}Xe/^{130}Xe$ are stored in P and are transferred into D. In P, isotopic shifts are due to decay of U- and Th- decay series nuclides, ^{40}K, ^{129}I, and ^{244}Pu over 4.5Ga. Radiogenic ^{136}Xe in P is dominantly from ^{244}Pu. In the well-outgassed D reservoir, additional isotopic shifts are due to decay of U- and Th- series nuclides and ^{40}K over a residence time of ~1.4Ga. Since ^4He, ^{21}Ne, ^{40}Ar, and ^{136}Xe are produced in proportions fixed by nuclear parameters, the resulting isotopic shifts are correlated. The model predicts that the shift in $^{21}Ne/^{22}Ne$ in D relative to that in P is the same as that for $^4He/^3He$ in the respective mantle reservoirs. This is compatible with the available data for MORB and hotspots. The minimum $^{40}Ar/^{36}Ar$, $^{129}Xe/^{130}Xe$, and $^{136}Xe/^{130}Xe$ ratios in P are found to be substantially greater than the atmospheric ratios. The range in Ne, Ar, and Xe isotopes measured in MORB are interpreted as reflecting contamination of mantle rare gases by variable proportions of atmospheric rare gases. Subduction is not significant for He and Ne, but may account for a substantial fraction of Ar and Xe in D. The rare gas relative abundances in P are different than that of the atmosphere and are consistent with possible early solar system reservoirs as found in meteorites. The $^3He/^{22}Ne$ and $^{20}Ne/^{36}Ar$ ratios of P are within the range for meteorites with 'solar' Ne isotope compositions. The $^{130}Xe/^{36}Ar$ ratio of the lower mantle is greater than that of the atmosphere, and may be as high as the ratio found for meteoritic 'planetary' rare gases.

In the model, atmospheric rare gas isotope compositions are distinct from those of the mantle. If the Earth originally had uniform concentrations of rare gases, degassing of the upper mantle would have provided only a small proportion of the nonradiogenic rare gases presently in the atmosphere. The remainder was derived from late-accreted material with higher concentrations of rare gases. However, radiogenic ^{129}Xe and ^{136}Xe abundances imply a substantial loss of rare gases up to ~10^8 years after meteorite formation either from the early Earth or from late-accreting protoplanetary materials. Rare gases must have been lost during accretion and the moon-forming impact, so that nonradiogenic rare gases in the atmosphere must have been supplied by subsequently accreted material with nonradiogenic Xe, possibly from comets. Fractionation of atmospheric Xe isotopes relative to other early solar system components occurred either on late-accreting materials or during loss from the Earth.

INTRODUCTION

Brown[1] originally argued that volatiles at the surface are derived from outgassing of the solid Earth, and subsequent measurements of isotopically distinctive He [2-5], Ne [6-9], and Xe [10,11] in terrestrial materials unambiguously document that nuclides originally trapped or formed early within the Earth continue to outgas. Models of the distribution of volatiles and the processes of outgassing have continued to evolve. Early models for the evolution of Ar isotopes postulated degassing of a single mantle reservoir to produce the atmosphere, with mantle Ar representing the residue of this process [12,13]. Based upon measurements of Ar in hotspot volcanics that were substantially less radiogenic than Ar in mid-ocean ridge basalts, Hart et al. [14] postulated that two mantle reservoirs evolved in an Earth originally uniform in Ar concentration; a primitive undegassed lower mantle sampled at hotspots and an upper mantle sampled by mid-ocean ridge basalts (MORB) which has served as the source of the atmosphere. This scheme followed models of crustal extraction from a limited portion of the mantle based upon Nd and Sr isotope studies [15,16]. Allegre et al. [17,18] have refined and extended the model to include all of the rare gases. The essential features of their model are: a) an Earth initially uniform in rare gas isotope compositions and concentrations; b) an upper mantle that has largely degassed early to form the atmosphere, so that subsequent *in situ* radiogenic production in the upper mantle has produced radiogenic rare gas isotope compositions; c) no other sources or losses of atmospheric rare gases, so that the atmosphere plus upper mantle has concentrations of all isotopes equal to those of the bulk Earth; d) a closed system lower mantle with high $^3He/^4He$ ratios as measured in ocean island basalts (OIB) and with bulk Earth concentrations of other rare gases, including radiogenic isotopes; and e) no subduction of rare gases or interaction between mantle reservoirs. Since the upper mantle is highly outgassed, the atmosphere-upper mantle system has approximately atmospheric rare gas isotope compositions. Measured Ar and Xe from OIB with atmospheric compositions were interpreted as representing the lower mantle reservoir, and confirming atmospheric bulk Earth ratios. Although Staudacher and Allegre [11] originally interpreted the ^{136}Xe excesses measured in MORB as due to the decay of ^{244}Pu, this was subsequently reinterpreted as due to the decay of ^{238}U [18] and the possible role of ^{244}Pu was not incorporated into their model. This model appeared successful at explaining the data available at that time.

As additional data has been acquired, objections to the above model have been raised:

1. Variations in $^{20}Ne/^{22}Ne$ ratios in mantle-derived materials have been found which cannot be ascribed to additions by nuclear processes and require a mantle component with a Ne isotope composition that is fundamentally different from that of the atmosphere[6-9]. The difference between the reservoirs may be due to initial heterogeneity in Ne or Ne losses from the atmosphere involving major fractionation. Both of these scenarios are at odds with an assumed uniform and closed system Earth.

2. The distinctive upper mantle Ne and Xe isotopic ratios must have been established early in Earth history. However, it is difficult to maintain these characteristics in a highly depleted reservoir underlain by an undegassed reservoir. The continuing rise of plumes through the upper mantle would be expected to contaminate the upper mantle signature [19].

3. Accretionary models of planetary formation imply substantial losses of rare gases in the atmosphere, both by 'atmospheric erosion'[20] and by the moon-forming impact[21,22]. In addition, the nonradiogenic Xe isotopes in the atmosphere are highly fractionated with respect to other solar system components, and it has been hypothesized that this is due to hydrodynamic losses from the atmosphere [23,24]. Under these circumstances, the assumption that the atmosphere represents the unaltered complement of the upper mantle is untenable.

4. Interaction between the mantle reservoirs has been postulated for ^3He [25] and for Pb [26]. Transfer of these elements is most plausibly by mass transfer [26,27], so that all of the rare gases must also accompany these elements. Although it was postulated[18] that He diffused alone into the upper mantle, a substantial diffusive flux of He does not appear to be a reasonable mechanism for maintaining the global ^3He flux at mid-ocean ridges.

5. Evidence in support of the residual upper mantle model included the measurement of atmospheric Ar and Xe isotope compositions in ocean island basalts together with high $^3He/^4He$ ratios, which were interpreted as reflecting the lower mantle Ar and Xe isotope composition [17,18]. However, it now appears that the measured ratios reflect atmospheric contamination either during transit through the crust, during eruption, or at the surface[28].

O'Nions and Oxburgh [25] first proposed that He in the upper mantle is open to flows from the underlying mantle, noting that the rate at which radiogenic ^4He is released at mid-ocean ridges is equal to that produced in the upper mantle, and arguing that the accompanying ^3He is derived from the lower mantle. Galer and O'Nions [26] argued that Pb is in a steady state in the upper mantle and is supported by a mass transfer from an undepleted mantle. They further noted that such a flow would also support other highly incompatible elements which are efficiently removed from the upper mantle. Kellogg and Wasserburg [27] presented a detailed analysis of a model for maintaining a steady state He inventory in the upper mantle by mass transfer inflow of lower mantle He with a high ^{3}He/^{4}He ratio at hotspots, radiogenic production of ^{4}He within the upper mantle, and outflows to the atmosphere of He at ridges and hotspots by melting. The lower mantle was assumed to have evolved essentially as a closed system. Porcelli and Wasserburg[29-31] extended the analysis to Xe and incorporated production of Xe by the short-lived nuclides ^{129}I and ^{244}Pu as well as ^{238}U, along with subduction of Xe. These workers showed that in this model upper mantle radiogenic ^{129}Xe from extinct ^{129}I has been stored in the lower mantle, along with excess ^{136}Xe derived from extinct ^{244}Pu. ^{136}Xe was found to be substantially augmented in the upper mantle by production from ^{238}U. Porcelli and Wasserburg [32] extended the model to include all of the rare gases and demonstrated the links between the isotopic shifts of ^{3}He/^{4}He, ^{21}Ne/^{22}Ne, ^{40}Ar/^{36}Ar, and ^{136}Xe/^{130}Xe. Ne with a high ^{20}Ne/^{22}Ne ratio was shown to be stored in the lower mantle.

This model contrasts with previous rare gas models by incorporating interactions between mantle reservoirs. The open system upper mantle requires long-term storage of rare gases in the lower mantle. This model has now been shown to be successful in explaining the available data for He, Ne, Ar, and Xe isotope variations, and provides a unified scheme for the distribution and transfer of rare gases in the Earth. The systematics of the steady state upper mantle model will be reviewed below.

One fundamental difference between the residual upper mantle model and the steady state mantle model is that the former starts with the assumption of a closed system crust-atmosphere that is equivalent to the lower mantle, and this sets the isotopic compositions of the lower mantle. In contrast, the steady state model presented here starts with the assumption that upper mantle isotope compositions are the result of *in situ* radiogenic production mixed with lower mantle rare gases and subducted atmospheric rare gases. The atmosphere is considered to be a separate and isotopically distinct reservoir. The present isotopic compositions of the upper mantle then limits the isotopic compositions and concentrations of the lower mantle as well as the contributions from subduction. The lower mantle concentrations and isotopic compositions calculated for the steady state upper mantle model have specific consequences for terrestrial volatile acquisition, which will be discussed below.

MODEL SYSTEMATICS

The steady state upper mantle model is shown schematically in Figure 1. The model assumes that the Earth initially had uniform concentrations of U, Th, Pu, I, ^{129}I, and K. The Earth is divided into three reservoirs; the 'undegassed' lower mantle (P), the extensively degassed upper mantle (D), and the atmosphere-crust. Each mantle reservoir is of fixed mass and homogeneous in rare gas isotopic composition and in rare gas and parent element concentrations. The lower mantle is assumed to have evolved isotopically approximately as a closed system with *in situ* decay of ^{129}I, ^{244}Pu, ^{238}U, ^{235}U ^{232}Th, and ^{40}K adding to the complement of initial rare gases. The upper mantle is open to interactions with both the atmosphere-crust and the P reservoir. Transfer of rare gases from the P reservoir into D occurs at hotspots, where a flow of material from P rises into D and a fraction degasses directly to the atmosphere, while the remainder is mixed into D. Species flows from P are determined by the concentrations of rare gases in the transferred masses. Rare gases also flow into the upper mantle from the atmosphere-crust by subduction. Degassing of the upper mantle occurs both at mid-ocean ridges and at hotspots, where some proportion of rising material from P entrains upper mantle material into plumes. It is assumed that there is no fractionation between rare gases during inter-mantle transfers or degassing. The concentrations of all the rare gases in the upper mantle are assumed to be in steady state, so that the flows of each species into D are equal to those out of D. The full derivations of the governing equations are given elsewhere [29,30,32].

For each rare gas isotope i in D, the equation relating the flows into and out of D are

$$\frac{d^i N_D}{dt}=\left[\sum_b {}^i P_b M_D {}^b C_D + {}^i C_{SUB}\dot{M}_{SUB} + {}^i C_P \dot{M}_{PD}(1-r)\right]-\left[{}^i C_D \dot{M}_{PD}ra + {}^i C_D \dot{M}_{MOR}\right]=0 \qquad (1)$$

The first three terms on the right hand side of eqn 1 are the inputs into D; ${}^i P_b$ is the production rate of i in D (per atom of parent) from parent nuclide b that has a concentration of ${}^b C_D$ in D, and M_D is the mass of D. ${}^i C_{SUB}$ is the concentration of i in material subducted into D at a rate of $\dot{M}_{SUB}$, ${}^i C_P$ is the concentration of i in P, $\dot{M}_{PD}$ is the total mass flow from P in plumes, and r is the fraction of plume material that is degassed directly to the atmosphere at ridges. The last two terms are outflows from D. The parameter a is the proportion of D material entrained in the plume flow, ${}^i C_D$ is the concentration of i in D and $\dot{M}_{MOR}$ is the rate at which D material is stripped of rare gases at mid-ocean ridges. In the steady-state upper mantle, $d^i N_D/dt=0$.

Equation 1 can be rearranged to

$$\frac{{}^i C_D}{{}^i C_P}=\frac{(1-r)\dot{M}_{PD}}{ra\dot{M}_{PD}+\dot{M}_{MOR}}\left(1+\delta^i_{SUB}+\delta^i_{PR}\right) \qquad (2)$$

where

$$\delta^i_{SUB}=\dot{M}_{SUB}{}^i C_{SUB}\Big/\left[{}^i C_P(1-r)\dot{M}_{PD}\right] \qquad (3)$$

is the fractional increase in ${}^i C_D$ due to the flow of i from subduction relative to the flow of i from P and so reflects the importance of subduction for isotope i. In addition,

$$\delta^i_{PR}=\sum_b {}^i P_b M_D {}^b C_D\Big/\left({}^i C_P(1-r)\dot{M}_{PD}\right) \qquad (4)$$

is the fractional increase in ${}^i C_D$ due to production by nuclear reactions in D relative to the flow of i from P. It is this term that governs the extent to which upper mantle compositions will be more radiogenic than those of the lower mantle. For the isotopes that have no radiogenic or nuclear contributions, i.e. ^{3}He, ^{20}Ne, ^{22}Ne, ^{36}Ar, and ^{130}Xe, then ${}^i P_b=0$ and $\delta^i_{PR}=0$. The term $\chi\equiv(1-r)\dot{M}_{PD}\big/\left[ra\dot{M}_{PD}+\dot{M}_{MOR}\right]$ is the ratio of the rate of mass flow into D to that of the mass flow from D that is outgassed, and is common to all expressions of ${}^i C_D/{}^i C_P$. Equation 2 is the key for the distribution of all rare gas isotopes, and the distribution of isotopic ratios is obtained by combining eqn 2 for each isotope. Note that the volume of ridge volcanism is much greater than that of hotspots, so that $\chi \approx (1-r)\dot{M}_{PD}\big/\dot{M}_{MOR}$.

The links between production of the isotopes of He, Ne, Ar, and Xe are the basis

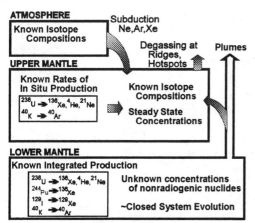

Figure 1. The steady state upper mantle model, with flows shown. See text for explanation.

for a unified model of rare gas isotope distributions. The relative production rates of the radiogenic rare gases are determined by the relative proportions of parent nuclides and known or estimated nuclear parameters. ^{4}He is produced by the decay of U- and Th- series nuclides as α particles. The Th/U ratio is well known for global reservoirs from Pb and Th isotope studies. ^{21}Ne is produced by the reaction ^{18}O$(\alpha,n)^{21}$Ne and to a lesser degree by the reaction ^{24}Mg$(n,\alpha)^{21}$Ne, and the relative production of ^{21}Ne to ^{4}He, $q = {}^{21}P/{}^4P$, is obtained from the estimated yields of these reactions. The parameter q is a constant based upon nuclear parameters and the bulk mantle composition. The relative production of ^{4}He to ^{40}Ar is determined by nuclear parameters and the mantle U/K ratio, which is well known. The production rate of ^{136}Xe relative to ^{4}He is determined by the fission yield of ^{136}Xe per decay of ^{238}U and ^{244}Pu, and is fixed by well-constrained nuclear parameters. The production rate of ^{136}Xe relative to ^{4}He is determined by the fission yield of ^{136}Xe per decay of ^{238}U and ^{244}Pu, and is fixed by well-constrained nuclear parameters. Since all δ^i_{PR} are proportional to nuclear production rates and parent element abundances, the isotopic shifts in all of the rare gases due to radiogenic production in D are correlated, although actual isotopic compositions also reflect the effects of subduction.

The total isotopic shifts in each rare gas isotope composition due to production within each mantle reservoir are proportional to the production rate of the radiogenic isotope and the rare gas residence time (i.e. the time of production). The isotopic shifts are inversely proportional to the abundance of the primordial isotopes. Therefore, isotopic shifts in each reservoir can be combined with known production rates to constrain the nonradiogenic rare gas relative abundances in D and P.

The concentrations of radiogenic (and nucleogenic) rare gas nuclides produced in 4.5Ga in P are listed in Table 1 for U=5.3x10^{13} atoms/g (21ppb) [33], Th/U=3.9, and (K/U)$_{wt.}$=1.27x10^4 (ref. 34). The abundance of Pu associated with terrestrial U at the time of meteorite formation is obtained from the ratio of (Pu/U)$_o$=0.0068 obtained from meteorite data[35]. Early losses of radiogenic Xe isotopes may have occurred[36,37]. Therefore, of the total fissiogenic ^{136}Xe produced by this Pu inventory, only a fraction (^{244}F) may have been accumulated by the Earth. The calculations presented here use ^{244}F=0.5. Note that production of ^{136}Xe in P is dominated by decay of ^{244}Pu (Table 1). The present production rates in D are also listed for U=8.0x10^{12} atoms/g, Th/U=2.6, and (K/U)$_{wt.}$=1.27x10^4.

In the lower mantle, the concentrations of nonradiogenic nuclides ^nC_P and radiogenic nuclides $^{i*}C_P$ are related to the isotopic shift $\Delta^{i/n}_P$ from the initial isotope ratio to the present ratio by

$$\Delta^{i/n}_P \equiv \left({}^iC_P/{}^nC_P\right) - \left({}^{io}C_P/{}^nC_P\right) = {}^{i*}C_P/{}^nC_P \tag{5}$$

where $^{io}C_P$ is the initial concentration of i. This is the key equation for calculating nonradiogenic isotope concentrations in P (^nC_P) from specification of the initial isotope composition, $^{i*}C_P$ (Table 1), and the present isotope ratio determined from the mixing relations in D.

The residence time for all of the rare gases in D is [27]

$$\tau_i = \frac{{}^iC_D M_D}{{}^iC_D(\dot{M}_{MOR} + ra\dot{M}_{PD})} \approx \frac{M_D}{\dot{M}_{MOR}} = 1.4\,\text{Ga} \tag{6}$$

TABLE 1

Nuclide	Present Production Rate in D	Total Production in P over 4.5Ga
^{4*}He	1.7x10^4 atoms/g-yr	1.02x10^{15} atoms/g
^{21*}Ne	1.5x10^{-3} atoms/g-yr	9.2x10^7 atoms/g
^{40*}Ar	4.5x10^3 atoms/g-yr	5.7x10^{14} atoms/g
^{136*}Xe	4.30x10^{-5} atoms/g-yr	2.5x10^7 atoms/g (from ^{244}Pu)
		1.9x10^6 atoms/g (from ^{238}U)

where $M_D=1\times10^{27}$g and $\dot{M}_{MOR}=7\times10^{17}$g/yr.

RARE GAS ISOTOPE COMPOSITIONS AND ABUNDANCES

The mechanics of the model are summarized in Figure 1. The balance of flows into and out of D is represented in eqns. 1-4. The key isotopic relationships (obtained by combining the equations for each isotope) are the mixing of inputs into D to produce the present isotopic compositions in D, which are constrained by the available MORB data. The rates of *in situ* production are known (Table 1). The flow of rare gases from the atmosphere in subducted materials of known isotope composition but unknown concentrations, must be balanced by the flow of rare gases from the lower mantle, with generally unknown isotope compositions but known concentrations of radiogenic nuclides. The isotopic compositions in the lower mantle and the concentrations of nonradiogenic nuclides there are coupled by the closed system evolution equation (eqn. 5). As shown below, the simplest case is for He, where it is argued that there is no subduction flux and the isotopic composition of both mantle reservoirs is well constrained. A flux of ^{4}He from P to D can be calculated, and this corresponds to a flux of material that excapes degassing at hotspots and mixes into D. By coupling the He fluxes with those of Ne, the isotopic distribution of the Ne isotopes are derived. For Ar and Xe, we argue subduction may be significant and there are no constraints available for the isotopic compositions of P. However, it will be shown that lower limits can be placed upon the lower mantle ^{40}Ar/^{36}Ar, ^{129}Xe/^{130}Xe, and ^{136}Xe/^{130}Xe ratios, and so upon the concentrations of Ar and Xe.

The model results for He were explored in detail by Kellogg and Wasserburg [27]. The isotopic composition of He in both mantle reservoirs can be estimated from available data. The (^{3}He/^{4}He)$_D$ ratio is taken from data for mid-ocean ridge basalts (MORB), which have a relatively uniform and well-constrained value of ^{3}He/^{4}He=1.1 x 10^{-5} (ref. 5). The value for P is taken from ocean island volcanics which have ^{3}He/^{4}He ratios that are greater than those in MORB and are generally considered to reflect contributions of He from a less degassed lower mantle[17,38]. The highest measured ratio of ^{3}He/^{4}He=4.5x10^{-5} from the Loihi hotspot [39] is taken as the value for P. An initial composition of ^{3}He/^{4}He=1.43x10^{-4} is taken from the 'planetary' component of meteorites[40]. The concentrations of He isotopes in P can be obtained immediately from eqn. 5 (Table 2) and are fixed.

From the He isotopic compositions of the mantle reservoirs and the rates of ^{4}He production, the undegassed mass flux from P to D can be calculated. From eqn. 2 for ^{3}He and ^{4}He (with $\delta^3_{SUB}=\delta^4_{SUB}=0$),

$$\left(^4\text{He}/^3\text{He}\right)_D = \left(^4\text{He}/^3\text{He}\right)_P\left(1+\delta^4_{PR}\right) \tag{7}$$

Using the above values for the He isotope ratios, $\delta^4_{PR}=3.1$. This value can be used with eqn. 4 to determine the undegassed mass flow from P, so that $(1-r)\dot{M}_{PD}=3.7\times10^{15}$g/yr. Note that since this is substantially less than geophysical estimates of $\dot{M}_{PD}=(2-20)\times10^{16}$g/yr, the fraction of rising plume material that degasses directly to the atmosphere must be large (r>0.84) [27]. The mass flow of non-degassed P material into D governs the transfer of all rare gas nuclides from P into D. Also, a value for $\chi\approx4.6\times10^{-3}$ is obtained and can be used with eqn 2 to obtain upper mantle rare gas concentrations.

Ne with ^{20}Ne/^{22}Ne and ^{21}Ne/^{22}Ne ratios greater than the atmospheric values of 9.8 and 0.029, respectively, have been reported for Kilauea volcanic gases [6], ultramafic xenoliths [41], and MORB [6,7,8,41,42]. ^{20}Ne/^{22}Ne and ^{21}Ne/^{22}Ne ratios in MORB are correlated (Figure 2). We interpret this as reflecting mixing between atmospheric Ne contamination and mantle Ne which has ^{20}Ne/^{22}Ne≥12.5 and ^{21}Ne/^{22}Ne≥0.065, following previous workers [43,44]. Ne measured in hotspots span a similar range in ^{20}Ne/^{22}Ne ratios as MORB but with corresponding ^{21}Ne/^{22}Ne ratios that are substantially lower [8,45-48] (see Fig. 2) and this range is also considered to be due to the effects of atmospheric contamination.

Solar system compositions with ^{20}Ne/^{22}Ne ratios greater than that of the atmosphere are shown in Fig. 2. 'Solar' Ne as measured in trapped solar wind[50,52], lunar soils[51], and gas-rich meteorites[49] have ratios of $^{21}Ne/^{22}Ne=0.032$, and this is taken here as the value for Ne initially trapped in the Earth. The measured $^{20}Ne/^{22}Ne$ ratios in these compositions are substantially different. The ratio for gas-rich meteorites of $^{20}Ne/^{22}Ne=12.5$ is similar to the value for D' taken from MORB analyses (Fig. 2). If this value pertains to Ne trapped within the Earth and so to Ne in P, then subduction of atmospheric Ne has not significantly changed the $^{20}Ne/^{22}Ne$ ratio of Ne in D from that in P. Therefore, $\delta_{SUB}^{22} \approx 0$. If the ratio for trapped solar wind of $^{20}Ne/^{22}Ne=13.6$ pertains to Ne trapped in P, then the minimum value for D is lower than that of P, and $\delta_{SUB}^{22} \leq 0.4$. In this case, Ne in D is still dominantly from P. For simplicity, we will use $\delta_{SUB}^{22} =0$ below.

The systematics of Ne are analogous to He, and

$$\left(^{21}Ne/^{22}Ne\right)_D = \left(^{21}Ne/^{22}Ne\right)_P\left(1+\delta_{PR}^{21}\right) \tag{8}$$

We define $\Delta_D^{21/22}$ as the isotopic shift in the $^{21}Ne/^{22}Ne$ ratio due to radiogenic addition of ^{21}Ne in D to Ne derived from P, so that $\Delta_D^{21/22} \equiv (^{21}Ne/^{22}Ne)_D - (^{21}Ne/^{22}Ne)_P$. Similarly, we define $\Delta_D^{4/3} \equiv (^{4}He/^{3}He)_D - (^{4}He/^{3}He)_P$. Combining eqns 7 and 8,

$$\left(\frac{\Delta_D^{21/22}}{\Delta_P^{21/22}}\right)=\left(\frac{\Delta_D^{4/3}}{\Delta_P^{4/3}}\right) \tag{9}$$

It follows that the ratio of the isotopic shift in D (the increase in $^{21}Ne/^{22}Ne$ in D relative to that in P), to the isotopic shift in P (the increase in $^{21}Ne/^{22}Ne$ in P relative to the initial value) is the same as the ratio of the corresponding shifts in $^{4}He/^{3}He$. Using the values for the isotopic compositions of He

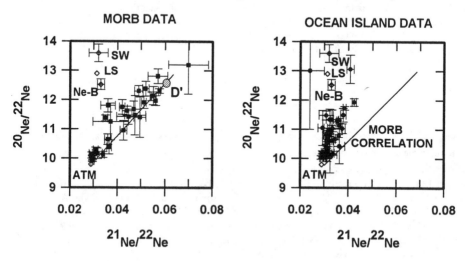

Figure 2. Ne isotope data for mid-ocean ridge basalts[6,8,41,42] and ocean islands (hotspots)[8,45-48]. Note the separate correlations of MORB and ocean island data extending from atmospheric values, reflecting variable atmospheric contamination to mantle Ne. The 'solar' values are from gas-rich meteorites (Ne-B)[49], lunar soils (LS)[51], and solar wind (SW)[50,52]. D' is the minimum value for D.

discussed above, the right hand side of eqn 9 is equal to 4.4. Therefore, the shift in ^{21}Ne/^{22}Ne generated in the upper mantle by the ^{18}O(α,n)^{21}Ne and ^{24}Mg(n,α)^{21}Ne reactions is 4.4 times larger than those generated by the same processes in the lower mantle. This is due to the extensive outgassing of D and the substantial concentration of U in D relative to the rare gases. If $(^{21}$Ne/^{22}Ne)$_D$ = 0.065 and $(^{21}$Ne/^{22}Ne)$_O$ = 0.032, then $(^{21}$Ne/^{22}Ne)$_P$ = 0.038 and $\Delta_P^{21/22}$ = 0.006. This value for $(^{21}$Ne/^{22}Ne)$_P$ is compatible with the ocean island basalt data (Figure 2). The concentrations of ^{22}Ne and ^{20}Ne in P and D can be calculated from eqns 2 and 5 (Table 2) and are fixed by the value for $\Delta_P^{21/22}$.

Measured ^{40}Ar/^{36}Ar ratios in MORB vary widely due to mixing of mantle Ar with variable proportions of atmospheric Ar contamination [17,53-55]. The highest value of 2.8x10^4 (ref. 55) is the minimum value for the MORB source and for computational purposes is taken here as the value for D. Hotspot volcanics with high ^{3}He/^{4}He ratios typically have measured ^{40}Ar/^{36}Ar ratios similar to atmospheric Ar [e.g. 17,47]. It has been argued [28] that Ar in these samples is highly contaminated with atmospheric Ar and does not reflect mantle compositions. Therefore, the ^{40}Ar/^{36}Ar value of P is considered unknown. The initial ^{40}Ar/^{36}Ar ratio in the solar system is ~10^{-3} (ref. 56); essentially all terrestrial ^{40}Ar is radiogenic.

Argon systematics are coupled to those of He by nuclear parameters. It has been argued that K and U have not been fractionated in global differentiation processes and so in the model K/U is the same in D and P. The difference between the ^{4*}He/^{40*}Ar production ratio in D (integrated over the last 1.4Ga) and ^{4*}He/^{40*}Ar in P (integrated over 4.5Ga) is due to the different half-lives of ^{238}U and ^{40}K. Using eqn 4 for δ_{PR}^{40} and for δ_{PR}^{4} (with a value calculated using the chosen He isotope compositions) so that the mass terms and parent element concentrations cancel, a value of δ_{PR}^{40}=1.9 is obtained. This indicates that the radiogenic production of ^{40}Ar in the upper mantle is 1.9 times greater than the flow of radiogenic ^{40}Ar from P into D.

The isotopic compositions of Ar in D and P are related by:

$$\left(^{40}Ar/^{36}Ar\right)_P = \left(^{40}Ar/^{36}Ar\right)_D \frac{1+\delta_{SUB}^{36}}{1+\delta_{PR}^{40}} = 9700(1+\delta_{SUB}^{36}) \qquad (10)$$

$(^{40}$Ar/^{36}Ar)$_P$ is much greater than $(^{40}$Ar/^{36}Ar)$_{ATM}$, with a minimum value of 9700 for δ_{SUB}^{36} = 0.

TABLE 2

	D	P	Initial	Atmosphere
^3C_i (atoms/g)	3.1 x 10^8	6.8 x 10^{10}	6.7 x 10^{10}	--
$^{20}C_i$ (atoms/g)	8.6 x 10^8	1.9 x 10^{11}	1.9 x 10^{11}	1.75 x 10^{12}
$^{36}C_i$ (atoms/g)	2.7 x 10^8	≤5.9 x 10^{10}	≤5.9 x 10^{10}	3.35 x 10^{12}
$^{130}C_i$ (atoms/g)	~6 x 10^5	≤1.1 x 10^8	≤ 9.8 x 10^7	3.77 x 10^8
^{3}He/^{4}He	1.1 x 10^{-5}	4.5 x 10^{-5}	1.43 x 10^{-4}	1.4 x 10^{-6}
^{20}Ne/^{22}Ne	12.5	12.5	12.5	9.8
^{21}Ne/^{22}Ne	0.065	0.038	0.032	0.029
^{40}Ar/^{36}Ar	2.8 x 10^4	≥ 9700	10^{-3}	296
^{136}Xe/^{130}Xe	2.45	≥ 2.34	2.11	2.176
^{129}Xe/^{130}Xe	7.35	≥ 7.35	6.05	6.496

Concentrations in D and P are calculated. Isotope ratios in D are inferred from MORB data;.those in P are calculated, except for He, which is inferred from OIB data. Initial isotope ratios are from meteorite data, and initial concentrations are equal to those of P. Concentrations for the atmosphere are atmospheric abundances divided by mass of the upper mantle (1x10^{27}g). See text for references.

$(^{40}Ar/^{36}Ar)_P$ is larger if a substantial proportion of Ar in D is derived from subduction. The concentration of ^{36}Ar in P can be readily obtained from the $(^{40}Ar/^{36}Ar)_P$ ratio (eqn 5). Note that since the amount of radiogenic ^{40}Ar in P is fixed, greater values of $(^{40}Ar/^{36}Ar)_P$ correspond to smaller concentrations of ^{36}Ar in P.

Using the $(^{40}Ar/^{36}Ar)_P$ ratio calculated by eqn. 10 as a function of δ_{SUB}^{36}, the concentration of ^{36}Ar in the lower mantle can be obtained from eqn 3. For the case of no subduction of Ar, $^{36}C_P = 5.9 \times 10^{10}$ atoms/g. Greater proportions of subducted Ar in D($\delta_{SUB}^{36} > 0$) gives greater values of $(^{40}Ar/^{36}Ar)_P$, so that the value of $^{36}C_P$ decreases. The concentration of ^{36}Ar in D is obtained by combining eqn.2 for ^{36}Ar and eqn. 10 (see ref. 30 for derivation);

$$^{36}C_D = \left(\frac{(1-r)\dot{M}_{PD}}{ra\dot{M}_{PD} + \dot{M}_{MOR}} \right) \left(\frac{^{40*}C_P \left(1 + \delta_{PR}^{40}\right)}{\left(^{40}Ar/^{36}Ar\right)_D} \right) \tag{11}$$

Note that the calculated value of $^{36}C_D$ is independent of δ_{SUB}^{36}. Using the values for each parameter as discussed above, $^{36}C_D = 2.7 \times 10^8$ atoms/g.

No isotopic variations have been observed or are expected between Kr in the various reservoirs, and so Kr will not be discussed further. Measured MORB $^{129}Xe/^{130}Xe$ and $^{136}Xe/^{130}Xe$ ratios [11,17,42,45,55] (Figure 3) are correlated [11] and extend from the composition of the atmosphere. This is interpreted here as reflecting mixing of atmospheric contaminant Xe with a mantle component. The highest ratios of $^{129}Xe/^{130}Xe = 7.35$ and $^{136}Xe/^{130}Xe = 2.45^{ref.55}$ are lower bounds (D') for the mantle component, and for computational purposes are taken here as the values for D. Xe from hotspot volcanics have atmospheric isotope compositions [e.g.17]; this has been interpreted as reflecting atmospheric contaminant Xe [28]. Therefore, the isotopic composition of Xe in P is considered unknown. Identifying an initial Xe isotope composition for the Earth is difficult due to the complexity of solar system components [37], although the specific value chosen is not critical to the model results[29].

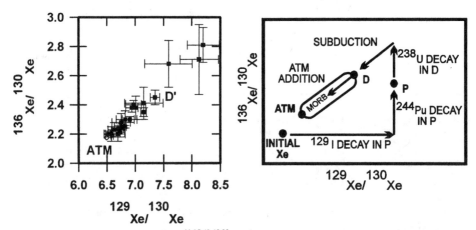

Figure 3. Xe isotope ratios for MORB [11,17,42,45,55]. The observed correlation [11] is interpreted as reflecting variable proportions of atmospheric contamination to a MORB Xe composition . The highest precise MORB analysis, D', is the minimum composition of D.

Figure 4. The general pattern of Xe evolution. ^{129}Xe and ^{136}Xe are increased in P by decay of ^{129}I and ^{244}Pu, producing the P composition. Xe transferred into D is mixed with ^{136}Xe from decay of U in D and with subducted Xe, resulting in the D composition. Varying degrees of contamination produces the MORB array.

In this study, we will use the most recent estimates for nonradiogenic, atmospheric Xe of $(^{136}Xe/^{130}Xe)_O=2.11$ and $(^{129}Xe/^{130}Xe)_O=6.05$ [37,57]. Measurements of Xe in CO_2 well gas samples suggest that a different, 'solar' nonradiogenic Xe composition may be present within the Earth[58]. The use of other nonradiogenic compositions for the lower mantle will not qualitatively change the conclusions drawn here (see ref. 29 for discussion).

A value for $\delta_{PR}^{136*}=0.46$ is obtained using eqn 4 for both δ_{PR}^{136*} and δ_{PR}^4, so that the mass terms and parent element concentrations cancel. This is the input of the radiogenic ^{136}Xe into the upper mantle produced by ^{238}U decay in D relative to the flow of radiogenic ^{136}Xe from P, which was produced by ^{244}Pu decay. The ^{136}Xe excess in D therefore is composed of plutogenic ^{136}Xe from P substantially augmented by uranogenic ^{136}Xe produced in D.

The isotopic compositions of D and P are related by:

$$\left(^{129}Xe/^{130}Xe\right)_P = \left(^{129}Xe/^{130}Xe\right)_D\left(1+\delta_{SUB}^{130}\right)-\left(^{129}Xe/^{130}Xe\right)_{ATM}\delta_{SUB}^{130} \qquad (12)$$

$$\left(^{136}Xe/^{130}Xe\right)_P = \frac{\left(^{136}Xe/^{130}Xe\right)_D\left(1+\delta_{SUB}^{130}\right)-\left(^{136}Xe/^{130}Xe\right)_{ATM}\delta_{SUB}^{130}}{1+\delta_{PR}^{136}} \qquad (13)$$

Equation 12 is a simple mixing relationship, where the fraction of ^{130}Xe in D from subduction is $X_{130} \equiv \delta_{SUB}^{130}/\left(1+\delta_{SUB}^{130}\right)$. Note that subduction of Xe lowers the $^{129}Xe/^{130}Xe$ ratio of Xe provided from P, so that $(^{129}Xe/^{130}Xe)_P \geq (^{129}Xe/^{130}Xe)_D$ and the ^{129}Xe excesses in the mantle must be stored in P. Similarly, equation 13 can be viewed as a mixing relation between Xe from P and subduction, with the addition of radiogenic ^{136}Xe produced in D. For greater values of δ_{SUB}^{130}, both $(^{129}Xe/^{130}Xe)_P$ and $(^{136}Xe/^{130}Xe)_P$ are greater and $^{130}C_P$ is smaller (eqn 3). The minimum values for the isotopic composition of P, and the maximum concentration of ^{130}Xe in P, are listed in Table 2. It has been shown earlier [29] that the very limited data available for subducting materials do not rule out derivation of a substantial proportion of upper mantle ^{130}Xe by subduction.

The concentration of radiogenic ^{129}Xe in P can be calculated from eqn 5 using values for $(^{129}Xe/^{130}Xe)_P$ calculated from eqn. 12. It can be shown that for all values of δ_{SUB}^{130}, $^{129*}C_P/^{136*}C_P \approx 3.8$ (ref. 29). This is 10^4 times less than the solar system value of $\sim 2 \times 10^4$ (see ref. 29), indicating that there has been substantial depletion in ^{129}I in the Earth prior to retention of radiogenic Xe.

The general pattern of mantle Xe isotopic evolution is shown in Figure 4. Starting with the initial Xe isotope composition, the $^{129}Xe/^{130}Xe$ and $^{136}Xe/^{130}Xe$ ratios in P are increased by the decay of ^{129}I and ^{244}Pu, respectively, to produce the present P composition. Small amounts of Xe are transferred into D, where due to a higher $^{238}U/^{136}Xe$ ratio the $^{136}Xe/^{130}Xe$ ratio is increased by spontaneous fission production of ^{136}Xe during the residence time of Xe in D. This Xe is also mixed with subducted atmospheric Xe, resulting in composition D. The observed range in MORB samples is attributed to variable proportions of atmospheric contamination added in the crust or at the surface.

MANTLE RARE GAS RELATIVE ABUNDANCES

From the concentrations in Table 2, $^3C_P/^{20}C_P = 0.36$. This is comparable with a solar wind value of $0.26^{ref.52}$ and an order of magnitude greater than the value of 0.031 for the 'planetary' component of meteorites [59]. This similarity of the light rare gas relative abundances in P with solar-type rare gases is compatible with Ne in the lower mantle having a 'solar' isotopic composition. Note that Honda et al.[43] and O'Nions and Tolstikhin[60] have also argued that the lower mantle has a solar $^3He/^{20}Ne$ ratio.

The lower mantle ^{20}Ne concentration is fixed, and the calculated lower mantle ^{36}Ar concentration is inversely proportional to $(1+\delta_{SUB}^{36})$, so that

$$^{20}C_P/^{36}C_P = 3.2(1+\delta^{36}_{SUB}). \tag{14}$$

The minimum value for $^{20}C_P/^{36}C_P$ is 3.2, which is much greater than the the atmospheric value of 0.52 and is within the range of meteorite compositions with 'solar' Ne isotope compositions (see ref. 32). The $^{20}C_P/^{36}C_P$ ratio is equal to the 'solar' ratio of 27^{56} for $\delta^{36}_{SUB}=7.4$. This corresponds to $X_{36}=\delta^{36}_{SUB}/(1+\delta^{36}_{SUB}) \approx 0.9$, where X_{36} is the fraction of ^{36}Ar in the upper mantle that is derived from subduction.

The calculated lower mantle ^{36}Ar/^{130}Xe ratio is dependent upon δ^{130}_{SUB} and δ^{36}_{SUB}, neither of which can be independently constrained. An equation for $^{36}C_P/^{130}C_P$ can be obtained from eqn. 2 for ^{36}Ar and ^{130}Xe (see ref. 32 for derivation), so that

$$\frac{^{130}C_P}{^{36}C_P} = \left[\delta^{130}_{SUB}\left(430 - {}^{36}C_{SUB}/{}^{130}C_{SUB}\right) + 530\right]^{-1} \tag{15}$$

In the case of no subduction ($\delta^{130}_{SUB}=0$), $^{130}C_P/^{36}C_P=1.9\times10^{-3}$. This value is similar to the meteoritic 'planetary' ratio of 1.56×10^{-3} (ref. 59) and substantially greater than either the atmospheric ratio of 1.12×10^{-4} or the 'solar' value of 1.67×10^{-6} (ref. 51). For the calculated $^{130}C_P/^{36}C_P$ to be much less than 1.9×10^{-3}, two conditions must be met in eqn. 15. Firstly, the ratio of $^{36}C_{SUB}/^{130}C_{SUB}$ must be less than 430, which requires that $^{36}C_{SUB}/^{130}C_{SUB} <0.05(^{36}$Ar/^{130}Xe)$_{ATM}$, so that atmospheric Xe in subducted materials must be substantially enriched with respect to atmospheric Ar. This is not unreasonable. Secondly, δ^{130}_{SUB} must be large. In figure 5, $^{130}C_P/^{36}C_P$ is plotted against X_{130} (the fraction of ^{130}Xe in D that is derived from subduction) where $X_{36}=\delta^{36}_{SUB}/(1+\delta^{36}_{SUB})$, for various values of $^{36}C_{SUB}/^{130}C_{SUB}$ normalized to the atmospheric ratio. For the calculated ratio of ^{130}Xe/^{36}Ar in the lower mantle to be as low as that of the atmosphere requires $X_{130}>0.95$, regardless of the ratio of Xe/Ar in subducted materials. This is extreme, and it appears that the ^{130}Xe/^{36}Ar ratio of the lower mantle is higher than the atmospheric ratio, and possibly as high as the chondritic ratio.

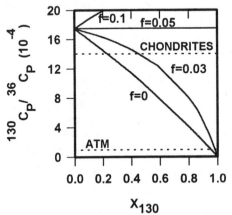

The rare gas abundance pattern in D is more constrained than that of the lower mantle, since the ^{36}Ar calculated concentration in D is not dependent upon the fraction of upper mantle Ar derived from subduction (eqn 11). The calculated ^{130}Xe concentration in D changes by a maximum of 10% for different values of X_{130} [29]. It follows that $^{20}C_D/^{36}C_D=3.2$ and $^{130}C_D/^{36}C_D\approx1.7\times10^{-3}$. These values are both an order of magnitude greater than the corresponding ratios for the atmosphere of 0.52 and 1.1×10^{-4}, respectively. This pattern of Ne and Xe enrichment over Ar with respect to the atmospheric rare gas relative abundances is consistent with measured MORB abundance patterns [18]. Since the upper mantle

Figure 5. The calculated ^{130}Xe/^{36}Ar ratio in P plotted against X_{130}, the fraction of ^{130}Xe in D derived from subduction. Curves are shown for different efficiencies of subduction of Ar relative to Xe, with f=$(^{36}$Ar/^{130}Xe)$_{SUB}/(^{36}$Ar/^{130}Xe)$_{ATM}$. The ^{130}Xe/^{36}Ar ratio calculated for P is similar to the CI chondrite value for $X_{130}<0.5$, and approaches the atmospheric value only for $X_{130}>0.95$.

rare gas abundance pattern is not sensitive to the amount of Ar and Xe subduction, it cannot be used to infer the amount of rare gas subduction or further constrain the lower mantle abundance pattern.

FORMATION OF THE ATMOSPHERE

The model treats the atmosphere as an independent reservoir with rare gas isotopes that are distinct from both mantle reservoirs. This is a direct consequence of the observation of distinct $^{20}Ne/^{22}Ne$ ratios in the mantle, and storage of ^{129}Xe and ^{136}Xe excesses in the lower mantle. Since the model begins with observations of the upper mantle and interaction between D and P, there are no initial assumptions about the atmosphere in relation to the solid Earth. However, from consideration of atmospheric and mantle rare gas systematics, requirements for both rare gas loss from the Earth and rare gas acquisition can be obtained. The correlation of $^{136}Xe/^{130}Xe$ with $^{129}Xe/^{130}Xe$ in MORB samples [11] indicates that the $^{129}I/^{244}Pu$ ratio of the lower mantle at the time of its formation was $\sim 10^{-4}$ of the early solar system value[29]. An I depletion of 10^{-2} has been inferred from degree of depletion of elements with volatile affinities[61], so that the additional depletion of 10^{-2} reflects early losses of daughter ^{129}Xe over $\sim$6 half-lives of ^{129}Xe, or $\sim 10^8 yr$. This implies a I-Xe 'age' of the Earth of $\sim 10^8$ years younger than that of meteorite parent bodies of 4.55Ga. It has been argued that the abundance of atmospheric radiogenic Xe implies a similar young age[36,37]. Such losses up to 4.45Ga may have occurred from the fully-formed Earth or from proto-planetary materials of a late-forming Earth.

Losses of radiogenic Xe would necessarily be accompanied by losses of nonradiogenic rare gases. Losses may have occurred from the proto-Earth atmosphere during accretion [20] and during impact of a Mars-size body to produce the moon[21,22]. The coincidence of the I-Xe timescale with the young ^{207}Pb-^{206}Pb age of the Earth and moon[62-64] suggests that Pb/U fractionation may have occurred during these events.

Most of the nonradiogenic Xe in the atmosphere must be derived from late-accreted material that is more Xe-rich than the lower mantle to generate an atmospheric $^{136}Xe/^{130}Xe$ ratio that is lower than that of the mantle[29]. The calculated concentrations of rare gases in the lower mantle are fully consistent with the hypothesis that a large proportion of rare gases presently in the atmosphere must be from late accreted material. If the Earth originally had a uniform concentration of rare gases, then degassing of the upper mantle would have provided only a small fraction of the atmosphere inventory. The remainder of the atmospheric nonradiogenic rare gases must be derived from material with a higher concentration of rare gases than the lower mantle. The source of the atmosphere cannot be from the late addition of CI chondrite material, which has Xe that is too radiogenic. A source rich in Xe relative to I is required, and our preferred source is cometary material (see also ref. 65 and 66). This material must arrive after catastrophic loss of rare gases from the Earth's atmosphere during the moon-forming impact.

The Xe in the earth's atmosphere has been isotopically fractionated with regard to any anticipated source. It is possible that this fractionation resulted from losses on the late-accreting bodies[67] or by hydrodynamic loss from the Earth's atmosphere after accretion[23,24]. This process may also be responsible for the lower atmospheric Xe/Ar ratio than that of the solid Earth (see ref. 24). Such losses which occurred from the Earth to generate the present isotopic composition must have followed late accretion of rare gases. Fractionating loss processes may also be responsible for the lower atmospheric $^{20}Ne/^{22}Ne$ ratio than the mantle, although this can also be explained by late accretion of Ne with a lower average $^{20}Ne/^{22}Ne$ ratio. The possible relationships between the inferred loss of the "original" volatile inventory along with radiogenic Xe and the inferred loss responsible for the present observed isotopic fractionation and abundance patterns in the atmosphere are open questions.

The above conclusions regarding the evolution of the terrestrial atmosphere are compatible with current understanding of planetary processes[68]. Comparisons can also be drawn with the rare gas systematics of Mars. It has been argued that Mars has lost a substantial proportion of its volatile inventory based upon such factors as geomorphological evidence for abundant water that is no longer observed[69] and the low abundances of rare gases and other volatiles measured in the martian atmosphere by Viking[70,71]. Evidence of SNC meteorites, which have been attributed to Mars, also support the idea of Mars as a planet that accreted rich in volatiles, but which has suffered substantial atmospheric loss[72]. Pepin[24] and Zahnle[73] have reviewed the martian data and possible evolutionary processes. Measurements from the SNC meteorites have obtained precise determinations of the isotopic compositions of martian rare gases [e.g.74]. Xe isotopes are fractionated with respect to solar or CI chondrite Xe compositions, and this has been attributed to fractionation during loss from the atmosphere by hydrodynamic escape[24]. These losses have resulted in a decoupling of the current atmosphere from the interior.

In summary, early losses of gases have occurred from both planets. The Earth appears to show late accretion of gas-rich material that has supplied volatiles to the terrestrial atmosphere. However, both planets have atmospheres that evolved by large losses that have caused isotopic and elemental fractionations, as well as by additions of radiogenic nuclides from the interior.

ACKNOWLEDGMENTS

This work was supported by NASA NAGW3337 and DOE DE-FG-88ER13851. This is Division Contribution 5477(874). Reviews by M. Honda and D. Patterson are particularly appreciated.

REFERENCES

1. H. Brown, in The Atmospheres of the Earth and Planets (ed. G. P. Kuiper), (Univ. of Chicago Press, IL, 1952) 258.
2. W. B. Clarke, M. A. Beg and H. Craig, Earth Planet. Sci. Lett. 6, 213 (1969).
3. B. A. Mamyrin, I. N. Tolstikhin, G. S. Anufriev, and I. L. Kamanskiy, Dokl. Akad. Nauk S.S.S.R. 184, 1197 (1969).
4. A. Krylov, B. A. Mamyrin, L. A. Khabarin, T. I. Mazina, and Yu. I. Silin, Geokhim. 8 , 1220 (1974).
5. J. E. Lupton and H. Craig, Earth Planet. Sci. Lett. 26, 133 (1975).
6. H. Craig and J. E. Lupton, Earth Planet. Sci. Lett. 31, 369 (1976).
7. R. J. Poreda and F. Radicati di Brozolo, Earth Planet. Sci. Lett. 69, 277 (1984).
8. P.Sarda, T. Staudacher, and C. J. Allègre, Earth Planet. Sci. Lett. 91, 73 (1988).
9. M. Honda, J. H. Reynolds, E. Roedder, and S. Epstein, J. Geophys. Res. 92, 12507 (1987).
10. W. A. Butler, P. M. Jeffery, J. H. Reynolds, and G. J. Wasserburg, J. Geophys. Res. 68, 3283 (1963).
11. T. Staudacher and C. J. Allègre, Earth Planet. Sci. Lett. 60, 389 (1982).
12. K. K. Turekian, Geochim. Cosmochim. Acta 117, 37 (1959).
13. M. Ozima, Geochim. Cosmochim. Acta 39, 1127 (1975).
14. R. Hart, J. Dymond, and L. Hogan, Nature 278, 156 (1979).
15. S. B. Jacobsen and G. J. Wasserburg, J. Geophys. Res. 84, 7411 (1979).
16. R. K. O'Nions, N. M. Evensen, and P. J. Hamilton, J. Geophys. Res. 84, 6091 (1979).
17. C. J. Allègre, T. Staudacher, P. Sarda, and M. Kurz, Nature 303, 762 (1983).
18. C. J. Allègre, T. Staudacher, P. Sarda, Earth Planet. Sci. Lett. 81, 127 (1986).
19. D. Porcelli, J. O. H. Stone, and R. K. O'Nions, Lunar Planet. Sci. XVII, 674 (1986).
20. A. G. W. Cameron, Icarus 56, 195 (1983).
21. W. K. Hartmann and D. Davis, Icarus 24, 504 (1975).
22. A. G. W. Cameron and W. R. Ward, Lunar Sci. 7, 120 (1976).
23. D. M. Hunten, R. O. Pepin, and J. C. G. Walker, Icarus 69, 532 (1987).
24. R. O. Pepin, Icarus 92 , 2 (1991).
25. R. K. O'Nions and E. R. Oxburgh, Nature 306, 429 (1983).
26. S. J. G. Galer and R. K. O'Nions, Nature 316, 778 (1985).
27. L. H. Kellogg and G. J. Wasserburg, Earth Planet. Sci. Lett. 99, 276 (1990).
28. D. B. Patterson, M. Honda, and I. McDougall, Geophys. Res. Lett. 17, 705 (1990).
29. D. Porcelli and G. J. Wasserburg, Geochim. Cosmochim. Acta, in press (1995).
30. D. Porcelli and G. J. Wasserburg, Lunar Planet. Sci. XXV, 1097 (1994)
31. D. Porcelli and G. J. Wasserburg, Abstr. 8th ICOG Conf. 255 (1994).
32. D. Porcelli and G. J. Wasserburg, Geochim. Cosmochim. Acta, submitted (1995).
33. R. K. O'Nions, S. R. Carter, N. M. Evensen, and P. J. Hamilton, The Sea 7, 49 (1981).
34. K. P. Jochum, A. W. Hoffmann, E. Ito, H. M. Seufert, and W. M. White, Nature 306, 431 (1983).

35. G. B. Hudson, B. M. Kennedy, F. A. Podosek, and C. M. Hohenberg, Proc. 19th Lunar Planet. Sci. Conf. 547 (1989).
36. G. W. Wetherill, Ann. Rev. Nuclear Sci. 25, 283 (1975).
37. R. O. Pepin and D. Phinney, unpublished manuscript (1978).
38. M. D. Kurz, W. J. Jenkins, and S. R. Hart, Nature 297, 43 (1982).
39. W. Rison and H. Craig, Earth Planet. Sci. Lett. 66, 407 (1983).
40. J. H. Reynolds, U. Frick, J. M. Niel, and D. L. Phinney, Geochim. Cosmochim. Acta 42, 1775 (1978).
41. T. K. Kyser and W. Rison, J. Geophys. Res. 87, 5611 (1982).
42. B. Marty, Earth Planet. Sci. Lett. 94, 45 (1989).
43. M. Honda, I. McDougall, and D. Patterson, Lithos 30, 257 (1993).
44. K. Farley and R. J. Poreda, Earth Planet. Sci. Lett. 114, 325 (1993).
45. H. Hiyagon, M. Ozima, B. Marty, S. Zashu, and H. Sakai, Geochim. Cosmochim. Acta 56, 1301 (1992).
46. M. Honda, I. McDougall, D. Patterson, A. Doulgeris, and D. Clague, Nature 349, 149 (1991).
47. M. Honda, I. McDougall, D. Patterson, A. Doulgeris, and D. Clague, Geochim. Cosmochim. Acta 57, 859 (1993).
48. R. J. Poreda and K. A. Farley, Earth Planet. Sci. Lett. 113, 129 (1992).
49. D. C. Black, Geochim. Cosmochim. Acta 36, 347 (1972).
50. J.-P. Benkert, H. Baur, P. Signer, and R. Wieler, J. Geophys. Res. 98, 13147 (1993).
51. P. Eberhardt, J. Geiss, H. Graf, N. Grogler, M. D. Mendia, M. Morgeli, H. Schwaller, and A. Stettler, Proc. Lunar Sci Conf. 2, 1821 (1972).
52. J. Geiss, F. Buehler, H. Cerutti, P. Eberhardt, and C. H. Filleaux, Apollo 16 Prelim. Sci. Rep. NASA SP-315, 14-1 (1972).
53. M. Ozima and S. Zashu, Earth Planet. Sci. Lett. 62, 24 (1983).
54. P. Sarda, T. Staudacher, and C. J. Allègre, Earth Planet. Sci. Lett. 72, 357 (1985).
55. T. Staudacher, P. Sarda, S. H. Richardson, C. J. Allègre, I. Sagna, and L. M. Dmitriev, Earth Planet. Sci. Lett. 96, 119 (1989).
56. A. G. W. Cameron, Space Sci. Rev. 15, 121 (1973).
57. G. Igarashi, Ph.D. Dissertation, Geophysical Institute, Univ. of Tokyo, (1988).
58. M.W. Caffee, G.B. Hudson, C. Velsko, E.C. Alexander Jr., G.R. Huss, and A.R. Chivas, Lunar Planet. Sci. XIX, 154 (1988).
59. E. Mazor, D. Heymann, and E. Anders, Geochim. Cosmochim. Acta 34, 781 (1970).
60. R. K. O'Nions and I. N. Tolstikhin, Earth Planet. Sci. Lett. 124, 131 (1994).
61. H. Wanke, Phil. Trans. R. Soc. Lond. A303, 287 (1981).
62. F. Tera and G. J. Wasserburg, Proc. 5th Lunar Conf. 2, 1571 (1974).
63. J. S. Stacey and J. D. Kramers, Earth Planet. Sci. Lett. 26, 207 (1975).
64. A. J. Gancarz and G. J. Wasserburg, Geochim. Cosmochim. Acta 41, 1283 (1977).
65. T. Owen, A. Bar-Nun, and I. Kleinfeld, Nature 358, 43 (1992).
66. M. Ozima and N. Wada, Nature 361,693(1993) and T. Owen and A. Bar-Nun, Nature 361,693(1993).
67. M. Ozima and K. Zahnle, Geochem. J. 27, 185 (1993).
68. D. Porcelli and G.J. Wasserburg, Lunar Planet. Sci. XXVI, 1129 (1995).
69. M. Carr, Icarus 68,187 (1986).
70. A.O. Nier and M.B. McElroy, J. Geophys. Res. 82, 4341 (1977).
71. T. Owen, K. Biemann, D.R. Rushneck, J.E. Biller, D.W. Howarth, and A.L. Lafleur, J. Geophys. Res. 82, 4635 (1977).
72. G. Dreibus and H. Wanke, Meteoritics 20, 367 (1985).
73. K. Zahnle, J. Geophys. Res. 98, 10899.
74. T.D. Swindle, M.W. Caffee, and C.M. Hohenberg, Geochim. Cosmochim. Acta 50, 1001 (1986).

PRIMITIVE XENON IN THE EARTH

G. Igarashi

Hiroshima University, Higashi-Hiroshima 724, Japan

ABSTRACT

The isotopic composition of primitive Xe common to the Earth's atmosphere and carbonaceous meteorites is determined using multivariate correlation analysis. Primitive Xe is defined as the intersection of the unfractionated, fission-free Earth's atmospheric Xe and the correlation line formed by Xe in carbonaceous meteorites in a multidimensional data space. The primitive Xe isotopic composition is similar to that of solar Xe rather than U-Xe. The isotopic variations of Xe in carbonaceous meteorites used in the present analysis are mostly accounted for by those in chemically separated phases, whereas U-Xe was determined using only those on bulk samples. The discrepancy between the present and previous studies may suggest that U-Xe itself is a multi-component mixture and is altered by chemical separation processes. However, since it is not guaranteed that bulk meteoritic samples really preserve unaltered primitive Xe, an alternative interpretation can be that unbiased primitive Xe is isotopically closer to solar Xe. The isotopic spectrum of the fission-like component in the Earth's atmospheric Xe, which is estimated through the procedure for determining the primitive Xe isotopic composition, does not agree satisfactorily with either of two possible parent nuclides, ^{244}Pu and ^{238}U, although it is closer to that of ^{244}Pu than ^{238}U. The amount of the fissiogenic Xe added to the Earth's atmosphere is tentatively estimated as ^{136}Xe$_{fission}$/^{136}Xe$_{total}$ = 2.8±1.3 %.

INTRODUCTION

It has long been known that the isotopic composition of Xe in the Earth is largely different from that in meteorites, which is one of the most important facts against the common source of volatiles to the Earth and meteorites. The difference in Xe isotopic composition probably reflects the complicated processes of the formation and early evolution of the Earth as well as the isotopic heterogeneity in the primitive solar nebula itself.

To make the problem simpler, it is desirable to fix a genetic standard, i.e., primitive Xe from which any Xe isotopic compositions observed in the Earth and meteorites can be reproduced through well-defined nucleogenic and/or physico-chemical processes. Unfortunately, no one has ever succeeded in isolating primitive Xe experimentally from now existing materials. It is inevitable to introduce some statistical treatments if one wishes to determine the primitive Xe isotopic composition.

Takaoka[1] made the first attempt to determine the isotopic composition of primitive Xe, which was defined as the intersection of two different linear correlation lines in 3-isotope plots between ^{130}Xe/^{136}Xe and ^{M}Xe/^{136}Xe (M = 128, 131, 132, and 134): a linear correlation was formed by Xe in carbonaceous chondrites and lunar soils, and another was formed by achondrites. Xe in the Earth's atmosphere did not lie on either of the correlation lines. Takaoka[1] argued that the terrestrial Xe could have been differentiated from primitive Xe through the addition of fissiogenic Xe followed by a mass fractionation process.

Pepin and Phinney[2] extended the 2-dimensional analysis proposed by Takaoka[1] to a multi-dimensional correlation analysis. By analyzing step-heated Xe data on seven carbonaceous meteorites using a multi-dimensional least-squares algorithm, Pepin and Phinney[2] determined 7-dimensional correlation equations (^{129}Xe was excluded because of its large mono-isotopic variations). The data distribution in 7-dimensional space was named "hyperplane". Xe in the Earth's atmosphere did not lie on the hyperplane. Pepin and Phinney[2] succeeded in fitting the atmospheric Xe to the hyperplane by correcting it for fractionation and contribution of ^{244}Pu fission. The intersection of the unfractionated, fission-free atmospheric Xe and the hyperplane was regarded as primitive Xe common to the Earth and meteorites. This is U-Xe. Although the manuscript[2] in which the analytical procedure to determine U-Xe is fully described is still unpublished, U-Xe has frequently and widely been referred to as primitive Xe.

The U-Xe isotopic composition is almost identical to the Takaoka's primitive Xe; both are distinctly different from any major Xe components such as solar Xe[3-5] and Xe-Q[6] that have actually been observed in meteorites and lunar soils. This may suggest that there had existed primitive Xe isotopically different from any Xe components observable in now surviving solar system materials. Is U-Xe observable or not?

Recently, some efforts to hunt achondritic meteorites for U-Xe have brought a peculiar component resembling U-Xe isotopic composition [7, 8]. These efforts are undoubtedly essential to investigate the origin of volatiles in terrestrial planets. On the other hand, plenty of evidence has been accumulated that solar type light noble gases (He and Ne in the solar wind) spread throughout the Earth's mantle. This strongly suggests that the Earth has originally captured volatile elements, at least light noble gases, which are isotopically identical to those in the present Sun. Therefore, it is natural to expect that the primitive Xe isotopic composition is also identical to solar Xe. If it is not the case, we have to accept that there had been unknown processes in the early evolution of the Sun having caused the Xe isotopic difference between the present Sun and the primitive solar nebula where the Earth was formed.

In the present paper, the author examines multi-dimensional variations of Xe isotopic compositions in carbonaceous meteorites and their relationship to the Earth's Xe by using an approach similar to those adopted by the previous studies [1, 2]. The main purpose of this study is to evaluate how the primitive Xe isotopic composition depends on the differences in the analytical method and data base. Because of the limit of the paper length, the author will focus on the statistical aspect in determining the primitive Xe isotopic composition.

CHARACTERIZATION OF XE IN CARBONACEOUS METEORITES

Three-dimensional views of isotopic variations of Xe in carbonaceous meteorites, which are compiled from available publications, are shown in Fig. 1. All data are re-normalized as $^{130}Xe = 100$, since variations due to the addition of H+L-Xe [2] are considered lowest, if any, in 130Xe. The literature sources for the data set are listed in Appendix. Of course this data set is not a complete data base. It contains a large number of Xe data on chemically separated phases besides those on bulk samples. Most of the Xe data on chemically separated phases show much larger isotopic variations than those on bulk samples. Consequently, the linear correlations seen in all the 3-dimensional spaces of Fig. 1 are essentially accounted for by those on chemically separated phases. This is the point most different from the Pepin and Phinney's study; they used a data base consisting entirely of compositions evolved from bulk samples by stepwise heating.

To characterize this multi-isotopic linear correlation, "Principal Component Analysis" is applied in this study instead of a least-squares analysis applied in the previous studies [1, 2]. The main purpose of Principal Component Analysis is to find new coordinate axes that provide a better perspective of multi-dimensional data variations. The following is a summary of Principal Component Analysis.

Let us consider a rotation of an original p-dimensional coordinate system $x = (x_1, x_2, \cdots, x_p)$ using a p x p dimensional orthodiagonal matrix $L = (l_{ij})$ to construct a new coordinate system $z = (z_1, z_2, \cdots, z_p)$,

$$z = xL. \tag{1}$$

Now we have n sets of p-dimensional data whose values are expressed as an n x p dimensional data matrix $X = (x_{ij})$ ($i = 1,2, \cdots, n; j = 1, 2, \cdots, p$) in the original coordinate system. The data matrix Z (n x p dimensions) in the new coordinate system will become

$$Z = XL. \tag{2}$$

Here, we can construct the new coordinate system under the following conditions,

$$V[z_k] = \lambda_k, \tag{3}$$

$$Cov[z_k, z_k{}^t] = 0, \tag{4}$$

$$\lambda_1 \geq \lambda_2 \geq \cdots \geq \lambda_p \geq 0, \tag{5}$$

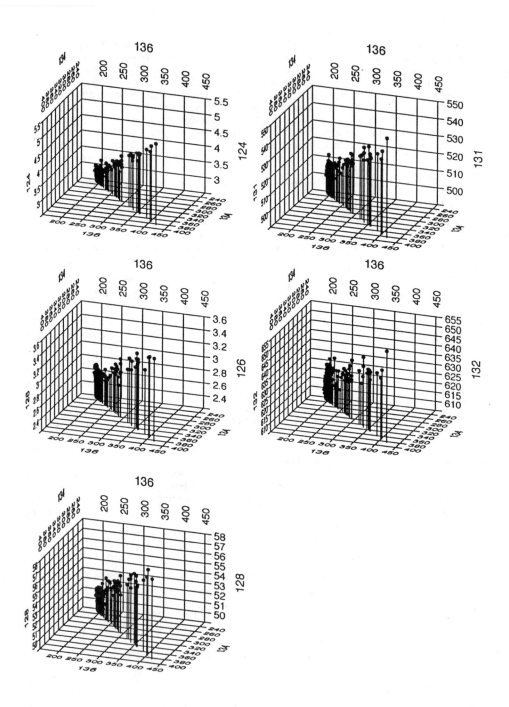

Fig. 1. Three dimensional diagrams of Xe isotope data on carbonaceous meteorites compiled for the analyses. All data are re-normalized as $^{130}Xe = 100$.

where $V[z_k]$ is a variance of z_k and $Cov[z_k, z_k^t]$ is a covariance between z_k and z_k^t (the superscript 't' denotes a transposition). Then the variance-covariance matrix Λ in the new coordinate system z can be calculated using equations (3), (4), and (5) as

$$\Lambda = Z^tZ/(n-1) = L^t(X^tX)L/(n-1) = L^tVL, \tag{6}$$

where V is a variance-covariance matrix in the original coordinate system x. Using $LL^t = I$ (I is a unit matrix), we obtain

$$VL = L\Lambda. \tag{7}$$

This is an eigenvalue problem about the matrix V; diagonal elements of Λ, $(\lambda_1, \lambda_2, \cdots, \lambda_p)$, are the eigenvalues and columns of L, $(l_1, l_2, \cdots, l_p)$, are the eigenvectors. Thus the data distribution is characterized by the eigenvalues and eigenvectors of the variance-covariance matrix of the data.

By solving equation (7), the new coordinate axis can be expressed by a linear combination of an eigenvector and the old coordinate axis as,

$$z_k = l_k^t x = l_{1k}x_1 + l_{2k}x_2 + \cdots + l_{pk}x_p. \tag{8}$$

This z_k is called the k-th principal component. The eigenvalues satisfy the relation

$$\lambda_1 + \lambda_2 + \cdots + \lambda_p \geq \sigma^2, \tag{9}$$

where σ^2 denotes the total variance of the data. Equation (9) means that σ^2 is decomposed into $\lambda_1, \lambda_2, \cdots, \lambda_p$.

Table 1 shows the result of Principal Component Analysis applied to the compiled data shown in Fig.1. In addition to eigenvalues and eigenvectors, contribution ratios of z_k to the total variance of the data, which are calculated as

$$c_k = \lambda_k/\sigma^2 \tag{10}$$

Table 1. Principal components of Xe in carbonaceous meteorites.

Number	1	2	3	4	5	6	7
Eigenvalue	3174.32	17.403	2.998	0.384	0.154	0.007	0.003
Contribution ratio (%)	99.34	0.54	0.09	0.01	0.00	0.00	0.00
Eigenvector							
$^{124}Xe/^{130}Xe$	0.0078	-0.0084	0.0087	-0.0169	0.0292	0.8632	-0.5035
$^{126}Xe/^{130}Xe$	0.0035	-0.0044	0.0045	-0.0099	0.0082	0.5033	0.8640
$^{128}Xe/^{130}Xe$	0.0258	-0.0133	0.1106	-0.2325	0.9653	-0.0342	0.0074
$^{131}Xe/^{130}Xe$	0.1411	0.4604	0.8500	-0.0748	-0.1174	-0.0029	-0.0009
$^{132}Xe/^{130}Xe$	0.1148	0.8633	-0.5088	-0.0908	0.0369	0.0092	-0.0008
$^{134}Xe/^{130}Xe$	0.5501	-0.0074	0.0023	0.8136	0.1809	0.0059	0.0019
$^{136}Xe/^{130}Xe$	0.8146	-0.2060	-0.0796	-0.5194	-0.1395	-0.0143	-0.0002

are also listed in Table 1. It has been found out that more than 99 % of the total variance can be expressed by the first principal component. Thus the single component structure of the data variation in 7-dimensional space has been revealed; i.e., the five linear correlations seen in Fig. 1 have been reduced to a single linear correlation in a 7-dimensional data space. The one-dimensional variations can be expressed by a single new variable of z_1. The eigenvector of the first principal component, which determines the direction of z_1, is identical to the isotopic composition of H+L-Xe [2]. It should be noted that the eigenvector of the second principal component has particularly large values of $^{131}Xe/^{130}Xe$ and $^{132}Xe/^{130}Xe$, which are the mixtures of s- and p-process nuclides. The second principal component might reflect variations due to s-process Xe. However, because of its small contribution ratio, the existence of the second principal component does not affect seriously the determination of the primitive Xe isotopic composition, and will be ignored here. Contribution ratios of the third to seventh principal components are still much smaller. They are probably less than the level of experimental errors in the data, and will not offer any information about meaningful Xe components.

DETERMINATION OF THE "PRIMITIVE EARTH XE"

To determine the isotopic composition of primitive Xe, we next have to find out the intersection of the carbonaceous meteorite data space and the "unfractionated, fission-free atmospheric Xe".

When a new data set $x' = (x'_j)$ is given, a projection of x'_j to the first principal component z_1 can be calculated as

$$x'^{*}_j = \sum_{k=1}^{p} x'_k \, l_{k1} \, k_{j1}. \tag{11}$$

Then distance, Δ, of x' from the carbonaceous meteorite data space is calculated as

$$\Delta^2 = \sum_{j=1}^{p} (x'_j - x'^{*}_j)^2. \tag{12}$$

To determine the intersection, it is better to minimize the distance normalized by the variance, s_j, of $(x'_j - x'^{*}_j)$,

$$\Delta^{*2} = \sum_{j=1}^{p} (x'_j - x'^{*}_j)^2/s_j^2, \tag{13}$$

where,

$$s_j = 1 - \lambda_1 l_{j1}^2. \tag{14}$$

The unfractionated, fission-free atmospheric Xe, x_j^{CA}, may be defined as

$$x_j^{CA} = (x_j^A - \alpha_j)/f_j \tag{15}$$

where x_j^A is the Xe isotopic composition of the Earth's atmosphere [9], α_j is a fission component, and f_j is a fractionation factor. In the present study, the spectrum of the fission component is not specified but fission yields of the light isotopes (M = 124, 126, 128) are assumed to be zero; then four free parameters are remained to be determined: $\alpha_j = (0, 0, 0, \alpha_4, \alpha_5, \alpha_6, \alpha_7)$. Two different types of fractionation functions, which could operate in the gain and loss of Xe, will be adopted; one is the gravitational differentiation in porous planetesimals [10, 11]

$$f^g_j = \exp\{\beta(\mu_j - \mu_{130})\}, \tag{16}$$

and another is the diffusive loss

$$f^d_j = \exp[\beta\{(\mu_{130}/\mu_j)^{1/2} - 1\}], \tag{17}$$

where μ_j is an atomic weight of a Xe isotope and β is a parameter to be determined.

Now, putting x_j^{CA} to x'_j of eq. (11) and (13), and minimizing Δ^{*2}, the isotopic composition of the intersection of the "unfractionated, fission-free atmospheric Xe" and the carbonaceous meteorite data space can be obtained. The intersection obtained with a fractionation function of the gravitational differentiation in porous planetesimals is given in Table 2. The degree of the fractionation of the atmospheric Xe that is needed to obtain the intersection is (3.1 ± 0.1) %/amu. Although results with a fractionation function of the diffusive loss are slightly different, the differences are smaller than the uncertainties listed in Table 2 that come from the scattering of the carbonaceous meteoritic Xe data.

Table 2. Estimated primitive Xe, fissiogenic Xe and fission-free Xe in the Earth's atmosphere.

	^{124}Xe	^{126}Xe	^{128}Xe	^{131}Xe ^{130}Xe =	^{132}Xe 100	^{134}Xe	^{136}Xe
Primitive Earth Xe	2.803	2.461	50.09	501.0	613.2	220.8	176.4
	±0.084	±0.065	±0.47	±1.6	±1.1	±1.7	±2.5
Fissiogenic Xe in				4.8	9.2	7.0	6.1
the atmosphere				±0.5	±0.7	±1.8	±2.8
(Relative Fission				0.79	1.51	1.15	=1.00
Yield)				±0.37	±0.70	±0.61	
Fission-free Xe in	2.337	2.180	47.146	516.5	651.5	249.3	211.5
the atmosphere	±0.007	±0.011	±0.047	±0.8	±0.9	±1.9	±3.0

DISCUSSION

Figure 2 plots δ-values defined as,

$$\delta_{130}^M = \{(^MXe/^{130}Xe)/(^MXe/^{130}Xe)_S - 1\} \times 1000 \tag{18}$$

where $(^MXe/^{130}Xe)_S$ is solar Xe observed in the Pesyanoe aubrite [4]. In Fig.2, the isotopic composition of the intersection is labeled as PE (Primitive Earth Xe) Xe. PE-Xe has an isotopic composition similar to solar Xe observed in the Pesyanoe aubrite [4] and lunar soils [3, 5], whereas primitive Xe [1] and U-Xe [2] have distinct isotopic ratios of ^{134}Xe/^{130}Xe and ^{136}Xe/^{130}Xe. The difference in the analyzed population of the meteoritic Xe data contributes largely to the discrepancy between the present and previous studies; the data variations used in the present analysis are essentially accounted for by those in chemically separated phases, whereas the data bases used in the previous studies consisted entirely of compositions evolved from bulk samples. That is, Xe isotopic compositions in chemically separated phases tend to shift the intersection closer to the solar Xe isotopic composition. The difference in the analytical method has a smaller effect on the resulting isotopic composition; by applying Principal Component Analysis to a smaller data population of meteoritic Xe, the author derived a primitive Xe composition closer to U-Xe [12].

Pepin and Phinney [2] pointed out that chemically separated phases contain a primitive component somewhat different from U-Xe, which they attributed to an alteration of U-Xe by chemical attacks. As they discussed, it is likely that U-Xe itself is a mixture of some sub-components and can be altered by physical and chemical processes. However, it is not guaranteed that bulk meteoritic samples really preserve unaltered primitive Xe. For example, in Pepin and Phinney's analysis, the hyperplane equation for the lightest Xe isotope, ^{124}Xe, did not show satisfactory fits to the data. They appeared to overcome this problem by rearranging the equation for ^{126}Xe to that for ^{124}Xe. This may suggest that the primitive Xe component has already been altered even in bulk meteoritic samples. Hence, there can be an alternative interpretation that unbiased primitive Xe is isotopically closer to solar Xe.

Three-isotope plots of Fig. 3 a) - c) will be helpful to grasp the relationship between PE-Xe and other major Xe components. Two processes, gravitational differentiation [10, 11] and diffusive loss, causing the Xe isotopic fractionation are adopted to determine the primitive Xe isotopic composition in the present study. Either process gives roughly the same result. There are some other processes proposed for the isotopic fractionation of Earth's Xe including hydrodynamic escape [13-15] and low energy ion implantation [16]. Further studies will be needed to examine whether the other processes lead to the same final results or not. The author's prejudice is that the gravitational differentiation [10, 11] is promising, since it can explain why Earth's Xe suffered such a severe isotopic fractionation leaving lighter noble gases including Kr almost intact, by taking account of the size distribution of the Earth-accreting planetesimals [17].

Michel and Eugster [7] found a primitive Xe component in a 1200 °C fraction of the step-heated Tatahouine diogenite, whose isotopic composition is similar to that of U-Xe. This may support the existence of a primitive Xe component isotopically distinct from solar Xe. However, as seen in Fig. 2 and Fig. 3, relatively large experimental uncertainties in the Xe data on the Tatahouine diogenite blur the difference from solar Xe. Furthermore, Eugster et al. [8] proposed the average primitive achondrite Xe (AVPA Xe) based on the Xe isotopic compositions extracted from the Lodran lodranite in addition to the Tatahouine diogenite. They argue that AVPA Xe is similar to primitive Xe[1] or U-Xe [2]. However, the difference between solar Xe and AVPA Xe does not appear to be persuasive. Further accumulation

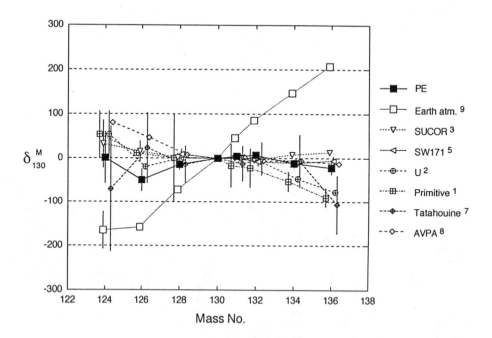

Fig. 2. The isotopic composition of Primitive Earth (PE)Xe determined as the intersection between the unfractionated, fission-free Earth's atmospheric Xe and carbonaceous meteorite data space. All spectra are normalized to solar Xe observed in the Pesyanoe aubrite [4]. For delta-values, see text.

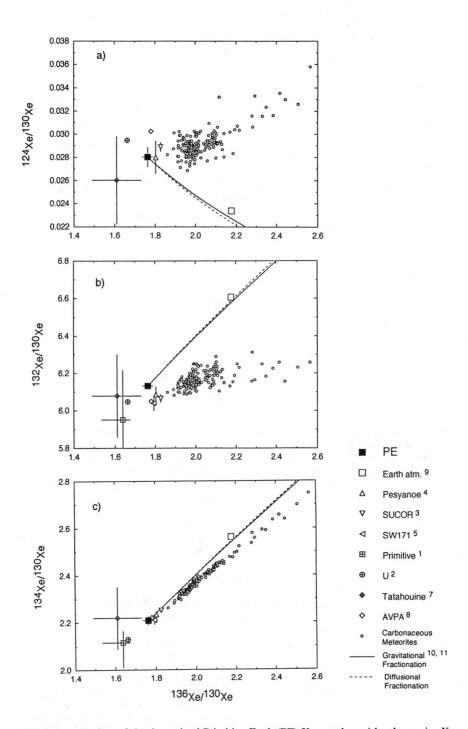

Fig. 3. Three-isotope plots of the determined Primitive Earth (PE) Xe together with other major Xe components and Xe in carbonaceous meteorites.

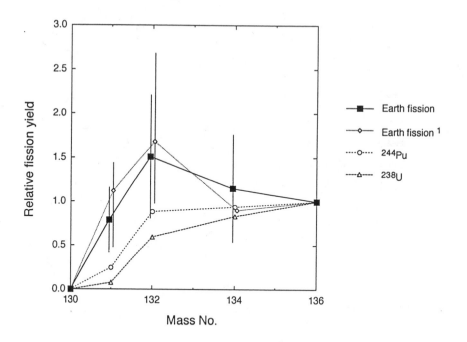

Fig. 4. The relative yields of the fission-like component in the Earth's atmospheric Xe.

of precise Xe isotope data on primitive achondrites will settle the problem whether or not we need to accept primitive Xe which is distinctly different from solar Xe.

Figure 4 shows the relative yields of the fission-like component in the Earth's atmospheric Xe. The obtained fission yields are similar to those obtained by Takaoka [1], but do not agree satisfactorily with those for two possible parent nuclides, ^{244}Pu [18] and ^{238}U [19], although they are closer to those of ^{244}Pu rather than ^{238}U. It is difficult to conclude, only from the statistical point of view, that the parent nuclide of the fissiogenic Xe is ^{244}Pu. The estimated amount of the fissiogenic Xe in the atmosphere also inevitably suffers from analytical uncertainties. The amount of fissiogenic Xe in the Earth's atmosphere is estimated as $^{136}Xe_{fission}/^{136}Xe_{total} = (2.8\pm1.3)$ %, which is smaller by about 40 % compared with the value (4.65±0.30) % obtained by Pepin and Phinney [2].

CONCLUSIONS

1) Published data on Xe isotopic compositions in carbonaceous meteorites, including those in chemically separated phases as well as bulk samples, are compiled and examined using a multi-dimensional correlation analysis. A simple structure of the data variations of carbonaceous meteoritic Xe is revealed by Principal Component Analysis; more than 99 % of the total variance of the data can be expressed by a single 7-dimensional principal component, whose isotopic composition is identical to H+L-Xe.

2) The isotopic composition of primitive Xe common to the Earth's atmosphere and carbonaceous meteorites is determined as the intersection of the unfractionated, fission-free Earth's atmospheric Xe and the multi-dimensional data space formed by carbonaceous meteoritic Xe. The determined primitive Xe, which the author refers to as Primitive Earth (PE) Xe, is close to solar Xe rather than U-Xe. The discrepancy between the present and previous studies is mainly due to the difference in the data population used in the analysis; the present study does not make any data selection and

consequently Xe data on chemically separated phases essentially account for the data variations, whereas previous studies determined U-Xe (primitive Xe) using only those on bulk samples.

3) The isotopic spectrum of the fission-like component in the Earth's atmospheric Xe is estimated through the procedure for determining the primitive Xe isotopic composition. It does not agree satisfactorily with either of two possible parent nuclides, ^{244}Pu and ^{238}U, although it is closer to ^{244}Pu rather than ^{238}U. It is difficult to specify the parent nuclide only from the statistical point of view. The amount of the fissiogenic Xe added to the Earth's atmosphere is tentatively estimated as ^{136}Xe$_{fission}/^{136}$Xe$_{total} = 2.8 \pm 1.3$ %.

ACKNOWLEDGEMENTS

The author thanks M. Ozima for invaluable discussion and encouragement, and R. O. Pepin and K. V. Ponganis for critical comments and suggestions.

REFERENCES

1.   N. Takaoka, Mass Spectroscopy 20, 287-302 (1972).
2.   R. O. Pepin and D. Phinney, (unpublished manuscript) (1978).
3.   F. A. Podosek et al., Earth Planet. Sci. Lett. 10, 199-216 (1971).
4.   J. S. Kim and K. Marti, Proc. Lunar Planet. Sci. 22, 145-152 (1992).
5.   R. Wieler and H. Baur, Meteoritics 29, 570-580 (1994).
6.   R. Wieler et al., Ceochim. Cosmochim. Acta 56, 2907-2921 (1992).
7.   Th. Michel and O Eugster, Meteoritics 29, 593-606 (1994).
8.   O. Eugster et al., In Noble Gas Geochemistry and Cosmochemistry (Terra Sci. Publ., Tokyo, 1994), pp. 1-9.
9.   J. R. Basford et al., Proc. Lunar Sci. Conf. 2, 1915-1955 (1973).
10.   M. Ozima and K. Nakazawa, Nature 284, 313-316 (1980).
11.   K. Zahnle et al., Geochim. Cosmochim. Acta 54, 2577-2586 (1990).
12.   G. Igarashi, in Japan-U.S. Seminar on Terrestrial Rare Gases (Dept. of Physics, Univ. California, Berkeley, 1986), pp. 20-23.
13.   D. M. Hunten et al., Icarus 69, 532-549 (1987).
14.   S. Sasaki and K. Nakazawa, Earth Planet. Sci. Lett. 89, 323-334 (1988).
15.   R. O. Pepin, Icarus 92, 2-79 (1991).
16.   T. J. Bernatowicz and B. E. Hagee, Geochim. Cosmochim. Acta 51, 1599-1611 (1987).
17.   G. Igarashi and M. Ozima, Proc. NIPR Symp. Antarctic Meteorites 1, 315-320 (1988).
18.   E. C. Alexander Jr. et al., Science 172, 837-840 (1971).
19.   G. W. Wetherill, Phys. Rev. 92, 907-912 (1953).

Appendix. Sources for the data set of Xe isotopic compositions in carbonaceous meteorites.

Name	Class	Reference
Orgueil	C1	O. Eugster et al., Earth Planet. Sci. Lett. 3, 249-257 (1967)
		P. M. Jeffery and E. Anders, Geochim. Cosmochim. Acta 34, 1175-1198 (1970)
		U. Frick and R. K. Moniot, Proc. Lunar Sci. Conf. 8, 229-261 (1977)
		R. O. Pepin and D. Phinney, (unpublished manuscript) (1978)
Cold Bokkeveld	C2	O. Eugster et al., Earth Planet. Sci. Lett. 3, 249-257 (1967)
		J. D. Macdougall and D. Phinney, Proc. Lunar Sci. Conf. 8, 293-311 (1977)
		J. H. Reynolds et al., Geochim. Cosmochim. Acta 42, 1775-1797 (1978)
		R. O. Pepin and D. Phinney , (unpublished manuscript) (1978)
Murchison	C2	P. K. Kuroda et al., J. Geophys. Res. 80, 1558-1570 (1975)
		J. D. Macdougall and D. Phinney, Proc. Lunar Sci.□ Conf. 8, 293-311 (1977)
		B. Srinivasan et al., J. Geophys. Res. 82, 762-778 (1977)

		J. H. Reynolds et al., Geochim. Cosmochim. Acta 42, 1775-1797 (1978)
		B. Srinivasan and Anders, Science 201, 51-56 (1978)
		L. Alaerts et al., Geochim. Cosmochim. Acta 44, 189-209 (1980)
		C. M. Jones et al., J. Geophys. Res. 90 supple, C715-C721 (1985)
		R. Wieler et al., Geochim. Cosmochim. Acta 56, 2907-2921 (1992)
Murray	C2	K. Marti, Earth Planet. Sci. Lett. 2, 243-248 (1967)
		P. K. Kuroda et al., J. Geophys. Res. 79, 3981-3992 (1974)
		R. O. Pepin and D. Phinney, (unpublished manuscript) (1978)
		J. H. Reynolds et al., Geochim. Cosmochim. Acta 42, 1775-1797 (1978)
		C. M. Jones et al., J. Geophys. Res. 90 supple, C715-C721 (1985)
Renazzo	C2	J. H. Reynolds and G. Turner, J. Geophys. Res. 69, 3263-3281 (1964)
Allende	CV3	D. Phinney, Ph. D. Thesis, Univ. of Minneapolis (1971)
		O. K. Manuel et al., Geochim. Cosmochim. Acta 36, 961-983 (1972)
		R. S. Lewis et al., Science 190, 1251-1262 (1975)
		R. S. Lewis et al., J. Geophys. Res. 82, 779-792 (1977)
		U. Frick and S. Chang, Meteoritics 13, 465-470 (1978)
		J. H. Reynolds et al., Geochim. Cosmochim. Acta 42, 1775-1797 (1978)
		B. Srinivasan et al., Geochim. Cosmochim. Acta 42, 183-198 (1978)
		R. V. Ballad et al., Nature 277, 615-620 (1979)
		U. Ott et al., Geochim. Cosmochim. Acta 45, 1754-1788 (1981)
		R. Wieler et al., Geochim. Cosmochim. Acta 56, 2907-2921 (1992)
		J. Matsuda et al., Geochim. Cosmochim. Acta 44, 1861-1874 (1980)
Grosnaja	CV3	J. Matsuda et al., Geochim. Cosmochim. Acta 44, 1861-1874 (1980)
Leoville	CV3	O. K. Manuel et al., Geochim. Cosmochim. Acta 36, 961-983 (1972)
Mokoia	CV3	M. W. Rowe, Geochim. Cosmochim. Acta 32, 1317-1326 (1968)
		D. Phinney, Ph. D. Thesis, Univ. of Minneapolis (1971)
		O. K. Manuel et al., Geochim. Cosmochim. Acta 36, 961-983 (1972)
Vigarano	CV3	J. Matsuda et al., Geochim. Cosmochim. Acta 44, 1861-1874 (1980)
Ferix	CO3	K. Marti, Earth Planet. Sci. Lett. 2, 243-248 (1967)
Lance	CO3	O. Eugster et al., Earth Planet. Sci. Lett. 3, 249-257 (1967)
		K. Marti, Earth Planet. Sci. Lett. 2, 243-248 (1967)
Kainsaz	CO3	L. Alaerts et al., Geochim. Cosmochim. Acta 43, 1421-1432 (1979)
Ornans	CO3	B. Srinivasan et al., J. Geophys. Res. 82, 762-778 (1977)
		L. Alaerts et al., Geochim. Cosmochim. Acta 43, 1421-1432 (1979)
Karoonda	C5	B. Srinivasan et al., J. Geophys. Res. 82, 762-778 (1977)
Dingo Pup Donga	Ureilite	D. D. Bogard et al., Geochim. Cosmochim. Acta 37, 547-557 (1973)
Havero	Ureilite	H. W. Weber et al., Earth Planet. Sci. Lett.13, 205-209 (1971)
		D. D. Bogard et al., Geochim. Cosmochim. Acta 37, 547-557 (1973)
Kenna	Ureilite	L. L. Wilkening and K. Marti, Geochim. Cosmochim. Acta 40, 1465-1473 (1976)
Novo Urei	Ureilite	K. Marti, Earth Planet. Sci. Lett. 2, 243-248 (1967)
		D. Phinney, Ph. D. Thesis, Univ. of Minneapolis (1971)

HE-ISOTOPIC INVESTIGATION OF GEOTHERMAL GASES FROM THE TABAR-LIHIR-TANGA-FENI ARC AND RABAUL, PAPUA NEW GUINEA

K.A. Farley, Desmond Patterson, and Brent McInnes

Department of Geological and Planetary Sciences, MS 170-25 Caltech, Pasadena, CA 91125

ABSTRACT

In order to investigate the behavior of slab-derived volatiles in the subduction environment, helium isotope ratios have been measured in geothermal gases from the Tabar-Lihir-Tanga-Feni (TLTF) chain in the Bismarck Archipelago of Papua New Guinea. As recorded by several geochemical tracers, these volcanos carry an exceptionally large slab-derived component, and therefore may provide new insights to the old question of volatiles in subduction zones. Geothermal gases from Lihir Island have homogeneous $^3He/^4He$ ratios of 7.18 ± 0.07 times the atmospheric ratio (R_A), while those from Ambitle Island (Feni Group) have lower ratios of 6.61 ± 0.13 R_A. These $^3He/^4He$ ratios are within the range defined by more-typical arc volcanos, but lie at the low end of the spectrum observed in arc volcanos erupted through purely oceanic crust. Although a small slab-derived signature ($^3He/^4He$ ratio lower than depleted mantle) exists in the TLTF gases, these data demonstrate that even in volcanos with a comparatively large slab component, He is overwhelmingly derived from the depleted mantle wedge. This observation further confirms the relative insensitivity of He isotopes to the presence of slab fluids.

He isotope ratios of 6.25 R_A were measured in geothermal gases from the Rabaul Caldera on New Britain Island. Coincidentally, these samples were taken six months prior to the major 1994 eruption at Rabaul. In conjunction with samples taken from the same locality 8 years earlier, these data allow us to test whether increasing He isotope ratios associated with fresh ascending magmas precede volcanic eruptions. Although some of the 1986 samples had much lower $^3He/^4He$ ratios (5 R_A) than observed in 1994, one did not. We thus find no strong evidence for a systematic rise in the He isotope ratio of the Rabaul fluids between 1986 and 1994. If a $^3He/^4He$ increase did precede the Rabaul eruption, then it occurred either prior to 1986 or sometime between our 1994 sampling and the eruption.

INTRODUCTION

Convergent margin processes play a critical role in the partitioning of volatiles between the Earth's interior and atmosphere by filtering elements from the subducting package, thereby restricting their return to the mantle. Elements which are tightly retained may be returned to the deep mantle along with the residual slab, while more loosely-bound volatiles may be released to the ambient mantle in a fluid or melt phase. In the latter case, the volatiles may be expelled to the Earth's surface by arc volcanism, or possibly by ejection of fluids into the trench or shallow forearc[1]. It is also possible that volatiles lost from the slab may be held within a subduction-modified (metasomatized, hydrated) mantle wedge. Identification of the ultimate fate of subducted volatiles has proven difficult because the composition of the downgoing material and the outward volatile flux at subduction zones are both poorly known. In addition, volatiles released by supra-subduction zone volcanos are often dominated by a component derived from the depleted upper mantle, making it difficult to uniquely identify the slab flux.

Helium isotopic measurements of arc and back-arc volcanic systems can potentially be used to address these issues because there is likely to be a large contrast between the $^3He/^4He$ ratio of the subducting and depleted mantle components. As recorded in mid-ocean ridge basalts (MORBs) the depleted mantle has a relatively uniform $^3He/^4He$ ratio of 8±1 times the atmospheric value[2,3] (8 R_A). Subducted material is likely to have much lower ratios. Helium within the subducting package is composed of atmospheric and radiogenic components, with a possible addition of extraterrestrial (ET) helium from micrometeorite fallout[4]. ET helium is characterized by very

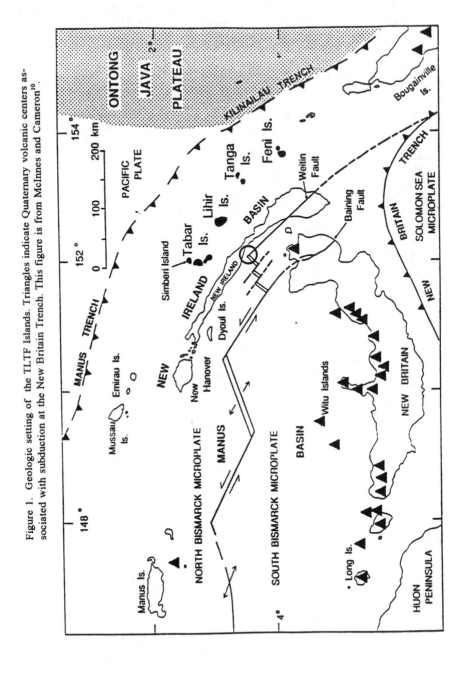

Figure 1. Geologic setting of the TLTF Islands. Triangles indicate Quaternary volcanic centers associated with subduction at the New Britain Trench. This figure is from McInnes and Cameron[10].

high $^3He/^4He$ ratios (>100 R_A), but recent experimental studies indicate that it readily diffuses out of the sediments and is likely lost in the very shallowest portions of the subduction system[5]. Thus entrapment of atmospheric helium ($^3He/^4He=1$ R_A) and radioactive decay of U and Th ($^3He/^4He<<1$ R_A) in the sediments and altered oceanic crust probably give the downgoing package a sub-atmospheric $^3He/^4He$ ratio.

He isotope ratios in arc-related geothermal gases range from the MORB value to < 1 R_A (refs 6-9). In most cases the low values arise from addition of crustal helium to the gases, or possibly as a consequence of crustal contamination of the magmas. In general, the uncontaminated magmas themselves are thought to have almost MORB-like $^3He/^4He$ ratios[8]. This demonstrates that in most cases the helium budget of arc volcanos is overwhelmingly dominated by a depleted mantle component derived from the mantle wedge, effectively precluding the use of He isotopes for exploring slab devolatilization in normal arc settings.

Recent chemical, isotopic, and geophysical evidence suggests that the unusual volcanos of the Tabar-Lihir-Tanga-Feni (TLTF) arc of the Bismarck Archipelago, Papua New Guinea (Figure 1) carry an exceptionally large component derived either directly from the slab or from mantle wedge severely-metasomatized by slab fluids[10,11]. These volcanos potentially offer a unique opportunity to investigate the fate of subducted volatiles with helium isotopes.

A further reason for studying the TLTF volcanos is that they host significant volcanigenic gold deposits, including the world-class Ladolam Mine on Lihir Island. On both Lihir and Ambitle (Feni Group) gold deposits occur within calderas which are still highly geothermally active[12]. The genesis of the TLTF ore deposits may be related to the presence of oxidizing fluids derived from the slab, possibly traceable with helium.

We have also analyzed geothermal gases from the Rabaul Caldera of Eastern New Britain. These samples were collected just prior to the largest eruption at Rabaul in 50 years[16].

GEOLOGIC SETTING AND SAMPLES

The Bismarck Archipelago lies in a tectonically complex portion of the western Pacific Ocean (Figure 1). Westward subduction of the Pacific plate beneath the Indo-Australian Plate at the Manus-Kilinailu (M-K) Trench occurred until approximately 10 Ma ago, when the buoyant Ontong-Java Plateau collided with the trench and precipitated subduction polarity reversal and migration. Subduction of the Indo-Australian plate now occurs along the New Britain Trench, producing numerous Quaternary volcanic centers on the island of New Britain, including Rabaul. Poorly-understood *extensional* processes within the New Ireland Basin, the former forearc region of the M-K Trench system, led to the emplacement of the TLTF volcanos over about the last 5 Ma. These volcanos were most likely produced by adiabatic decompression melting of mantle wedge modified by tens of millions of years of subduction at the M-K trench[10]. It is also possible that these volcanos carry a component derived directly from the fossil slab underlying the wedge[10].

Both Lihir and Ambitle Island in the Feni Group have active geothermal systems (Figures 2). On Lihir, activity is largely confined to high and moderate temperature hotsprings in an area of about 15 km^2 contained in the breached Luise Caldera. All major thermal areas in the caldera were sampled, including submarine springs in the harbor. A single low temperature spring in the Kinami Caldera, about 8 km south of Luise, was also sampled. Hotsprings on Ambitle occur predominantly within the central caldera structure, and are probably being driven by a quartz trachyte body emplaced approximately 2300 years ago[17]. Additional minor activity on Ambitle occurs on the outer flanks of the caldera. Sample locations are further detailed in Table 1.

The Rabaul Harbor samples were collected from two shallow submarine springs between the airport and the base of Tavurvur Volcano ("Bubbly Beach", Figure 2). These gases were collected simply to test the field gear, so no effort was made to sample Rabaul thoroughly. The composition of active gases from this general locality have been described by other workers [16].

Samples were collected in pinch-seal copper tubes and were analyzed for $^3He/^4He$ and He and Ne abundances at Caltech using techniques which will be fully documented elsewhere. Estimat-

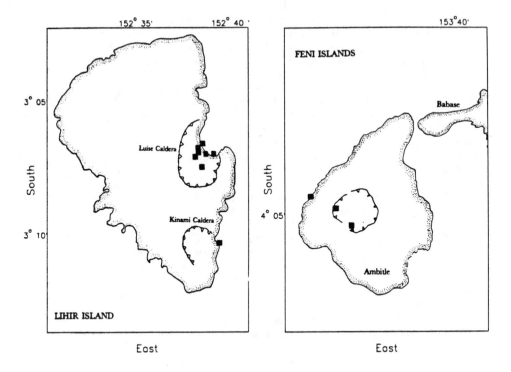

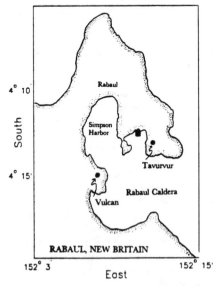

Figure 2. Locations of geothermal samples (filled boxes) from a) Lihir, b) Ambitle, and c) Rabaul. In c) the loci of the 1994 Rabaul eruption at Tavurvur and Vulcan vents are shown.

ed uncertainties on the He/Ne ratio and He abundance are about 5%, while the uncertainty on the He isotopic ratio is ~ 1%.

RESULTS

The Table and Figure 3 present the results of these analyses. All but three samples yielded high He/Ne ratios, ranging from 16 to 1300 times higher than the atmospheric ratio. Such high ratios imply very little contamination by atmospheric gases and adjustments to the $^3He/^4He$ ratio for this effect are very minor. The three samples with He/Ne ratios <10 *may* have suffered much larger degrees of atmospheric addition, for which no reliable correction can be made. These samples are essentially ignored in the following discussion. For the sake of consistency with previously published work, all tabulated $^3He/^4He$ data have been corrected for air addition by assuming contamination with air saturated water[6].

Excluding one locality, all gases in Luise Caldera have $^3He/^4He$ ratios which are identical within analytical uncertainty, at 7.18±0.07 R_A. The exception is sample 94-LSUB-3, with a $^3He/^4He$ ratio of 6.8 R_A. The single sample from Kinami is isotopically indistinguishable from the Luise fluids, and is notable for its factor of 100 enrichment in He. The Ambitle samples have distinctly lower $^3He/^4He$ ratios than those from Lihir, between 6.5 and 6.75 R_A. Like Lihir, the samples are essentially within error of each other, averaging 6.61±0.13 R_A. The Rabaul sample yielded a $^3He/^4He$ ratio of 6.27 R_A, substantially lower than both Ambitle and Lihir.

DISCUSSION

Lihir and Ambitle

The helium isotopic homogeneity of the Luise samples implicates a single source for the helium in all sampled thermal areas within the caldera. It further suggests that addition of crustal helium (which would be variable, depending on the history of each sampled fluid) is probably insignificant. The $^3He/^4He$ ratio of the magmatic body driving the Luise Caldera circulation is therefore close to 7.2 R_A. The sole outlier from this value was collected from near the edge of the caldera in very diffusely bubbling beach sands, and possibly has accumulated more-radiogenic helium from surrounding country rocks.

A magmatic body with this same He isotopic ratio apparently lies within Kinami Caldera, or alternatively, there may be subsurface flow allowing fluids from the Luise magmatic body to vent at Kinami. The high He abundance of the Kinami fluid may bear on this question. Given its high $^3He/^4He$ ratio, the He enrichment is not a consequence of accumulation of crustal helium, but is more likely the result of a process which preferentially concentrates helium in the gas phase. High-temperature, near-source geothermal fluids are dominated by the non-conservative gases CO_2 or CH_4 (ref 13). If these species are removed by chemical and/or biological reactions, the result will be enrichment of helium in the residual gas. Such an effect is more likely to occur away from the high temperature source of the fluid, and might be expected if the Kinami gas originates in the distal Luise Caldera. Further confirmation of this hypothesis requires analysis of the bulk composition of the geothermal gases.

As at Lihir, the homogeneity of $^3He/^4He$ ratios on Ambitle suggests a single magmatic source, with a ratio of about 6.6 R_A.

The helium isotopic ratio of these magmatic systems is well-within the range observed in arc volcanos worldwide (Figure 3), suggesting that helium may not record an unusually large slab component in the TLTF lavas. However, the worldwide arc range in Figure 3 includes samples erupted through crust of a variety of ages and compositions, from young oceanic to old continental. When arc volcanos erupting through purely oceanic crust are considered separately, the $^3He/^4He$ ratios cluster near the high end of the range (> 7 R_A). For example, $^3He/^4He$ ratios in the Marianas and Izu Bonin arcs range from 7.22 to 8.10 R_A (refs 6,9). The same observation can be made com-

Sample Number	Location	T (°C)	^{3}He/^{4}He (R$_A$)	(He/Ne) (He/Ne)$_A$	He (ppm)	(^{3}He/^{4}He)$_c$ (R$_A$)
Ambitle Island						
94-AMB-1	Kabang Prospect- creek-bed acid hotspring	48	6.22	16	3.1	6.50
94-AMB-2B	"	"	6.50	18	2.9	6.75
94-AMB-3B	Caldera Saddle acid hotspring	92	6.56	640	17	6.57
94-AMB-4B	Bubbling beach near Danlam	65	6.48	8.9	1.4	7.02
Lihir Island -						
Luise Caldera						
94-LUI-1	Ladolam - acid hotspring	92	7.23	830	10	7.23
94-LUI-2A	Ladolam - creek-bed bubbling vent	25	7.14	830	14	7.14
94-LUI-3A	Lower Kapit - neutral hotspring	80	7.08	180	5.0	7.11
94-LUI-4	Lower Kapit - muddy acid pool	65	7.17	33	5.1	7.33
94-LUI-5A	Kapit Beach - acid hotspring in alteration zone	92	7.11	1300	4.8	7.11
94-LUI-6	Kapit Beach - bubbling sand	99	7.10	1300	5.2	7.10
94-LUI-7	Between Kapit Beach and Lower Kapit, acid hotspring	88	7.07	28	4.2	7.25
94-LUI-8A	Upper Kapit, N. slope acid hotspring	88	7.15	240	8.0	7.17
94-LUI-9B	Upper Kapit, S. slope	42	7.17	200	7.4	7.19
94-LUI-10A	100m NW. of Lihir Mining Camp - bubbling beach sand	95	6.42	0.23	0.18	--
94-LUI-10B	-duplicate-	"	6.64	1.46	0.33	--
94-LSUB-1B	Luise Harbor, offshore of Kapit Village, 10 m depth submarine		7.17	260	5.1	7.19
94-LSUB-3A	Lihir Mining Camp Boat Launch, bubbling sand	69	6.73	48	5.9	6.82
94-LSUB-3B	-duplicate-	"	6.68	52	5.2	6.77
Kinami Caldera						
94-KIN-1	Bubbling vent in reef offshore of Kinami Village, 1 m depth	41	7.04	140	750	7.08
Rabaul Harbor						
94-RAB-1	"Bubbly Beach", shallow submarine vent in lava, near Tavurvur volcano	78	6.26	4.9	0.18	7.29
94-RAB-2	"Bubbly Beach", shallow submarine vent in sand	52	6.24	130	0.50	6.27

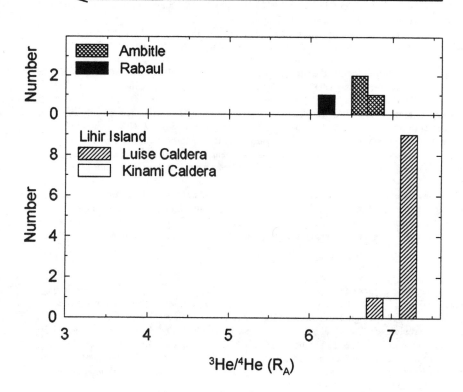

Figure 3. Corrected helium isotopic ratios in the TLTF and Rabaul gases. Shown for comparison are the range of 3/4 ratios in arc volcanos worldwide, and in arc volcanos erupting purely through oceanic crust. Data for other arcs are from refs 6-9, 13 and H. Craig, unpublished.

pletely within the Bismarck Archipelago (Figure 1). $^3He/^4He$ ratios within the purely oceanic portions of the New Britain arc have $^3He/^4He$ ratios >8 R_A, whereas those underlain by the older arc crust of New Britain Island (including Rabaul) have $^3He/^4He$ ratios ranging downward from 6.2 R_A (H. Craig, personal communication, and our Rabaul data). On the bases of similar arguments, it has been suggested[13,18,6,8] that most parental arc magmas are characterized by MORB-like $^3He/^4He$ ratios (8 R_A), and that the lower $^3He/^4He$ ratios commonly observed may simply be the product of contamination by helium within the uppermost crust.

Because the TLTF samples are erupting through oceanic crust, and for the reasons described above, we believe our $^3He/^4He$ ratios represent uncontaminated magmatic values. This supposition is supported by preliminary helium analyses of phenocrysts from primitive basalts from these islands, which match the geothermal gases extremely well[14]. If it is true that uncontaminated arc magmas are characterized by MORB-like He, then the TLTF arc may indeed have a small slab-derived helium component that is more strongly manifest at Ambitle than at Lihir. Although speculative, the along-strike variation within the TLTF chain may arise because the sub-Lihir mantle has been convectively flushed by rising MORB-like magmas associated with the actively-spreading Manus Bark-arc basin, to the southwest (Figure 1). A tectonic association between Lihir and the Manus Basin is supported by structural features on and near Lihir (including several recently discovered seamounts) which lie on an extension of the Manus spreading center through New Ireland[15].

In any case, as in most arc settings, He in the volcanos of the TLTF chain is dominantly derived from the depleted mantle wedge. Thus even in a locality where a slab component is thought to be unusually large, He does not provide a particularly good tracer for slab volatiles. This likely reflects the fact that the downgoing package is very poor in helium relative to the wedge.

Rabaul

It has been suggested that the $^3He/^4He$ ratio of volcanic gases may rise when fresh magma, uncontaminated by crustal helium, enters an existing magma chamber[18]. As a consequence, rising $^3He/^4He$ ratios may signal an increased likelihood of impending volcanic eruption. In conjunction with previous data, our Rabaul samples allow us to test this prediction. Our samples from Rabaul were taken on April 16, 1994, shortly prior to a major eruption which commenced on September 21. This eruption was focused through several intracaldera vents, including one (Tavurvur) within a few hundred meters of the sampled spring (Figure 2c). This same general area was sampled by H. Craig in March of 1986.

We collected two gas samples which yielded identical raw $^3He/^4He$ ratios of 6.25 R_A. While one of these had a low He/Ne ratio (5 times atmospheric), the agreement between the two samples leads us to suspect that there has been no significant addition of atmospheric helium to either sample. In 1986, one sample was collected from a vent on the summit of Tavurvur, while two samples were taken in the immediate vicinity (± 50 meters) of our sampling location. The Tavurvur sample yielded a $^3He/^4He$ ratio of 5.1 R_A. Of the samples taken very close to ours, one yielded a $^3He/^4He$ ratio of 4.55 R_A, while the other had a low He/Ne ratio (3 times atmospheric) but a much higher raw $^3He/^4He$ ratio of 6.2 R_A (H. Craig, personal communication). The corrected $^3He/^4He$ ratio of the latter sample cannot be reliably calculated, but because air contamination can only lower $^3He/^4He$, the raw ratio provides a solid lower limit for this sample. We find the similarity in the raw ratios in the two 1994 samples and this 1986 sample to be striking.

Two of the 1986 samples indicate an increase in the $^3He/^4He$ ratio between 1986 and 1994, but the third does not. While not conclusive owing to the limited spatial and temporal resolution of the sampling, these observations do not provide strong evidence for a systematic increase in the $^3He/^4He$ ratio of Rabaul geothermal fluids in the 8 years preceding the September 1994 eruption. This observation suggests either 1) magmas injected into the Rabaul magma chamber did not have an isotopic contrast with the magmas/fluids already present; or 2) injection occurred either prior to 1986, or between the 1994 sampling and the eruption. The possibility of emplacement prior to 1986 is given some support by the fact that an intense seismic crisis (consistent with rising magma) was observed within Rabaul Caldera between 1983 and 1985 (ref 19). A post-eruption sampling of Rabaul fluids would be useful to further constrain this issue.

CONCLUSIONS

Although the volcanos of the TLTF arc are believed to carry an unusually large slab signature, geothermal fluids from Lihir and Ambitle Islands have $^3He/^4He$ ratios only slightly lower than mid-ocean ridge basalts and fluids sampled from other purely oceanic arc volcanos. As at most arcs, He in the TLTF fluids is derived overwhelmingly from the mantle wedge and not from the slab, and therefore is not a useful tracer of slab volatiles. On both islands the geothermal gases are being driven by very late-stage magmas which are compositionally different from those responsible for stratovolcano construction. Thus analyses of phenocryst-borne helium are necessary to determine if this conclusion applies to all phases of volcanic activity on these islands.

Geothermal gases sampled from the Rabaul Caldera in 1986 and 1994 show no systematic change in He isotopic ratio preceding the large September 1994 eruption.

ACKNOWLEDGEMENTS

We wish to thank the Lihir Management Company and Father Tony Gendusa for providing invaluable assistance in collecting these samples. We also thank H. Craig for allowing us to cite his unpublished data from the Bismarck Archipelago. H. Hiyagon provided a helpful review. This work was supported by NSF OCE-9402159 to KAF.

REFERENCES

1.   Fryer, P., Pacific Science 44, 95-114 (1990).

2.   Craig, H. & Lupton, J. E., Earth Planet. Sci. Lett. 31, 369-385 (1976).

3.   Kurz, M. D. et al., Earth Planet. Sci. Lett. 58, 1-14 (1982).

4.   Ozima, M. et al., Nature 311, 449-451 (1984).

5.   Hiyagon, H., Science 263, 1257-1259 (1994).

6.   Poreda, R. & Craig, H., Nature 338, 473-478 (1989).

7.   Hilton, D. R. & Craig, H., Nature 342, 906-908 (1989).

8.   Gasparon, M., Hilton, D. R. & Varne, R., Earth Planet. Sci. Lett. 126, 15-22 (1994).

9.   Tsunogai, U. et al., Earth Planet. Sci. Lett. 126, 289-301 (1994).

10.   McInnes, B. I. & Cameron, E. M., Earth Planet. Sci. Lett. 122, 125-141 (1994).

11.   Kennedy, A. K., Hart, S. R. & Frey, F. A., J. Geophys. Res. 95, 6929-6942 (1990).

12.   Wallace, D. A. et al., Australian B. of Min. Res. Geology and Geophys. Report 243 (1983).

13.   Giggenbach, W. F., Sano, Y. & Wakita, H., Geochim. Cosmochim. Acta 57, 3427-3455 (1993).

14.   Patterson, D. B., Farley, K. A. & McInnes, B., EOS Trans. Am. Geophys. Union 75, 740 (1994).

15.   Herzig, P. et al., EOS Trans. Am. Geophys. Union 75, 513-515 (1994).

16.   Obzhirov, A.I., Geomarine Lett. 12, 54-59 (1992).

17.   Licence, P.S., Terrill, J.E. & Fergusson, L.J. Pacific Rim Congress 87, 273-278 (1987).

18.   Sano, Y., & Wakita, H., J. Geophys. Res. 90, 8729-8741 (1985).

19.   Mori, J., & Mckee, C. Science 235, 193-194 (1987).

HALOGEN GEOCHEMISTRY OF MANTLE FLUIDS IN DIAMOND

R. Burgess and G. Turner
*Department of Earth Sciences, University of Manchester, Oxford Rd., Manchester,
M13 9PL, U.K.*

ABSTRACT

Argon and halogens (Cl, Br and I) have been measured, using the ^{40}Ar-^{39}Ar stepped heating method, in diamonds from Jwaneng, Orapa (both in Botswana) and Zaire. The samples analysed included cubic (coated) stones and polycrystalline diamonds of eclogitic association. Both these types of diamond contain H_2O, CO_2, carbonate and silicate inclusions. Coated stones have relatively constant ^{40}Ar*/Cl and Br/Cl, show limited variation in I/Cl, and have normal mantle δ^{13}C values (-5 to -7‰). This contrasts with polycrystalline diamonds which, although having similar Br/Cl values to coated stones, possess significantly higher and more variable ^{40}Ar*/Cl and I/Cl values coupled with lower δ^{13}C values ($\leq$ -20‰). The origin of polycrystalline diamonds with high I/Cl-low δ^{13}C is tentatively considered in terms of the subduction of organic carbon and iodine in pelagic sediment. Coated stones have Br/Cl, I/Cl and δ^{13}C values that are similar to depleted upper mantle (MORB source). Mantle fluid trapped in the coated stones is enriched in halogens and ^{40}Ar by about a factor of 5000 relative to present-day upper mantle values. However, the estimated halogen content of the source from which the fluid derived is 7 ppm Cl, 25 ppb Br and 0.1-2.5 ppb I. These values are strikingly similar to those estimated previously for the source of MORB and therefore indicates that the halogens, like other volatile elements (e.g., noble gases, C and N), are homogeneously distributed throughout large portions of the upper mantle.

INTRODUCTION

The origin and evolution of volatile elements in the Earth's principle reservoirs are important problems in geochemistry. While it has long been known that volatile elements, e.g., CO_2, noble gases, halogens, N, S, etc. are predominantly concentrated in the external reservoirs of the Earth (atmosphere, oceans and crust), it is less clear whether transport of these elements from the deep Earth occurred during an initial period of catastrophic degassing, or by progressive degassing of the Earth's interior by volcanism throughout geological time. Models of catastrophic degassing have been invoked to explain the high ^{40}Ar/^{36}Ar value of MORB source ($\geq$ 25,000) compared to the atmosphere (296) [1-4]. These models require almost total extraction of ^{36}Ar (>99%) from the upper mantle during an early phase of Earth degassing, followed by growth of ^{40}Ar from radioactive decay of ^{40}K. Supporting evidence that an early phase of degassing affected the Earth comes from ^{129}I-Xe systematics. The atmosphere contains a large contribution of ^{129}Xe derived from ^{129}I in the upper mantle (e.g., ref 5). The relatively short half-life of ^{129}I (17 Ma) means that this isotope would have been extinct within about 100 Ma of nucleosynthesis. Therefore, severe degassing must have occurred before all the ^{129}I in the Earth had decayed. It also appears that lower mantle was much less affected by the early degassing episode as indicated by elevated ^{3}He/^{4}He in hot spot volcanics of ≥ 1.7x10^{-5} compared to MORB value at ≈ 1.1x10^{-5} (see ref 6 for data sources)

Noble gases and halogens show similar geochemical behaviour, both are normally incompatible elements in silicate melts and will tend to partition into the liquid or fluid phase during partial melting or crystallisation. Indeed, it appears that I may be considerably less compatible than Xe in mantle minerals[7]. The abundances of halogens in the major reservoirs of the Earth shows a marked progression from chlorine which is concentrated in seawater, to Br which is roughly equally split between seawater and crust, through to iodine which is concentrated mainly in the crust. This distribution is a direct consequence of the increasing degree of association with organic matter in order of Cl<Br<I. All halogens show relatively low abundance in the upper mantle. The Cl and Br contents of the oceans may be accounted for in terms of a continuous flux from the mantle, albeit at a decreasing rate throughout geological time[8]. The high concentration of iodine in the crust raises the possibility of using this element as an indicator of sediment recycling in the mantle.

The efficiency with which volatile elements like noble gases and halogens (Cl, Br and I) can be transported back into the mantle by subduction processes is largely unknown. Recycling of Xe into the upper mantle has been considered in geodynamic models[5,9], but the quantitative transport of sediment-derived noble gases into the mantle by subduction has been challenged[10]. There is evidence from subduction zone volcanics that incompatible elements (LIL, LREE) are returned to the mantle above the slab by de-watering of the oceanic crust and sediments (e.g., ref. 11). However, the mechanisms for return of elements beyond the limit of arc volcanism are less clearly defined. Previously it has been proposed that diamonds with low $\delta^{13}C$ values ($\leq$ -20‰) have inherited their isotopic signature from organic matter subducted in pelagic sediments (see ref. 12 for a review), although alternative explanations include either mantle-related carbon isotope fractionation processes[13,14], or primordial carbon isotope heterogeneity in the Earth[15]

To investigate the nature of volatiles in the mantle we undertook a detailed study of the noble gas (Ar) and halogen (Cl, Br and I) composition of mantle fluids trapped in diamonds. Diamonds were chosen because their inherent physical and chemical properties make them robust containers of pristine mantle samples ($\geq$150 km). Moreover, since diamonds have formed at various periods throughout time (1-3 Ga[16]), their study also holds the potential of obtaining "fossil" isotopic and chemical compositions of the ancient mantle. Thus, in future studies of diamonds it will be important to establish links between age measurements and chemistry in the same samples. The use of ultra-low blank laser ^{40}Ar-^{39}Ar dating methods represent the best hope for achieving this aim and considerable progress is currently being made[17-19].

EXPERIMENTAL METHODS

Using extensions of the ^{40}Ar-^{39}Ar method it is possible to measure noble gases and halogens simultaneously in the same sample, so that their relative chemical behaviour and distribution can be compared. During nuclear irradiation of samples, thermal neutron reactions convert F, Cl, Br and I to stable noble gas isotopes[20].

$$^{19}F(n,\gamma)^{20}F(\beta)^{20}Ne$$

$$^{37}Cl(n,\gamma)^{38}Cl(\beta)^{38}Ar$$

$$^{79}Br(n,\gamma)^{80}Br(\beta)^{80}Kr$$

$$^{127}I(n,\gamma)^{128}I(\beta)^{128}Xe$$

In each of these reactions the β decay times are on the order of minutes and therefore have completely decayed prior to analysis. The reaction on ^{19}F was not utilised in this study, it has a relatively small thermal neutron cross section which, given a typical thermal neutron dose of 10^{18} n cm^2, limits its use on our instrument to about 1 nanogram F (this does not necessarily preclude F measurements in future diamond studies). Noble gases were extracted by stepped heating of the diamonds in a double vacuum furnace capable of attaining a temperature of 2200°C which is significantly above the diamond graphitisation point (2000°C).

Two kinds of fluid-bearing diamond were analysed. Cubic diamonds with a central (mainly inclusion-free) core surrounded by a fibrous coat containing sub-micrometer inclusions (from hereon these are referred to as coated stones). Studies of coated stones have included the mineralogy and chemistry of the inclusions[21-23]; argon and chlorine relations[24]; noble gases[25] and C and N isotopes[13,14,26,27]. Samples for the present study consist of 10 brown and green stones from an unidentified source in Zaire (Congo tops) and 3 green cubes from Jwaneng. The second type of diamond investigated were black polycrystalline stones from Jwaneng (2 samples) and Orapa (5 samples), both in Botswana. Polycrystalline diamond (framesite) consists of aggregates of randomly-oriented diamond crystals. Crushed residues of polycrystalline stones were found to contain macroscopic garnet and pyroxene inclusions indicating an eclogitic association. Infra-red absorption spectra indicate that both coated and polycrystalline diamonds contain water, CO_2 and carbonate[22,23] (M. Schrauder, personal communication).

RESULTS

Noble gases were released by stepped heating diamond fragments between 600 and 2150°C in six temperature intervals; typical release patterns for halogens are shown for two diamonds in Fig. 1. Two major releases are observed; for Br and Cl (and Ar) the major releases occurs at 2000°C, corresponding to graphitisation, with a minor peak at about 1200°C. Iodine shows a less dominant release at high temperature, and a more prevalent peak at ≤900°C. The 1200°C release of Cl was noted previously[24] where the suggested cause was mechanical rupture of the largest inclusions. This may also explain the low temperature iodine release, although it is not obvious why this occurs at 200-300°C lower temperature. However, we note that halogen and argon ratios are indistinguishable in gases released at both low and high temperature.

Coated and polycrystalline stones contain excess ^{40}Ar (^{40}Ar*) with most samples giving anomalously old apparent ages >4.5 Ga. ^{40}Ar* shows a good correlation with Cl and Br. This is illustrated for ^{40}Ar* and Cl in Fig. 2 for coated stones from Jwaneng and Zaire. The ^{40}Ar*/Cl value of $(8.5 \pm 0.9) \times 10^{-4}$ molar (M) obtained from Fig. 2. is indistinguishable from that obtained previously from coated stones from Zaire[24]. ^{40}Ar*/Cl values for polycrystalline diamonds are higher and show significantly more variation $(3-76) \times 10^{-4}$ M.

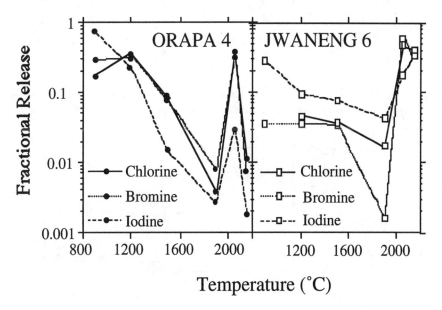

FIG.1 Fractional release of Cl, Br and I versus temperature for stepped heating of a polycrystalline stone (Orapa 4) and a coated stone (Jwaneng 6). Halogens are measured as noble gas isotopes (^{38}Ar$_{Cl}$, ^{80}Kr$_{Br}$ and ^{128}Xe$_{I}$) produced during nuclear irradiation.

Br and Cl show a good correlation in both cubic and polycrystalline stones (Fig. 3) with a Br/Cl value of $(1.5 \pm 0.2) \times 10^{-3}$ M which is, within error, equal to both the seawater and MORB ratios of 1.55×10^{-3} M and about 1.4×10^{-3} M (ref. 8 and Table I) respectively. Figure 3 shows that I and Cl are not as well-correlated, although coated diamonds show lower and less variable I/Cl $[(2-10) \times 10^{-5}$ M] than polycrystalline diamonds $[(30-300) \times 10^{-5}$ M]. Both types of diamonds have I/Cl values considerably higher than that of seawater (I/Cl = 9.5×10^{-7} M), but only the values from coated stones encompasses the MORB value of I/Cl = 3×10^{-5} M (see Table I for data sources).

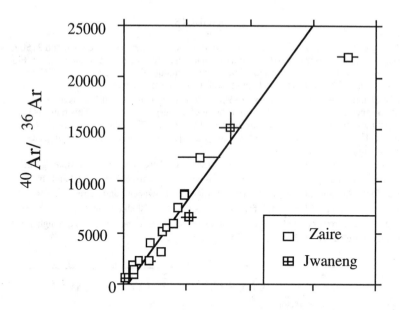

FIG. 2 ^{40}Ar/^{36}Ar versus Cl/^{36}Ar for coated stones form Zaire and Jwaneng. Each point represents the total release from an individual diamond. The best-fit line gives ^{40}Ar*/Cl = 8.5 x 10^{-4} M.

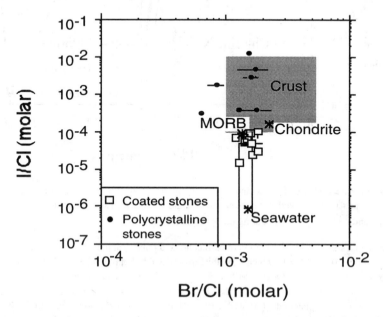

FIG. 3   I/Cl versus Br/Cl for coated and polycrystalline stones. Each point represents the total release from an individual diamond. The shaded area is the range characteristic of crustal values[28-31].

DISCUSSION

Halogen content of the sub-continental mantle

The halogen content of the mantle fluid trapped in coated diamonds are estimated to be 2-5 wt.% Cl, 60-120 ppm Br and 1-3 ppm I and the concentration of ^{40}Ar is (1-3) x 10^{-2} ml/g. This represents about a factor 5000 enrichment relative to present-day upper mantle values (MORB source). To achieve this level of enrichment the fluid phase must be present at a level of about 200 ppm (c.f. ref. 24). The halogen abundances present in the mantle region from which the fluids derive can be estimated assuming an upper mantle ^{40}Ar concentration of 4 x 10^{-6} ml STP/g (ref. 1) and using the measured ^{40}Ar*/Cl value in diamond of 8.5 x 10^{-4} M (Fig. 2). The estimated values are 7 ppm Cl, 25 ppb Br and 0.1-2.5 ppb I. In Table I these values are compared with abundances in the MORB source, crust and seawater. From the values given in Table I, it can be seen that there is close correspondence between the halogen contents of the sub-continental mantle (from diamonds) and sub-oceanic mantle (MORB source)[32,33]. Two important implications of this finding are: (1) The apparent uniformity of halogen abundances and ratios throughout large regions of the upper mantle; a result which is consistent with observations made for other volatile elements (e.g., noble gases, C and N). (2) The halogens are not significantly fractionated with different degrees of partial melting (this is also partly an implicit assumption made in the calculation to obtain the values given in Table I). More specifically, it probably rules-out significant involvement of Cl-bearing minerals (e.g., amphiboles, phlogopite or hydroxychlorides) during formation of the fluids.

TABLE I Halogen abundances in the main terrestrial reservoirs

	Cl (ppm)	Br (ppb)	I (ppb)
Diamond[§]	8	25	(0.1-2.5)
MORB[¶]	7	20	0.8
Crust	130	2700	820
Seawater	19000	67000	64

[§]Assumes upper mantle [^{40}Ar] = 4 x 10^{-6} ml/g (ref 1); ^{40}Ar*/Cl = 8.5 x 10^{-4} M.
[¶]MORB data: Cl and Br from ref. 32; I from ref. 33.

Recycling of halogens in the mantle

The fate of subducted halogens is of considerable importance for understanding the origin of these elements in external reservoirs, especially the oceans, which represent the main reservoir for Cl and to a lesser extent Br[8]. According to mass balance calculations, Cl and Br are not subducted, but released as volatile species from island arcs[8]. The similarity between the Br/Cl values of seawater, MORB and diamonds (Table I), means that it is unlikely that evidence for recycled material can be distinguished using this ratio. However, it may be possible to trace sediment subduction by use of iodine because of the strong affinity that this element has for organic matter. In this regard, iodine determinations carried-out in conjunction with C isotope studies may represent a particularly promising approach.

Figure 4 shows δ^{13}C versus I/Cl for coated and polycrystalline stones from Zaire, Jwaneng and Orapa. The coated stones show a restricted range in δ^{13}C (-5 to -7 ‰) that are typical of mantle carbon values[14,26,27] and relatively low I/Cl values that are similar to MORB. Two of the polycrystalline stones (one each from Orapa and Jwaneng), give mantle δ^{13}C values and high I/Cl values. However, the remaining polycrystalline diamonds are more depleted in ^{13}C and have δ^{13}C values as low as -20‰. These diamonds have I/Cl values that are up to two orders of magnitude higher than those of coated diamonds.

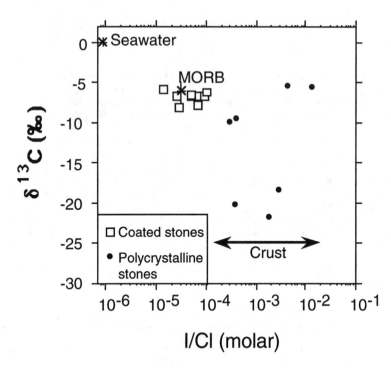

FIG. 4 δ^{13}C versus I/Cl for polycrystalline and coated stones. Crustal I/Cl values are from data sources given in Fig. 3

There are essentially three hypotheses regarding the origin of diamonds with unusually low δ^{13}C values and the halogen data will be briefly discussed in the context of each of these.

(1) Primordial carbon isotope heterogeneities[15]. However, as is shown in Fig. 3, the halogen ratios in the diamonds are distinct from those of meteorites[34].

(2) Fractionation of C isotopes during diamond growth[13,14]. If the carbon isotope signatures of the polycrystalline diamonds were generated by isotopic fractionation during growth then either the same process may be responsible for fractionating I from Br and Cl, or the isotopic fractionation may have occurred in a region of the mantle with unusually high I/Cl.

(3) The lack of a viable isotopic fractionation process capable of producing diamonds with δ^{13}C values as low as -20‰ has led to speculation concerning the involvement of subducted organic matter in their genesis. Supporting evidence for a subduction origin of some diamonds comes from S and Pb isotope compositions of sulphide inclusions in eclogitic diamonds that are within the range of crustal values[35]. If diamonds formed from subducted organic carbon then they be expected to have high I/Cl values (i.e., significantly higher than MORB) because marine sediments are enriched in I. Therefore, C isotope and I/Cl signatures of polycrystalline diamonds (Fig. 4), which indicate a separate formation event from coated stones, are consistent with a crustal origin. If this interpretation is substantiated by further measurements of C isotopes and I in diamonds, this would be the first indication of deep recycling of halogens (I) in the mantle.

For sediment recycling to be viable, significant amounts of I must be transferred to the deep mantle (>150 km) without significant fractionation from the associated organic matter. Geodynamic models for C indicate significant subduction of this element (e.g., ref. 36), and even relatively mobile species like B and Be, which are only loosely adsorbed onto clay minerals, may be transported into deep regions of the mantle. What then is the situation for I, can this element be subducted and in what amounts? These questions are best addressed by considering the quantity of iodine that can potentially be subducted in pelagic sediments (layer 1 of oceanic crust), since this is the dominant reservoir for I. Sediments are efficiently subducted into the mantle at a rate of about 1 km^3 yr^{-1} (ref. 37). The amount of I available for subduction in pelagic sediments can be estimated from the following sediment criteria: organic C content of 0.27 wt.%[38], I/C of ~10^{-4} M (ref. 39) and density of 2.0 g cm^3. This gives about 6 x 10^9 g I yr^{-1} available for subduction. Unknown proportions of I from the sediments in the subducting slab are: (a) removed by hydrous fluids, (b) returned to the crust *via* island arc volcanism, or (c) stored in the shallow mantle above the slab (the mantle wedge). Organically-bound I should be relatively insoluble in hydrous fluids and therefore, unlike many other mobile and incompatible elements (e.g., B, Be, Cs, Ba and K), is more likely to be subducted to greater depths. Analysis of a limited number of IAB indicates that they are only slightly enriched in I compared to MORB[33]. In contrast, hotspot volcanics have higher I contents than both MORB and IAB, although a large variation has been measured (7-363 ppb)[33], and this may indicate I transport to the relatively deep levels of plume involvement.

CONCLUSIONS

Mantle fluids in diamonds are of interest because they have the potential to record "fossil" geochemical signatures of the ancient mantle. Noble gases and halogens show a strong geochemical coherence due to their incompatible behaviour. These two groups of elements can be measured simultaneously using extensions of the ^{40}Ar-^{39}Ar method. Moreover, use of stepped heating enables separation of different noble gas and halogen carriers (e.g., atmospheric or crustal contamination). Mantle fluids in coated diamonds are enriched in Ar and halogens by about 5000 times over the present day upper mantle values. However, the concentrations of these elements in the source from which the fluids derived is estimated to be close to that of the sub-oceanic upper mantle (MORB source). Coated stones show halogen and Ar and carbon isotope ratios consistent with origin from a primitive upper mantle source. In contrast, the enhanced I/Cl values and lower $\delta^{13}C$ signature of polycrystalline diamonds, indicates a separate diamond-forming event which may have involved subducted crustal material.

ACKNOWLEDGEMENTS

We would like to thank J. Harris and J. Milledge for sample provision; D. Blagburn and B. Clementson for technical assistance; J. Saxton for software development and some initial sample analyses and D. Mattey for kindly providing C isotope data. This work benefited from discussions with J. Harris and M. Schrauder. We thank K. Farley and D.Patterson for reviews and K. Farley for editorial assistance.

REFERENCES

1. C.J. Allègre, T. Staudacher and P. Sarda, Earth Planet. Sci. Lett. **81**, 127 (1986).
2. Y. Zhang and A. Zindler, J. Geophys. Res. **94**, 13719 (1989).
3. I.Y. Azbel and I. Tolstikhin, Geochim. Cosmochim. Acta **54**, 139 (1990).
4. M. Ozima and K. Zahnle, Geochem. J. **27**,185 (1993).
5. I. Ya. Azbel and I.N. Tolstikhin, Meteoritics **28**, 609 (1993).
6. L.H. Kellog and G.J. Wasserburg, Earth Planet. Sci. Lett. **99**, 276 (1990).
7. D.S. Musselwhite, M.J. Drake and T.D. Swindle, Nature **352**, 697-699 (1991)
8. J.-G. Schilling, C.K. Unni and M.L. Bender, Nature **273**, 631 (1978).
9. D. Porcelli and G.J. Wasserburg, U.S.G.S. Circular **1107**, 255 (1994).
10. T. Staudacher and C.J. Allègre, Earth Planet. Sci. Lett. **89**, 173 (1989).
11. A.D. Saunders, M.J. Norry and J. Tarney, Phil Trans. R. Soc. Lond. **A335**, 377 (1991).
12. M.B. Kirkley, J.J. Gurney, M.L. Otter, S.J. Hill and L.R. Daniels, Appl. Geochem. **6**, 477 (1991).
13. S.R. Boyd, F. Pineau and M. Javoy Chem. Geol. **116**, 29 (1994).
14. M. Javoy, F. Pineau and H. Delorme Chem. Geol. **57**, 41 (1986).
15. P. Deines, J.W. Harris, P.M. Spear and J.J. Gurney, Geochim. Cosmochim. Acta **53**, 1367 (1989).

16.   S.H. Richardson, J.J. Gurney, A.J. Erlank and J.W. Harris, Nature **310**, 198 (1984).
17.   R. Burgess, G. Turner, M. Laurenzi and G. Turner, Earth Planet. Sci. Lett. **94**, 22 (1989).
18.   R. Burgess, G. Turner and J.W. Harris, Geochim. Cosmochim. Acta **56**, 389 (1992).
19.   R. Burgess, S.P. Kelley, G. Turner and J.W. Harris, U.S.G.S. Circular **1107**, 43 (1994).
20.   G. Turner, J. Geophys. Res. **70**, 5433 (1965).
21.   G.D. Guthrie, D.R. Veblen O. Navon and G.R. Rossman, Earth Planet. Sci. Lett. **105**, 1 (1991).
22.   O. Navon, I.D. Hutcheon, G.R. Rossman and G.J. Wasserburg, Nature **335**, 784 (1988).
23.   M. Schrauder and O. Navon, Geochim. Cosmochim. Acta **58**, 761 (1994).
24.   G.Turner, R. Burgess and M.P. Bannon, Nature **344**, 653 (1990).
25.   M. Ozima and S. Zashu, Earth Planet. Sci. Lett. **105**, 13 (1991).
26.   S.R. Boyd, D.P. Mattey, C.T. Pillinger, H.T. Milledge, M. Mendelson and M. Seal, Earth Planet. Sci. Lett. **86,** 341 (1987).
27.   S.R. Boyd and C.T. Pillinger, Chem. Geol. **116**, 43 (1994).
28   D.E. White, J.D. Hem and G.A. Waring, U.S.G.S. Prof. Paper 440-F (1963).
29   J.K. Bohlke and J.J. Irwin, Geochim. Cosmochim. Acta **56**, 203 (1992)
30   J.J. Irwin and E. Roedder, Geochim. Cosmochim. Acta **59**, 295 (1995).
31   J.J. Irwin and J.H. Reynolds, Geochim. Cosmochim. Acta **59**, 355 (1995).
32.   J.-G. Schilling, M.B. Bergeron and R. Evans, Phil. Trans. R. Soc. Lond. **A297**, 147 (1980).
33.   B. Déruelle, G. Dreibus and A. Jambon, Earth Planet. Sci. Lett. **108**, 217 (1992).
34.   G. Dreibus, B. Spettel and H. Wänke, in *Origin and Distribution of the Elements*, edited by L.H. Ahrens,(Pergamon, Oxford, 1979), p. 33.
35.   C.S. Eldridge, W. Compston, I.S. Williams, J.W. Harris and J.W. Bristow, Nature, **353**, 649 (1991).
36.   M. Javoy, F. Pineau and C.J. Allègre, Nature **300**, 171 (1982).
37.   R. von Huene and D.W. Scholl, Rev. Geophys. **29**, 279 (1991).
38.   R. Chester and S.R. Ashton, Chemical Oceanography, J.P. Ripley and R. Chester, eds., (Academic Press, London 1976), **6**, p.281.
39.   H. Elderfield and V. Truesdale, Earth Planet. Sci. Lett. **50,** 105 (1980).

THE ORIGIN AND EVOLUTION OF THE TERRESTRIAL ALKALI ELEMENT BUDGET

K. Lodders and B. Fegley, Jr.
Dept. of Earth and Planetary Sciences, Washington University
St. Louis, MO 63130-4899

Abstract

The origin and evolution of the terrestrial alkali element inventory is investigated in the framework of the accretion and differentiation history of the Earth. We predict that a significant percentage of the Earth's bulk alkali element inventory is in the core (30% for Na, 52% for K, 74% for Rb, and 92% for Cs). These predictions agree with independent estimates from nebular volatility trends and (for K) from terrestrial heat flow data. Vaporization and thermal escape during planetary accretion are unlikely to produce the observed alkali element depletion pattern. However, loss during the putative giant impact which formed the Moon cannot be ruled out.

Introduction

The depletion of Na, K, Rb, and Cs in the Earth's upper mantle and crust relative to their abundances in primitive, undifferentiated meteorites (chondrites) is a long standing problem in geochemistry[1,2]. The cosmochemical classification of the elements based on condensation temperatures in the solar nebula divides the alkali elements into moderately volatile (Na) and volatile (K, Rb, Cs) elements with increasing volatility from Na to Cs, while elements such as Al, Sr, and U are classified as refractory elements[3]. Elemental abundance ratios of moderately volatile and volatile elements to refractory elements (e.g., Na/Al, K/U, Rb/Sr) in chondrites reflect volatility related fractionations if these ratios are smaller than those found in CI-chondrites (which have primordial, solar composition except for the atmophile elements H, O, C, N and noble gases). Geochemically, the alkali elements are normally regarded as lithophile elements and therefore it is commonly assumed that the observed alkali element abundances in the Earth's mantle and crust represent the entire alkali element budget of the Earth. As can be seen from Table I, the Na/Al, K/U, and Rb/Sr ratios of the bulk silicate Earth are generally much lower than the respective chondritic ratios, which indicates that the silicate Earth underwent a massive depletion of alkali elements.

One requirement for modeling the alkali depletion process is that this process should integrate into the framework of accretion and differentiation of the Earth. In the next sections we will argue that two processes are responsible for producing the observed depletions, namely, core formation and vaporization during the hypothesized giant impact which formed the moon.

Accretion and Differentiation of the Earth

We adopt a heterogeneous accretion scenario for the Earth that is similar to models proposed by Wänke[4] and Ringwood[5]. An important feature of these models is the absence of core-mantle equilibrium. The model used here and the models of Wänke and Ringwood involve two stages: (a) initial accretion of highly reduced matter and core formation; and (b) accretion of more oxidized matter and a moon-forming impact to the Earth after core formation.

The underlying concept in this study is that chondritic matter is representative of the material which formed larger planetary objects. This is the difference between our model and the models of Wänke and Ringwood, who postulate that the initially accreting component is devoid of elements more volatile than Na. As illustrated in Table 1, none of the known chondrite groups match the hypothetical component A of Wänke and Ringwood. We discuss further below why we use EH-chondrites as the initially accreting component.

Table I. Some Compositional Data for Chondrites, the Aubrite Parent Body (APB) and the Earth and Moon.

Type	Total Fe wt%	Silicates Fe/(Fe+Mg) mole %	Fraction Fe as metal Fe_{met}/Fe_{tot}	metal & sulfide, wt%	$\delta^{17}O^a$ $^o/_{oo}$	$\delta^{18}O^a$ $^o/_{oo}$	Na/Al	K/U	Rb/Sr	Cs/Rb
Carbonaceous Chondrites										
CI	19.04	45	0	3	~ 8.8	~ 16.4	0.576	68300	0.295	0.081
CM	21.00	43	0	8.7	~ 4	~ 12.2	0.347	36400	0.168	0.074
CO	24.80	33	0-0.2	~ 10.2	~ -5.1	~ -1.1	0.287	26500	0.114	0.055
CV	23.50	35	0-0.3	14.4	~ -4	~ 0	0.189	18200	0.082	0.076
Ordinary Chondrites										
LL	18.50	27	0.11	9.4	3.9	4.9	0.588	60800	0.279	0.058
L	21.50	22	0.29	13.3	3.5	4.6	0.573	63500	0.279	0.090
H	27.50	17	0.58	23.2	2.7	4.2	0.566	65000	0.290	0.041
Enstatite Chondrites										
EL	22.00	0.26	0.74	28.7	2.9	5.6	0.552	73500	0.305	0.040
EH	30.62	0.98	0.62	36.6	2.7	5.2	0.860	88900	0.439	0.068
Differentiated Bodies										
APB	(30.62)	0.072-0.72	0.68	38.2	2.78	5.34	0.736^c	68700^c	0.257^c	0.053^c
Earth	32.04	11	0.82	32.5^b	2.78	5.38	0.138^c	10500^c	0.028 - 0.035^c	0.022 - 0.036^c
Moon	10.60	19	~ 0	2.5	2.78	5.38	0.019	2500	0.009	0.045

(a) bulk isotopic composition relative to SMOW (standard mean ocean water); (b) Earth's core also contains oxygen; (c) element ratio from silicate portion only. Data sources: [6-12, and references therein].

A generally accepted and likely accretion scenario invokes condensation of chondritic matter from the nebula gas and accumulation of solids into larger planetesimals. These planetesimals accrete further to form meteorite parent bodies and also the terrestrial planets. While the primary compositional signature from condensation processes is relatively well preserved in chondrite parent bodies (although chondrule formation and thermal metamorphism have partially erased the nebular signature), metal-silicate fractionation during core-mantle differentiation in the terrestrial planets efficiently sorted the initially more homogeneously distributed metal, sulfide, and silicate phases so that siderophile and chalcophile elements fractionated from the lithophile elements. These fractionation processes strongly depend on the oxidation state of the differentiating parent body.

We now address the question of which type of chondritic planetesimals accreted to form the Earth. From Table I we can see that chondrites display a wide range in oxidation state, as indicated by their molar FeO/(FeO+MgO) ratios in silicates, and their total metal and sulfide content. This is also shown in Figure 1. The oxidation state of terrestrial silicates falls between that of ordinary (LL, L, H) chondrites and enstatite (EH and EL) chondrites. The chondrites roughly plot along a line and the corresponding data for the Earth fall right into this trend. This plot also may indicate that there is

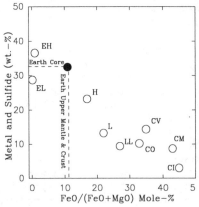

Figure 1. Comparison of the silicate oxidation state and metal-sulfide content in chondrites with corresponding data for the Earth. The terrestrial core size is easily explained if accretion started from EH-chondritic matter.

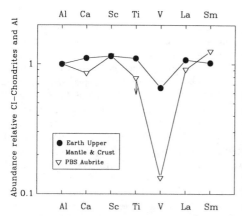

Figure 2. Refractory elemental abundances in the Earth's upper mantle and in the Pena Blanca Spring aubrite[11,13]. The depletion of V is explained by partitioning into metal-sulfide under highly reducing conditions.

indeed a relationship between chondritic compositions and bulk composition of the Earth. If the Earth formed from chondritic matter, it is necessary to supply enough metal and sulfides to account for the terrestrial core size of about one third of the Earth's mass. To fulfill this requirement, we have to turn to EH-chondritic matter, which is the only type of chondritic matter which can supply enough metal and sulfide. However, the current oxidation state of the Earth's silicates is higher than that of enstatite chondrites and thus at least two components, a reduced and an oxidized one, are necessary. This heterogeneous accretion has been discussed in the models by Wänke[4] and Ringwood[5], which show that about 2/3 of the Earth accreted from highly reduced matter and about 1/3 oxidized matter.

There are additional indications that the Earth's accretion started from highly reduced enstatite chondritic matter. One is the (well known) common oxygen isotopic signature of enstatite chondrites and the Earth-Moon system (Table 1). No other chondrite group plots along the terrestrial O-isotope fractionation line[10]. The only other meteorites matching the isotopic signature are the differentiated enstatite-rich meteorites known as aubrites, for which a link to enstatite chondrites is established (see below).

Another connection of terrestrial accretion to enstatite meteorites comes from the depletion of refractory V in the bulk silicate Earth[14,15]. Figure 2 shows the abundance of refractory elements in the Earth's upper mantle and in the Pena Blanca Spring aubrite. The depletion of V in both systems is obvious, although the depletion in the Earth's mantle is smaller than that in aubrites. Under the current oxidation state of the Earth's mantle V behaves as a lithophile element. However, in a highly reduced environment V becomes chalcophile and siderophile[16] and can be removed from silicates by a metal-sulfide melt. This partitioning explains the depletion of V in aubrites[13] and it is plausible to assume that as long as the Earth accreted reduced matter, removal of V into the core also occurred.

After about 2/3 of the Earth had accreted from reduced matter, more oxidized planetesimals accreted to the Earth and oxidized the mantle to its current state[4,5]. If the moon were formed by the impact of a Mars-sized body with the Earth and subsequent recondensation of evaporated mantle and impactor material[17], then this impact most likely occurred during or after accretion of the oxidized component to the Earth. The reason for postulating this late impact (after core formation) is due to the fact that the moon lacks a large core, implying that no metal was available in the recondensing matter. However, while it is relatively easy to identify the reduced chondritic component which initially accreted to the Earth, the task of finding the oxidized component among chondrites is more difficult. For modeling purposes, it is easiest to assume CI-composition, as done by Wänke and Ringwood in their models[4,5,14,15], but in that case, the requirement that the oxygen isotopic signature of the Earth-Moon system is preserved is violated. In addition, the elemental contribution of the Mars-sized impactor is certainly not negligible, but no attempts have yet been made to estimate the composition of this supposedly differentiated object. However, we will see below that despite these obstacles, we can model the evolution of the terrestrial alkali element budget by using the accretion scenario discussed so far.

ALKALI ELEMENT REMOVAL INTO THE EARTH'S CORE

As discussed above, the early Earth probably was highly reduced. From the mineralogy and chemistry of highly reduced meteorites (the enstatite chondrites and achondrites) we can see that we have to modify our views about the commonly accepted geochemical behavior of the elements. For example, the high degree of reduction in enstatite meteorites resulted in the partial occurrence of Si metal alloyed with FeNi, and led to the formation of exotic sulfides such as oldhamite (CaS), niningerite ((Mg,Fe)S), alabandite ((Mn,Fe)S and Ti-bearing troilite[18]. Two minerals of special interest in this study are djerfisherite (K_3(Cu,Na)(Fe,Ni)$_{12}$(S,Cl)$_{14}$) and caswellsilverite (NaCrS$_2$), which contain about 10 wt% K and 16 wt% Na respectively. Their occurrence in enstatite meteorites underlines the chalcophile nature of the alkali elements under reducing conditions. As a consequence, some fraction of the alkali elements may have partitioned as sulfides into the Earth's core during its early differentiation.

Alkali element partitioning into the Earth's core was modeled by assuming that alkali element partitioning during core formation on the aubrite parent body (APB) is analogous to that on the early Earth[12]. Aubrites (enstatite achondrites) represent samples of the silicate portion of the highly reduced, differentiated aubrite parent body. These brecciated meteorites consist mainly of almost iron free pyroxene (enstatite)[19]. Other major minerals are forsterite, diopside and albite. It has been shown earlier[13] that the bulk APB is most likely EH-chondritic in composition and that the silicate portion of the APB (which is sampled by the aubrites) formed as a result of core-mantle differentiation under very reducing conditions. If, as seems likely[12,13], the APB and the early Earth differentiated from EH-chondritic matter, then the APB provides a natural laboratory to study how core-mantle differentiation proceeds under highly reducing conditions. Because we can determine the compositions of the primitive EH-chondrites and the differentiated aubrites by measuring samples available from our meteorite collections, we can estimate the composition of the core formed by differentiation of EH-chondritic matter.

The calculated concentrations of the alkali elements in the core of the APB are obtained from the general mass-balance equation :

$$C(EH) = X(silicates) \cdot C(silicates) + X(core) \cdot C(core) \qquad (1)$$

where C stands for concentration by weight and X for the mass fraction of silicates (= 0.618) and core (0.382) in the APB. Details of these calculations are described elsewhere[12] and are only briefly repeated here. The composition of the entire APB is assumed to be EH-chondritic (Table II, column 1) and the composition of the silicate portion of the APB is computed from element correlations for analyses of aubrites (column 2). The calculated elemental abundances in the core of the aubrite parent body are listed in column 3. These results show that the APB core contains 2590 ppm Na, 550 ppm K, 3.4 ppm Rb, and 0.31 ppm Cs. Using equation (1) and the compositions of caswellsilverite and djerfisherite in column 4, we can also calculate how much caswellsilverite and djerfisherite are required on the entire APB to remove the alkali elements as sulfides into the core. For Na, about 0.63 wt% caswellsilverite is necessary and for K and Rb we need about 0.22 wt% djerfisherite.

Table II. Alkali Element Mass-Balance in the Aubrite Parent Body
and Alkali Element Concentration in Sulfide Minerals.

	EH-Chondrites (= bulk APB)	APB silicates (= aubrites)[b]	APB Core	Djerfisherite (D) & Caswellsilverite (C)
	(1)	(2)	(3)	(4)
Al	8200±200	13300±500	0	0
Na	7050±60	9800±500	2590±1000	15.7±1 wt% (C)
K	850±60	1030±120	550±260	9.5±1 wt% (D)
Rb	3.1±0.4	2.9±1.1	3.4±2.1	660 (D)
Cs	0.21±0.06	0.15±0.10	0.31±0.24	?

(a) data in ppm, if not noted otherwise, for data sources see [12]; (b) concentration of APB silicate from element correlations of aubrite samples; (c) core composition calculated from data in column (1) and (2) and equation (1).

The amounts of each mineral required are fairly low and agree with the trace abundances of these minerals that are observed in enstatite meteorites[20,21]. In addition to the low abundances of djerfisherite and caswellsilverite, the appearance of these minerals as solidified relics of sulfide melts in aubrites makes it plausible to assume that they are indeed alkali element carriers into the core.

If we assume that elemental fractionations during core formation on the APB and the early Earth are similar, then the Earth's core could contain the same alkali element abundances as the core of the APB. Indeed the concentration of alkali elements in the Earth's core calculated here agree with independent published estimates based on nebular volatility trends[11] (see also discussion below). One could argue that the analogy between the APB and early Earth is only valid as long as the Earth is comparable in size to asteroidal objects like the APB, because higher pressures and temperatures prevailing in the growing Earth may have altered the partitioning behavior of alkali elements between sulfides and silicates. However, Stevenson[22] presented arguments that metal-sulfide/silicate equilibration most likely occurred in the top layers of a differentiating body and that the dispersed metal-sulfide bodies from an incoming undifferentiated object equilibrated with the surrounding silicates until the metal-sulfides bodies accumulated into larger blobs which sank to the core. Because the pressure regime in the upper region of a larger planet is probably not very different than that in asteroidal sized bodies, we can expect that elemental partitioning between metal, sulfide, and silicates under comparable redox conditions and temperatures will lead to similar results.

Even if we assume that metal-sulfide blobs equilibrated with surrounding silicates in deeper portions of the Earth, we expect that the partitioning of the alkali elements into sulfides still took place. The terrestrial occurrence of djerfisherite in diamond[23] shows that djerfisherite is stable at higher pressure in deeper regions of the mantle. In addition, sulfide/silicate partition experiments at 4-6 GPa and 1525-2585°C show that K partitions more strongly into sulfides at high pressures than at lower pressures[24]. However, it is important to keep in mind that alkali removal into the core will only occur as long as highly reducing conditions are present during accretion and core formation. Once accretion proceeds with more oxidized matter, the redox state of the terrestrial silicates will not allow the continued formation of djerfisherite and caswellsilverite, and the alkali elements will remain lithophile.

At this point we want to mention that enstatite meteorites also contain oldhamite (CaS) which is the main Ca and REE carrier in these meteorites. It may appear from Figure 2 that some Ca was lost from the Pena Blanca Spring sample, indicating possible removal of CaS into the core. However, this depletion is an artifact from the normalization to Al, which is about 10% lower in the Pena Blanca Spring sample than in CI chondrites. Presently, no definitive conclusion can be drawn regarding possible Ca depletions in the silicate portion of the APB and possible CaS loss to the APB core. Comparison of the Ca abundances in the silicate portion of the APB and enstatite chondrites is hampered by the inhomogeneous distribution of CaS in aubrites and the lack of any element correlation of Ca with other major elements. Thus, no estimate of the total Ca abundances in the APB silicates can be made. It is also not very likely that CaS is as efficiently removed into the core as are the alkali-bearing sulfides because of its high melting point of ~2720 K and its low density (ρ=2.5 g/cm^3), which is lower than that of silicates. Formation of CaS solid-solutions with FeS, (which is likely to have occurred for the alkali element bearing sulfides and FeS) probably did not occur because the denser Fe/FeS melts (ρ> 4.5 g/cm^3) are removed at lower temperatures, before a possible FeS-CaS eutectic can form. The Fe-FeS eutectic is at ~1270 K and the FeS-CaS eutectic is at 1393 K (at a composition of 80 wt.-% FeS and 20 wt.-% CaS[25]). Given the fact that CaS is about 5 times less abundant than FeS in EH chondrites, the temperature where a solid solution between CaS and FeS would occur are even higher. Therefore, liquid Fe-FeS melts would migrate towards the core before CaS can be incorporated into FeS, leaving CaS behind in the silicate.

ALKALI ELEMENT REMOVAL BY THE MOON-FORMING IMPACT

After the reducing stage of accretion, the bulk silicate Earth would be similar in composition to aubrites, if the analogy between the early Earth and the APB is made. As can be seen from Table I, the elemental ratios (Na/Al, K/U, Rb/Sr) in the silicate portion of the APB (the aubrites) are significantly higher than those presently observed in terrestrial silicates. Because continuing accretion of a more oxidized component probably added even more alkali elements to the silicate portion of the Earth, some mechanism is required to reduce the alkali element abundances to their current state.

Vaporization is one possible mechanism to remove alkali elements from the early Earth. However, evaporation from a magma and thermal escape from the Earth during accretion and differentiation is unlikely for elements as heavy as the alkalis[26]. To loose alkalis, temperatures must first rise high enough to evaporate alkali elements and their compounds. Second, the gaseous vapors must leave the gravitational field of the (growing) Earth. Vaporization of the alkalis can be calculated using the MAGMA code[27], while thermal (Jeans) escape is governed by the ratio of the thermal velocity (v_{th}) of the gaseous species to the planetary escape velocity (v_{esc}). Two cases (v_{esc}/v_{th} = 1 and 5) were considered at the Fe-FeS eutectic temperature of 1270 K where planetary differentiation begins[26]. The results show that alkali loss is negligible from Mars sized and larger bodies under these conditions. Thus, thermal escape of the alkali elements from the Earth after core formation can be ruled out.

However, another efficient process involving vaporization which could have removed volatile elements from the Earth may have been the putative Moon-forming impact. If there were no loss of volatiles during the impact, one would expect that lunar Na/Al and K/U ratios would be similar to those in the terrestrial silicates, because the common oxygen isotope systematics suggest that the terrestrial and lunar silicates have a common parent. However, the lunar volatile to refractory element ratios are significantly lower than those found in the Earth's silicates (Table I) indicating that alkali elements were lost from the Earth-Moon system. Details of this volatile loss mechanism have not yet been modeled and require further investigation.

ALKALI ELEMENT BUDGET IN THE EARTH

We derive the alkali element abundances for the bulk Earth (core + mantle + crust) by combining our estimated abundances in the Earth's core with published estimates of alkali element abundances in the silicate portion of the Earth[11]. Terrestrial silicate abundances were derived from upper mantle and crust data[11]. We follow the generally made assumption that the abundances determined for the upper mantle and crust are representative for the entire silicate portion of the Earth and that no differences for alkali element abundances are found in the lower mantle.

Table III. Alkali Element Abundances in the Earth and in Chondrites (ppm).

	Earth' Mantle & Crust [11]	Core (this work)	Core[11]	Bulk Earth (this work)	Bulk Earth[11]	CI-Chondrites[6]	Ordinary Chondrites[7]	EH-Chondrites[12]
Na	2932 (2330-3640)	2590 (± 1000)	1400 ?	2820	2450	5000	6400-7000	7050±610
K	232 (175-300)	550 (±260)	210 ?	340	225	558	780-825	850±60
Rb	0.6 (0.48-0.73)	3.4 (±2.1)	1.1 ?	1.5	0.76	2.3	2.9-3.1	3.1±0.4
Cs	0.013 (0.007-0.025	0.31 (±0.24)	0.14 ?	0.11	0.055	0.187	0.12-0.18	0.21±0.6

Bulk Earth abundances calculated from equation (1) and assuming a core mass fraction of 0.325 for the Earth.

The data for the silicate portion, core, and the bulk Earth are given in Table III. Alkali element abundances for the bulk Earth of 2800 ppm Na, 340 ppm K, 1.5 ppm Rb, and 110 ppb Cs are obtained using the mass balance equation (1). A large fraction of the terrestrial alkali elements is hidden in the core (30% Na, 52% K, 74% Rb, and 92 % Cs). However, the calculated bulk Earth abundances of the alkalis are only 40% (Na), 43% (K), 50% (Rb) and 80% (Cs) of those found in ordinary chondrites suggesting that the process which established the alkali abundances in the terrestrial silicates was dominated by vaporization.

Our derived alkali abundance estimates for the core are consistent with independent results obtained by Kargel and Lewis[11]. These authors calculated the volatility trend for the Earth and compared their predictions with observed abundances in the silicate Earth. The difference between calculated bulk Earth abundances and the silicate portion was assigned to the core. Their core data are also shown in Table III and we see that their data are about 1.5-3x smaller than the core abundances determined here. However, within the uncertainties of the data both studies are consistent.

We can also compare our calculated K abundance with K abundance estimates from heat-flow data. Recently Breuer and Spohn[28] calculated that the present day heat flow can be matched if about 20 ppb U are in the mantle and 400-800 ppm K are in the core. Their estimated K abundance agrees well with the 550±250 ppm K derived in this study.

Acknowledgments. We thank Dr. J. Larimer for helpful comments on the manuscript and Dr. T. Ahrens for the invitation to contribute this paper. This work was supported by grants from the NASA Origins of Solar Systems and Planetary Atmospheres Programs (B. Fegley, P.I.).

REFERENCES

1. H. C. Urey, Proc. *Natl. Acad. Sci.* **41**, 127-144 (1955).
2. P.M. Hurley, *Bull. Geol. Soc. Amer.* **68**, 379-382 (1957).
3. H. Palme, J. W. Larimer and M. E. Lipschutz, in: Meteorites and the early solar system (Univ. Arizona Press, AZ., 1988), pp. 436-461.
4. H. Wänke, *Phil. Trans. R. Soc. Lond.* **A303**, 287-302 (1981).
5. A. E. Ringwood, *Proc. R. Soc. Lond.* **A395**, 1-46 (1984).
6. E. Anders and N. Grevesse, *Geochim. Cosmochim. Acta* **53**, 197-214 (1989).
7. J. T. Wasson and G. W. Kallemeyn, *Phil. Trans. R. Soc. Lond.* **A325**, 535-544 (1988).
8. D. W. G. Sears and R. T. Dodd, in: Meteorites and the early solar system (Univ. Arizona Press, AZ., 1988) pp. 3-31.
9. S. T. Taylor, in: Origin of the Moon (Lunar and Planetary Institute, Houston TX., 1986), pp. 125-143.
10. R. N. Clayton, *Ann. Rev. Earth Planet. Sci.* **21**, 115-149 (1993).
11. J. S. Kargel and J. S. Lewis, *Icarus* **105**, 1-25 (1993).
12. K. Lodders, *Meteoritics* **30**, 93-101 (1995).
13. K. Lodders, H. Palme and F. Wlotzka, *Meteoritics* **28**, 538-551 (1993).
14. H. Wänke, G. Dreibus and E. Jagoutz, in: Archean Geochemistry (Springer Verlag Berlin, Germany, 1984) pp. 1-24.
15. A. E. Ringwood, in: Origin of the Earth (Oxford Univ. Press, N. Y., 1990), pp. 101-134.
16 M. J. Drake, H. E. Newsom and C. J. Capobianco, *Geochim. Cosmochim. Acta* **53**, 2101-2111, (1989).
17. W. Benz and A. G. W. Cameron, in: Origin of the Earth (Oxford Univ. Press, N. Y. 1990), pp. 61-67.
18. K. Keil, *Geophys. Res. Lett.* **73**, 6945-6976 (1968).
19. T. R. Watters and M. Prinz, *Proc. Lunar Planet. Sci. Conf.* **10th**, 1073-1093, (1979).
20. L. H. Fuchs, *Science* **153**, 166-167 (1966).
21. A. Okada and K. Keil, *Amer. Mineral.* **67**, 132-136 (1982).
22. D. J. Stevenson, *Science* **241**, 611-619 (1981).
23. G. P Bulanova, O. Y. Shestakova and N. V. Leskova, *Dok. Akad. Nauk SSSR* **255**, 430-433 (1980).
24. E. Ohtani, K. Ohnuma and E. Ito, in: High-Pressure Research (Terra Publ., Tokyo, 1992), pp. 341-349
25. R. Vogel and T. Heumann, *Archiv. Eisenh.* **15**, 195-199 (1941).
26. K. Lodders, *Meteoritics* **29**, 492-493 (1994).
27. B. Fegley, Jr. and A. G. W. Cameron, *Earth Planet. Sci. Lett.* **82**, 207-222 (1987).
28. D. Breuer and T. Spohn, *Geophys. Res. Lett.* **20**, 1655-1658 (1993).

FORMATION OF CARBON SPECIES IN TERRESTRIAL MAGMAS

A. A. Kadik

V.I.Vernadsky Institute of Geochemistry and Analytical Chemistry, Russian Academy of Sciences, Kosigin St.19, Moscow V 334, Russia

ABSTRACT

Experiments on C(graphite) + basaltic melt and C(graphite) + Mg_2SiO_4 + melt, C(graphite) + Mg_2SiO_4 + (CO-CO_2) vapor + melt at 15 -40 kbar, 1400-1700^{0}C have shown that carbon may be soluble in forsterite and pyroxene in concentrations of 10-100 ppm, and reactions with graphite lead to formation in the melt of 100-1000 ppm CO_2 concentrations. Carbon is an incompatible element in the melt + crystals system. The redox state of the mantle and interactions among primary carbon, melt and crystals are suggested to play an important role in the formation of carbon species in terrestrial basalts.

INTRODUCTION

There is general agreement that primary carbon and the redox state of the mantle play an important role in the formation of carbon species in terrestrial basalts, as well as in the composition of gases evacuated by these liquids to the upper layers of the Earth. Over the past fifteen years most attention has been focused on the formation of carbon species in terrestrial magmas as a result of CO_2 solubility in melts. This model proposes melting of the upper mantle in the presence of some amount of fluid phase or carbonate at depth. In contrast it is assumed that accretional degassing, melting and vaporization of the solid at impact led to depletion of the original volatiles of the Earth, and the presence of a primary fluid phase in the mantle is rather doubtful[1,2]. In this case the knowledge of free carbon solubility in melt and crystals may be critical to understanding the carbon contents of magmas.

It is assumed that free carbon should react with magma in the process of partial melting of the upper mantle to produce CO_3^{2-}(carbonate ion) in the melt[3,4]

$$\text{·C(graphite)} + O^{2-}\text{(melt)} + O_2 = CO_3^{2-}\text{(melt)} \quad (1)$$

or

$$[CO_3^{2-}]_{melt} = K\,[O^{2-}]_{melt}\,fO_2 \quad (2)$$

Eq.(2) shows that CO_3^{2-} or CO_2 content in a melt in equilibrium with graphite is a function of oxygen fugacity (fO_2). This means that the evolution of the redox state of the mantle should be important for the formation of carbon species in magmas .

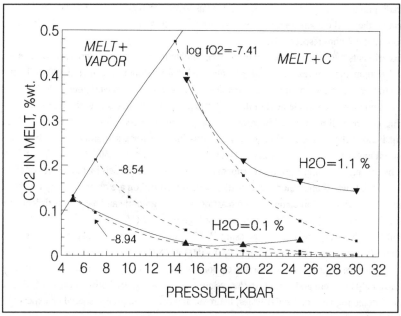

Fig.1 Experimental determination of graphite solubility in basaltic melt at 5-30 kbar and 1400° C. Dissolved carbonate ion contents are expressed as CO_2, wt %. C(graphite) + CO + CO_2 (CCO) equilibrium imposes a limit on the carbon stability and on the maximum value of oxygen fugacity in the basalt-graphite system. Carbon solubility at given fO_2 (dashed curves) - calculation based on Holloway's model[3].

Another mechanism that may control the formation of C-bearing components in magmas is the carbon solubility in minerals (Ol,Px,Gar) in concentrations of a hundred ppm by participating in point defect equilibria[5,6]. The source of the carbon species in the melt in this case is the fractionation of carbon between melt and carbon-bearing crystals. The mechanism and scale of this process are poorly understood.

In this study we provide: 1) experimental determination of the carbon solubility in melt and crystals at high pressure, 2) data on the upper mantle oxidation state and the free carbon distribution with depth, 3) estimation of carbon species formation in magmas during fluid- absent melting of the carbon-bearing mantle.

INTERACTION BETWEEN CARBON, CRYSTALS AND MELT: EXPERIMENTS.

The experiments[4] were made on C(graphite)-melt-crystal equilibria for C(graphite) + basaltic melt + CPx at 15-30 kbar and 1350-1400°C and on C(graphite) + Mg_2SiO_4 + melt, C+Mg_2SiO_4 +(CO-CO_2) vapor + melt for CaMgSi_2O_6-Mg_2SiO_4 composition at 25-40 kbar, 1400-1700°C. The solubility of carbon in crystals and silicate melts was determined by registration of the beta activity from ^{14}C and by the infrared determinations of species in glasses. Analytical treatment of autoradiograms by electron microprobe allowed us to see the distribution of carbon and to determine the local carbon content[7]. The scanning areas and the carbon detection limit were 25 microns and 0.2 ppm respectively.

Experiments have shown that carbon may be soluble in forsterite and pyroxene in concentrations of 10-100 ppm and that reactions with graphite lead to formation in the melt of 100-1000 ppm CO_2 concentrations. The distribution coefficient D = (C in melt/C in crystal) is about 10-100. Thus carbon is an incompatible element in the melt + crystals system and during the fluid-absent partial melting magma will be enriched in carbon in comparison with crystalline residue.

Fig.1 shows the results of experimental examinations of carbon solubility in basaltic melt with the initial amount of H_2O = 0.1 % and 1.1 % wt. at 1 - 30 kbar and 1250 - 1400°C. The fO_2 was not fixed by the oxygen buffer during experiments and was controlled itself by the system of given composition. It was found that structural sites of dissolved carbon are determined by the formation of the CO_3^{2-} carbonate ion complex. The reaction between graphite and basaltic melt leads to the formation of 0.04-0.40 % wt. CO_2 in the melt at high pressure (15 - 30 kbar, 1350 - 1700°C). In the presence of (CO, CO_2) vapor at low pressure (1 - 5 kbar, 1250°C.) the reactions between graphite and basaltic melt lead to the formation of 0.01 - 0.03 wt.% CO_2.

The influence of dissolved H_2O on the carbon concentration in melt was found. It increases with increasing H_2O content and the fO_2 in the system (Fig.1). Reactions between carbon and melt species interacting with molecular water, hydroxyl and carbonate groups should be expected in the fluid-absent region:

$$C(graphite)+2OH^-(melt)+2O=H_2O(melt)+CO_3^{2-}(melt) \quad (3),$$

where O represents a distinguishable reactive oxygen species e.g. bridging, non-bridging or free. Reaction (3) has an equilibrium constant defined by a ratio involving the activities of melt species:

$$[CO_3^{2-}] = K \, [OH^-]^2 \, [O]^2 \, /[H_2O] \qquad (4).$$

Eq.(4) provides a simple way to think about the influence of the ratio of dissolved molecular H_2O to hydroxyl groups on the reaction of carbon with melt. Melts with a higher ratio of OH^- species to molecular water and higher oxygen activity would have a higher concentration of carbonate ion in the melt.

LINES OF EVIDENCE ON FREE CARBON DISTRIBUTION IN THE UPPER MANTLE.

There has been a considerable debate over the redox state of the mantle and the distribution of free carbon with depth[8-12]. Data on the oxygen fugacities (fO_2) recorded by mantle rocks and experiments indicate that 1) the evolution of the upper mantle is characterized by a wide range of redox conditions mainly in the range between the quartz-magnetite-fayalite (QMF) and the iron-

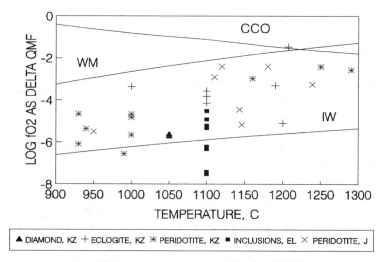

Fig. 2. The oxidation state of diamond-bearing peridotates[9,12,14,16] , eclocates[17] and mineral inclusions in diamonds[11] . fO$_2$ is given in log unites relative to the quartz-magnetite-fayalite (QMF) buffer.
(Kadik, Zharkova et al, Kadik, Sobolev et al - KZ[14,16,17], Eggler and Lorand -EL[11], Jaques et al -J[12])

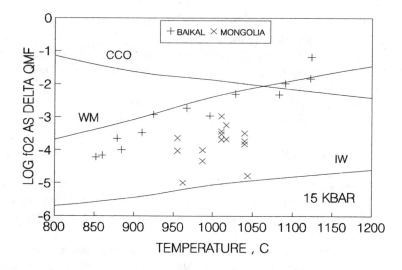

Fig.3. The oxidation state of spinel lherzolite xenoliths of the upper mantle beneath continental rift system, Mongolia[15] and Baikal[18].

wustite buffer (IW) and oxidation could be considered as the main evolutionary trend , 2) primitive and less modified spinel lherzolites may be in equilibrium with free carbon (graphite)[8-10], 3) carbon may be stored in mantle minerals (olivine, pyroxene, garnet) in concentrations of 10-100 ppm[5,9,13].

For the demonstration of these conclusions the following methods were used:
1) electrochemical measurements of the intrinsic fO_2 of coexisting minerals (Ol,OPx,CPx,Spl) from diamond-bearing peridotite and diamond-bearing eclogite nodules and slightly depleted spinel lherzolalite nodules from Mongolia and the Baikal region[9,14-17], 2) Raman spectroscopic determination of graphite particles in minerals[9] , 3) carbon measurements in olivines, pyroxenes and garnets by $^{12}C(d,n)^{13}N$, $^{12}C(d,n)^{13}N$ and $^{12}C(d,p)^{13}C$ nuclear reactions[9,13] .

Diamond-bearing peridotite and diamond-bearing eclogite nodules from kimberlites are supposed to be largely disintegrated representatives of the ancient lithosphere and astenosphere. The observed evolution of oxygen fugacities is characterized by a wide range of redox conditions mainly in the range between the wustite-magnetite (WM) iron-wustite buffer[11,12,14,16,17] (Fig.2). The overwhelming majority of samples fall well below the fO_2 defined by the C-CO-CO$_2$ buffer (CCO), the theoretical upper fO_2 permitted in the presence of elemental carbon. A separate fluid phase, if it is stable, would be saturated by carbon . Highly reduced diamond-bearing peridotites are interpreted as a relict from an earlier, lower-fO_2 regime in the outer carbon-bearing layers of the primitive Earth .

The spinel lherzolite nodules from the alkaline basalts of Mongolia are rather close in composition to the primitive upper mantle. The crystals are of pure gem quality and are free from fluid or other inclusions. Measurements indicate that the upper mantle beneath the continental rift systems is characterized by a wide range of redox conditions mainly in the range between the WM and IW buffers[9,15,18] (Fig.3). Most of the slightly depleted lherzolites are moderately reduced (around WM and WM-2). At 900-1100 °C, 10-20 kbar the range of oxygen fugacities correspond to the graphite stability in the C-O-H system.

This conclusion is confirmed by the presence of rare and extremely fine-grained C crystals in minerals and carbon dissolved in olivines and pyroxenes (5-100 ppm). Raman spectroscopy has been shown to be a sensitive indicator of graphite crystallinity in rocks and it indicates that graphite from the spinel lherzolites of Mongolia[9] is fully crystalline with a well defined, single first-order peak at 1581 cm^{-1} .

The carbon measurements in minerals by$^{12}C(d,n)^{13}N$, $^{12}C(d,n)^{13}N$ and $^{12}C(d,p)^{13}C$ nuclear reactions[9,19] show that carbon is dissolved in olivines, pyroxenes and garnets at 5-100 ppm (Fig.4). This concentration is below or very close to carbon saturation of the minerals at high pressure[4,9] (Fig.4).

The range of fO_2 values observed in peridotites from the oceanic upper mantle[8] is given in Fig.5. In general, upper mantle fO_2's are too high for graphite to be stable, but the most reduced samples derived from hot-spot peridotites and part of the rapidly quenched basalts consists of carbon-saturated fluids.

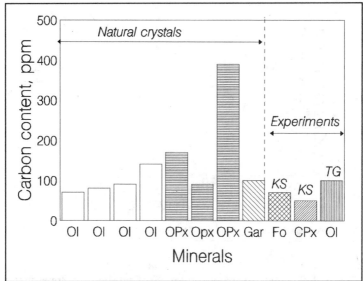

Fig.4. The carbon contents in minerals of lherzolite xenoliths (Mongolia) determined by [12]C (d,n) ^{13}N niclear reactions (Minaev,Shilobreeva and Kadik - SK[13]), and experiments (Kadik and Shilobreeva-KS[4], Tingle and Green-TG[5]).

FLUID-ABSENT MELTING OF CARBON-BEARING MANTLE.

Based on the experimental data it may be suggested that the carbon-melt-crystal reactions and oxidation during fluid - absent melting of carbon bearing mantle could control the formation of C-bearing species in basaltic magmas.

Analysis requires the consideration of three situations in the region of magma formation: 1) melting in the presence of the very low content of fluid phase that is more traditional in petrology (melt+fluid+crystal equilibrium), 2) fluid-absent melting in the presence of some free carbon (carbon saturated equilibrium of melt+crystal + graphite) and 3) fluid-absent melting of carbon-bearing minerals
(carbon unsaturated equilibrium of melt + carbon-bearing minerals).

According to experiments, the reaction between graphite and basaltic melt in the presence of some H_2O (0.1-0.5 % wt.), at 15-30 kbar pressure and redox condition in the range between fO_2= WM and fO_2= CCO may lead to the formation of 0.1-0.5 % wt.CO_2 in magma. This CO_2 content is observed generally in basaltic glasses and expected in primary magmas. Thus we conclude that the fluid-absent melting in the presence of free carbon could explain the formation of carbon-species in basaltic magmas. Oxidation of graphite-bearing mantle with increasing oxygen fugacity in the range of IW buffer up to CCO, which is typical for evolution of the upper mantle, creates a condition for the increase of carbon solubility in magmas (Fig.6).

Carbon contents in mantle minerals (Fig. 4) and carbon-silicate-liquid partition coefficient (D = C(melt)/ C(crystal)) can be used to model the carbon content in melt and residual crystals during partial melting of carbon-bearing minerals in the absence of the free carbon and fluid phases. The

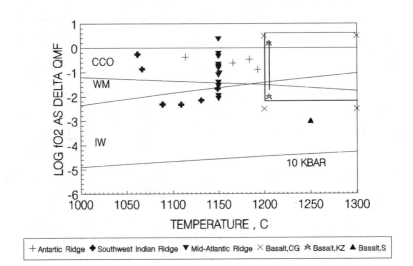

Fig.5. The oxidation state of spinel peridotites of sub-oceanic upper mantle (Bryndzia and Wood[8]) and basalts (Christie et al - CG[19], Kadik and Zharkova - KZ[9], Sato-S[20]).

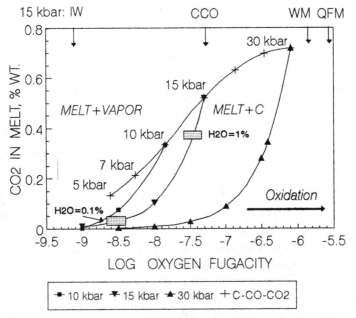

Fig.6. The effect of fO$_2$ on graphite solubility in basaltic melt according to experiments (Fig.1) and Holloway's model[3].

source of the carbon species in the melt in this case is partitioning of carbon between melt and carbon-bearing crystals. A simple assumption of batch and fractional melting indicates a large fraction of carbon in the melt during the initial stage of melting (several 100's of ppm). Melting from 10 to 15 % leads to formation of 300-400 ppm C or 0.1-0.15 wt % CO_2 in melt. This carbon content in melt is also consistant with CO_2 content observed in basaltic glasses (Fig.7).

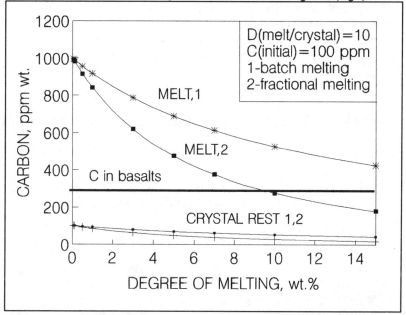

Fig.7. The carbon content in melt and residual crystals during partial melting of carbon-bearing minerals.

CONCLUSIONS.

The experimental determination of carbon solubility in melt and crystals at high pressure and data on the upper mantle oxidation state as well as the free carbon distribution with depth lead to the conclusion that fluid - absent melting of carbon-bearing peridotites could resulted in the formation of carbon species in mantle magmas in an amount comparable with carbon concentration in basaltic melts (several 100 ppm).

ACKNOWLEDGMENTS.

This research is supported by grant number MA 8000 of International Sciences Foundation.

REFERENCES

1. G.Arrhenius, Earth,Moon and Planets 37, 187 (1987)

2. H.Wanke, G.Dreibus and E.Jagoutz, In A.Kroner et al. (eds.), Archaean Geochemistry (Springer-Verlag, Berlin Hedelberg, 1984), 1.

3. J.R. Holloway, Eur.J.Mineral, **4**, N 1, 1054,
(1992).

4. A.A.Kadik, S.N.Shilobreeva, Mineral.Mag., 58A, 460 (1994).

5. T.N. Tingl, H.W.Green II, Geology, 15, 324 (1987).

6. E.B.Watson, Geophys. Res. Lett., 13, N 6, 529 (1986).

7. V.G. Senin, S.N. Shilobreeva, A.A.Kadik, Nucleonica, 36, 45 (1991).

8. L. Bryndzia and B.J. Wood, Am.J.Sci., 290, 1093, (1990).

9. A.A.Kadik, Proc.Indian Acad.Sci. 99,N 1, 141 (1990)

10. Balthaus C., Contrib. Mneral. Petrol., 114, 31,
(1993).

11. D.H.Eggler, J.P.Lorand, In: Kimberlites and related rocks, (1993,),in
press.

12. A.L.Jaques, H.St.C. O'Neill, C.B.Smith et al,
Contrib.Mineral.Petrol., 104, 255 (1990).

13. V.M.Minaev, S.N.Shilobreeva, A.A.Kadik, J.Radioanal.Nucl.Chem., in press (1995).

14. A.A.Kadik,N.V.Sobolev, H.V.Zharkova, N.P.Pohkilenko, Geochem. Intern., 27, N3, 41
(1990).

15. A.A.Kadik, H.V.Zharkova, V.I.Kovalenko, D.A.Ionov, Geochem. Intern.,
26, N1, 12 (1989).

16. A.A.Kadik, H.V. Zharkova, E.S.Efimova, N.V.Sobolev, Doklady Nauk, Russia, 328, N3, 389
(1993).

17. A.A.Kadik, H.V.Zharkova, Z.V.Specius, Doklady Nauk,USSR, 320, 440 (1992).

18. A.A.Kadik, H.V.Zharkova et al, Doklady Nauk, Rassia, in press (1994).

19 D.M.Christie, J.S.E. Carmichael, Ch.H. Langmuir, Earth and Planet.Sci.Lett., 79, 397 (1986).

20. M.Sato, Geol. Soc. Amer. Mem., 135, 289 (1972).

2. SOLAR SYSTEM VOLATILES

SOLAR SYSTEM FORMATION AND THE DISTRIBUTION OF VOLATILE SPECIES

Jonathan I. Lunine, Wei Dai and Fatima Ebrahim
Lunar and Planetary Laboratory, University of Arizona, Tucson, AZ 85721

ABSTRACT

Dynamical and chemical processes acting on *ice-forming* volatiles (i.e., water ice and more volatile species) within the solar nebula can lead to temporal and spatial dependencies which are complex and depend on transport in gaseous and solid phases. We examine some of these processes and then use the current ice inventories of Triton and Pluto to show that significant processing of these bodies has occurred relative to more primitive reservoirs such as comets.

INTRODUCTION

Spacecraft exploration and advanced ground-based spectroscopy have enabled the volatile budgets of the solar system's distant icy bodies to be quantified in recent years. Such an endeavor is important because observations of molecular clouds provide information on abundances of highly volatile species such as carbon monoxide, ammonia and methane[1]. Understanding of processes which led to the formation of outer solar system solid objects, and the giant planets, is then constrained by such observations only if we can assemble a compositional record of similar quality for the solar system. Of relevance here are comets, Titan, Triton and Pluto, and the newly discovered trans-Neptunian objects (which seem dynamically to be consistent with the predicted Kuiper disk of remnant planetesimals).

In this short paper we use new calculations and previous results to show the following:

1. Dynamical and chemical processes acting on volatiles within the solar nebula will lead to variations in space and time which are complex and depend on both gas and grain transport. Here we use the example of infall of grains from the nascent molecular cloud, which leads to a complex pattern of heating, sublimation and chemical reactions within the grain.

2. Following a suggestion by Stevenson[2], who showed with a simple model that volatile loss might be most severe from objects much smaller than Triton and Pluto, we model the formation of atmospheres around small primitive bodies, the source of which is outgassed volatiles driven by accretional or solar heating. We conclude that indeed substantial puffy atmospheres could form (as opposed to simple condensation of volatiles onto the surface) for objects of Kuiper Belt size or larger, but this is sensitive to the abundance of hydrogen available in the early solar nebula and in icy grains.

3. Triton and Pluto both have nitrogen-dominated surface volatile budgets, which is distinct from primitive reservoirs in which carbon monoxide should dominate over nitrogen. Both Triton and Pluto underwent significant modification of their volatile budgets during their early history. The question of what constitutes a reliably "primitive" reservoir in the outer system remains a difficult one. Comets do have a volatile budget which is consistent with what is known of gas and grains in molecular clouds, and hence could set constraints on the degree of alteration of grains during formation and infall into the solar nebula. However the radial location at which comets formed remains uncertain. Kuiper belt objects may be a better constrained sample in this regard, but likely underwent significant volatile loss during formation.

UPDATES ON GRAIN HEATING

Lunine et al.[3] quantified the heating of pre-existing molecular cloud grains during infall into the solar nebula and passage through postulated, mild accretion shocks at the surface of such a disk. The physics of the heating was based on Hood and Horanyi[4]. Here we have modified and improved the calculation by more naturally and explicitly including both the process of sublimation of volatiles and the heat released or taken up by other phase changes in the heated grains.

The picture is as follows: small icy grains fall inward with the collapsing gas of a protoplanetary clump. Such grains grow significantly during infall (perhaps to as much as 0.1 millimeters for fluffy particles[5]). In consequence, the grains decouple from the gas upon reaching the mild shock associated with the solar nebula disk surface. As a result grains end up with a velocity relative to the gas, leading to gas dynamical heating, sublimation, chemical modification of the grain, and eventual recondensation of sublimated constituents onto the grain in the saturated environment of the outer solar nebula.

Here we have extended the published results[3] to model two possible grain assemblages, for (in this case) a micron-sized, non-fluffy grain. The first assumes that the most volatile ices are arranged in an outer layer, concentric with the water ice and less volatile organics and silicates, so that volatiles individually sublime off of the grain beginning at very low temperatures (figure 1). The initial mass distribution is given in the table. The second assumes that a fraction (assumed here for concreteness to be 0.75) of the volatiles species is volumetrically adsorbed into amorphous water ice, evidence for which is present in experiments[6]. Here release of some volatiles is delayed until the amorphous to cubic transition at around 140 K.

Table: Assumed interstellar grain composition

Grain Component	Mass Fraction
$Silicates + MetalOxides$	0.20
$Carbon$ (Graphite)	0.06
Nonvolatile Complex Organic Refractory	0.19
H_2O	0.20
CH_4	0.007
CO	0.03
CO_2	0.02
N_2	0.015
C_2H_4	0.002
H_2CO	0.0005
NH_3HCO	0.0001
NH_3	0.002
Other $Molecules$ + $Radicals$	0.23

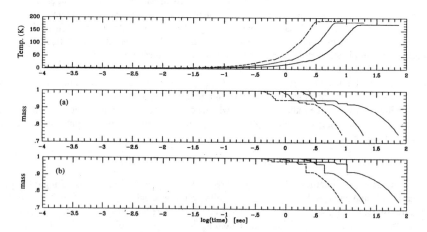

Fig. 1. Model of grain heating during infall to the solar nebula. Plotted versus time are grain temperature and grain mass for velocities of 4, 5 and 6 km/s. Two different mass versus time curves are shown. Panel (a) assumes all frosts are free to sublimate separately; panel (b) assumes 3/4 of the volatiles species are adsorbed (i.e. trapped) volumetrically in amorphous ice, and cannot be released until the amorphous-to-crystalline transition at roughly 140 K. Space limitations prevent labeling of each of the volatiles lost; the initial model grain composition is given in the table.

The results show that significant modification of molecular cloud grains is possible associated with infall and sublimation. While much of the material will be recondensed or adsorbed onto the grains after the latter cool[3], fractionation of volatiles according to the nature of the re-adsorption process will take place. This provides a straightforward set of predictions for volatile abundances in comet grains or other outer solar system grains, since rapid condensation and retrapping can be readily simulated in the laboratory[6]. Such predictions are work in progress, and the use of analytic laboratory devices on comets (via the European Space Agency mission *Rosetta*, for example) will be able to directly confront such predictions.

MASSIVE EARLY ATMOSPHERES ON PRIMITIVE OBJECTS

Stevenson[2] showed that when simple energy balance considerations are applied to volatile release and escape during accretion of icy objects, an object radius exists (roughly several hundred kilometers) at which the loss of volatiles is potentially a maximum. To some extent the location (and existence) of such a maximum is a function of the details of the outgassing and escape; we are exploring these issues. Here we show that one part of the escape process, formation of relatively warm, puffy atmospheres, is possible given the existence of sufficient molecular hydrogen.

We use an atmospheric radiative transfer model previously applied to a massive early atmosphere on Triton[7]. The physics is that of a grey atmosphere with collision-induced opacities supplied by molecular nitrogen, methane, carbon monoxide and molecular hydrogen. For a given surface heat flux (due to sunlight, accretional or tidal heating), the equilibrium surface temperature and pressure are computed assuming the atmosphere is in equilibrium with a layer of volatile ices which have been outgassed onto the surface. This is something of an oversimplification but not an unreasonable one.

Figure 2 shows the results. Here the base temperature of a volatile-driven atmosphere is plotted, for three different values of the hydrogen mole fraction, as a function of the surface heat flux. Marked on the figure are heat fluxes for insolation at 30 AU, and for accretional heating where the accretion time is given in years. For a Triton-sized object (not shown), elevated surface temperatures are possible for hydrogen mole fractions of 0.01% and higher, and for accretion times approaching 10^8 years (which is long). For comet-sized bodies up to hundreds of kilometers (which is the size of "Kuiper belt" objects so far detected[8]), insolation is sufficient to raise an atmosphere (in the presence of 0.01% hydrogen), but accretional heating on reasonable timescales is not.

The question of the source of enough hydrogen to raise such atmospheres is an important one. The background solar nebula would have, in the Uranus/Neptune region, a hydrogen gas pressure of 0.01-1 microbars. The puffy atmospheres, which tend to run away to a value corresponding to the critical pressure of the volatile components, have pressures of 10-100 bars; hence the mole fraction of hydrogen in such atmospheres is too low to sustain them if the source is the solar nebula gas itself.

Photolysis of organics is a possible source at later times when the nebula is optically thin to solar UV (or external stellar UV) radiation; this requires low gas pressures. Using a photochemical model for present-day Triton[9], a hydrogen mole fraction due to photolysis of 0.01% could be obtained.

Finally, trapping of significant amounts of molecular hydrogen is possible in clathrate hydrate ices[10], since the hydrogen is small enough to occupy cages already filled by methane, nitrogen and carbon monoxide. From a statistical mechanical model[10], one finds that relative to the primary volatile species trapped in the clathrate, a molecular hydrogen mole fraction of 0.01-1% is possible for double occupancy of voids. If double occupancy is disallowed, the fraction drops to 0.0001-0.01%. The double-occupancy values are adequate to sustain an atmosphere raised from such a volatile mix. Although adsorption in amorphous ice may have been preferred to clathration in the cold (50 K) grains of the outer solar nebula, the void sizes of the two ice structures are likely comparable and hence hydrogen might be abundant in amorphous ice as well. Experiments performed at 12 K indicate large amounts of molecular hydrogen in amorphous ice[18]. This source is attractive because it is available any time during accretion, whether the gas is present or not.

Given sufficient hydrogen, insolation alone may be enough to start the devolatilization process for small bodies as the nebula becomes optically thin. Objects exceeding hundreds of kilometers in radius may raise massive atmosphere due just to accretional heating, and 1000-km sized objects certainly will.

Such atmospheres enhance the escape rate by making gases less gravitationally bound relative to thin, cold atmospheres. The effectiveness of the escape process in removing and (most interestingly) fractionating volatiles as a function of radius is work in progress.

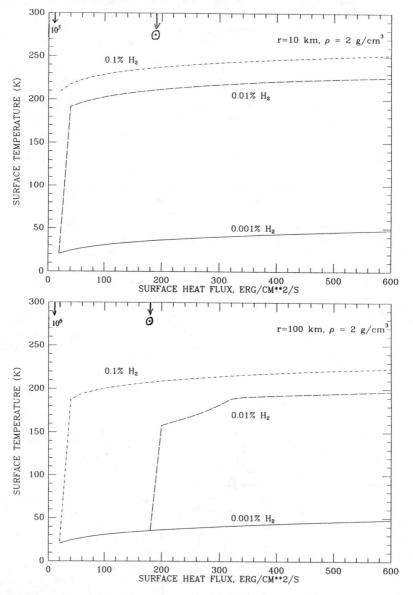

Fig. 2. Surface temperature versus surface heat flux for (top) 10 km and (bottom) 100 km radius objects. Each curve is labelled with the percentage of the atmosphere which is H_2. In both panels the lowest curve (0.001% hydrogen) corresponds to an optically thin atmosphere; elevated temperatures indicated optically thick (in the thermal infrared) atmospheres. Also shown are average surface heat flux due to sunlight (30 AU, albedo = 0.5; solar symbol) and accretion over the timescale indicated in years.

VOLATILE BUDGETS OF TRITON AND PLUTO

Enough data now exist to assemble initial volatile budgets for Triton and Pluto, both of which are thought to represent the upper end of Kuiper belt type objects and both of which likely were formed in solar orbit[11]. These budgets can be compared with those in more "primitive" reservoirs such as comets, and in theoretical primordial reservoirs.

Figure 3 shows the comparisons of nitrogen and carbon volatile budgets for the case of Triton (the results are similar for Pluto). For the surface of present-day Triton, we take an upper limit to the thickness of the frost deposits based on viscous spreading limits[12]. For the lower limit we use the fact that nitrogen is still present in the southern hemisphere and we compute the rate of transport over the season from last solstice to the present. Carbon monoxide is assumed dissolved in nitrogen and we use Raoult's law along with current atmospheric limits on abundance. Methane upper limits are given by assuming separate volatile transport of its frost; a lower limit on the original methane surface abundance is based on the photolysis rate integrated over the age of the solar system. The carbon dioxide number is derived directly from spectroscopic data[13].

The values for Triton in figure 3 are compared with "theoretical primordial" Tritons built up of comets[14] and from solar nebula and protosatellite nebula grains. These theoretical grain compositions are computed as described in McKinnon et al.[15]

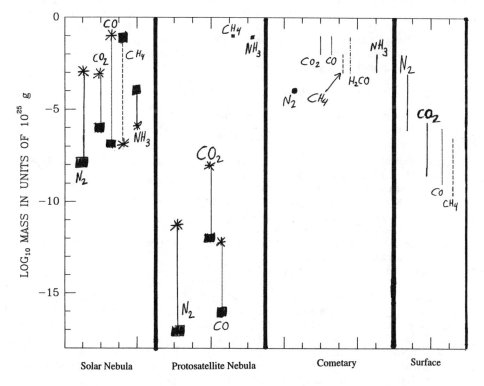

Fig. 3. Volatile budget for Triton. Surface values are derived from observations as summarized in the text. The comet models uses abundances from Mumma et al.[14]. Solar nebula and protosatellite nebula models are described in McKinnon et al.[15] In these models one must choose between grain catalysis versus no grain catalysis[17]. The squared ends represent the grain-catalyzed case, favoring lower temperature, reduced species; the starred ends correspond to no grain catalysis. Comparing values for different species, the reader must stay consistent in selecting either the starred or squared ends.

Comparing the theoretical bulk abundances with the surface values, we see that the current surface does not contain the bulk of the original volatile content which Triton was capable of possessing. The remainder has been lost or is still in the interior; tidal models suggest substantial lost. Although the *relative* abundances among the carbon species is not very different for the actual surface and theoretical bulk values, the carbon-to-nitrogen ratio is quite different. In particular, molecular nitrogen is grossly overabundant compared with the carbon species on Triton today.

Clearly volatile loss or selective outgassing has led to significant evolution of the volatile budget of Triton relative to primitive reservoirs. Carbon monoxide could have been hydrothermally processed into other species in Triton's interior[16], though whether such a model works for the other species remains to be determined. Loss of species through an early massive atmosphere and escape episode is also possible, with the current molecular nitrogen being at least in part a product of primordial ammonia. Even rough agreement between such a model and the current surface inventory remains to be demonstrated, but it appears from figure 3 that the "best" primitive inventory to start out with is comets.

SUMMARY

We have touched very quickly on aspects of the history of volatile species in the outer solar system and their implications for physical processes in the solar nebula. The next leap in understanding solar system ices will come from *in situ* analysis of collected material. In particular, understanding the way in which volatiles are trapped in the ice provides constraints on the thermal history of the grains and on the initial volatile composition of the material from which the grain was formed. However, the prospects for further progress are not limited by space flight programs; sensitive Earth- based spectroscopy, laboratory simulations and theoretical models are all at a relatively immature stage, and new discoveries on all these fronts are anticipated in the years ahead.

ACKNOWLEDGEMENTS

This work is supported by the NASA Origins Programs.

REFERENCES

1. E.F. Van Dishoeck, G.A. Blake B.T. Draine and J.I. Lunine, In Protostars and Planets III, eds. E.H. Levy and J.I. Lunine, (University of Arizona Press, Tucson, 1993), p. 163.
2. D.J. Stevenson, Lunar Planet. Sci. Conf. 24, 1355 (1993).
3. J.I. Lunine, S. Engel, B. Rizk and M. Horanyi, Icarus 94, 333(1991).
4. L.L. Hood and M. Horanyi, Icarus 93, 259 (1991).
5. S.J. Weidenschilling and T.V. Ruzmaikina, Astrophys. J. 430, 713 (1994).
6. A. Bar-Nun, I. Kleinfeld and E. Kochavi, Phys. Rev. B 38, 7749 (1988).
7. J.I. Lunine and M.J. Nolan, Icarus 100, 221 (1992).
8. D.C. Jewitt and J.X. Luu, Astron. J., in press.
9. J.R. Lyons, Y.L. Yung. and M. Allen, Science 256, 204 (1992).
10. J.I. Lunine and D.J. Stevenson, Astrophys. J. Suppl. 58, 493 (1985).
11. P. Goldreich, N. Murray, P.Y. Longaretti and D. Banfield, Science 245, 500 (1989).
12. R.H. Brown, and R.L. Kirk, J. Geophys. Res. 99, 1965 (1994).
13. D.P. Cruikshank, T.L. Roush, T.C. Owen, T.R. Geballe, C. deBergh, B. Schmitt, R.H. Brown and M.J. Bartholomew, Science 261, 742 (1993).
14. M.J. Mumma, P.R. Weissman and S.A. Stern, In Protostars and Planets III, eds. E.H. Levy and J.I. Lunine, (Univ. of Arizona Press, Tucson, 1993) p. 1177.
15. W.B. McKinnon, J.I. Lunine and D. Banfield, In Neptune and Triton, ed. D. Cruikshank. (University of Arizona Press, Tucson), in press.
16. E.L. Shock and W.B. McKinnon, Icarus 106, 464 (1993).
17. R.G. Prinn and B. Fegley, Jr., In Origin and Evolution of Planetary and Satellite Atmospheres, eds. S.K. Atreya, J.B. Pollack and M.S. Matthews, (University of Arizona Press, Tucson, 1989), p. 78.
18. R.W. Dissley, M. Allen and V.G. Anicich, Astrophys. J. 435, 685 (1994).

COMETS, IMPACTS AND ATMOSPHERES
II. ISOTOPES AND NOBLE GASES

Tobias Owen
University of Hawaii, Honolulu, HI 96822

Akiva Bar-Nun
Tel-Aviv University, Tel-Aviv, Israel

ABSTRACT

We suggest that impacts by icy planetesimals during the first billion years of solar system history played a major role in the origin of planetary atmospheres. In the outer solar system, the effect of these planetesimals is evident in the super-solar values of C/H and the variations of D/H in giant planet atmospheres. Laboratory experiments on the trapping of noble gases in ice forming at various temperatures provide an illuminating interpretation of noble gas abundances in the atmospheres of the inner planets, again implicating icy planetesimals. This model can be tested by searching for noble gases in comets, by investigating abundances and isotopic ratios in Jupiter's atmosphere, and by further studies of the SNC meteorites and the atmospheres of the inner planets.

INTRODUCTION

This is a summary of the talk given at the Symposium, which was based in part on a paper currently in press [1]. The central hypothesis is that impacts by icy planetesimals (comets) have been largely responsible for the inventories of volatile elements that we find on the inner planets today. An obvious corollary of this model is that impacts can also remove volatiles, especially if a planet is small, like Mars. It is the balance between delivery and removal that ultimately determines how much of an atmosphere a given body will possess.

Other authors have considered the effects of early massive hydrodynamic and Jeans escape of gases and gas loss by sputtering from Mars [2,3]. We are concentrating on an effort to determine the *source(s)* of planetary volatiles, searching for underlying relationships in the isotopic and elemental abundances we find in atmospheric gases today. In the companion paper[1], we have focussed on carbon and nitrogen. Here the emphasis is on hydrogen and the noble gases. Ultimately, all these different strands must be woven together, but it is already obvious that more data are needed before any such synthesis can be accomplished.

2. THE OUTER PLANETS

Icy planetesimals have also contributed to the generation of the atmospheres of the giant planets. Following the early work of Mizuno [4], the current paradigm for the formation of these planets begins with the accumulation of a core of icy and rocky material. This core will develop an atmosphere of its own, derived from the accretional degassing of its constituent materials.

As the core grows to its final size of approximately 10 Earth masses, it attracts gas directly from the surrounding solar nebula. Thus the final atmosphere of the planet will consist of a mixture of the early atmosphere produced by the core (plus late-accreting planetesimals) and the envelope of gravitationally captured solar nebula gas.

If this model is correct, one would expect the heavy volatile elements except neon (see §4) in the atmospheres of these planets to be enriched as a result of the "extra" gases coming from the cores. Furthermore, this enrichment should increase from planet to planet as the mass of the core increases relative to the total planetary mass, a prediction that is consistent with the observations. The progression from Jupiter to Neptune takes C/H from 2.6 to approximately 25 while the ratio core mass/total mass changes from 0.03 to 0.75 [5,6].

We can make an additional test of the Mizuno model by studying the relative abundances of deuterium in these atmospheres. The observations of D/H demonstrate the existence of two distinct reservoirs of deuterium in the outer solar system: one in hydrogen gas and the other in hydrogen compounds [7,8,9]. These reservoirs have been isolated from each other because most of the mass of the hydrogen-containing compounds was in the condensed phase in the solar nebula. In fact, most of this mass is in the form of water ice and compounds trapped in this ice, i.e., the very icy planetesimals we are discussing.

On Jupiter and Saturn, where the core masses are small fractions of the total mass, D/H is approximately 2×10^{-5}, close to the value found in local, interstellar hydrogen gas [10]: 1.65 (+0.07, -0.18) $\times 10^{-5}$. On Uranus and Neptune the higher values of D/H reflect the increased contribution of gases coming from the cores (Figure 1). As H_2O is the most abundant constituent of these cores, this interpretation implies that D/H in the interstellar H_2O that was present in the cloud that formed the solar system must have been considerably enriched over the value found in interstellar hydrogen gas. Observations of H_2O in interstellar clouds support this inference. In the "hot cores" of interstellar clouds where significant evaporation of icy grains occurs (dust temperature ~200K), Jacq *et al.* have found $2 \times 10^{-4} \leq HDO/H_2O \leq 10^{-3}$ (see Figure 1) [11]. Evidently one may expect water in the solar system to have values of D/H in this range, as the H_2O in meteorites, comets and the

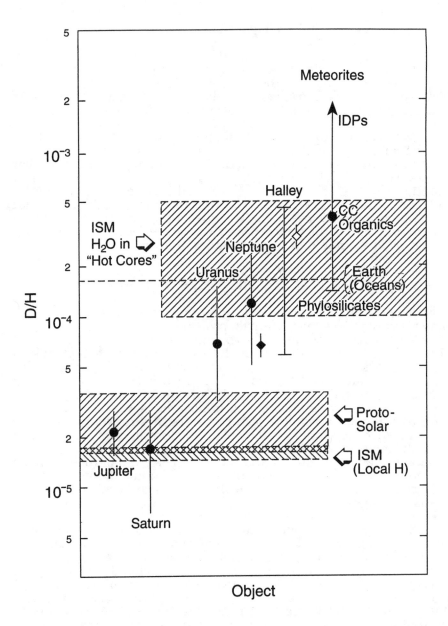

Fig. 1 This figure shows the values of D/H determined in the atmospheres of the giant planets compared with values measured in the interstellar medium (ISM), meteorites, comets and the Earth's oceans. The diamonds for Neptune and Halley represent the most recent measurements. The hatched region designated "protosolar value" comes from analysis of He in meteorites and the solar wind. (See [55] Fig. 18 for complete references).

Earth's oceans indeed does (Figure 1).

The most recent determination of D/H in the water from Halley's Comet [12] leads to a value of $D/H = 3.2 \pm 0.3 \times 10^{-4}$. This result adds support to the idea that comets may indeed be composed of grains from the interstellar medium that have not been significantly altered by their passage through the solar nebula.

We expect the same formation processes leading from interstellar clouds to planets to occur in other planetary systems as well. Ices in such systems should accordingly exhibit a similar enrichment in D/H. Thus in cases such as β Pictoris where material is falling into the star [13], it would be interesting to look for indications that $D/H \sim 2 \times 10^{-4}$ in that material, if such a difficult observation proves feasible [6]. The question remaining for our own system is whether the meteorites or the comets were the dominant source of the water and other volatiles that we find on the inner planets today.

3. THE INNER PLANETS: PROBLEMS WITH METEORITES

The traditional approach to the problem of the origin of inner planet atmospheres has been to propose one or more classes of chondritic meteorites as the volatile carriers that augmented whatever gases were trapped in the rocks composing the bulk of these planets [14,15,16]. These meteorites are thought to have supplied a late-accreting veneer of volatile-rich material that degassed upon impact and through subsequent processing on the planets, ultimately producing the atmospheres we observe today. On Earth alone, this simple picture has been strongly modified by the existence of liquid water, plate tectonics, and life, but a reconstruction of the Earth's volatile inventory reveals nearly the same abundances of carbon and nitrogen (per gram of rock) that we find today in the atmosphere of Venus [17] and about the same proportion of C/N that we see in the current atmospheres of both Venus and Mars [18].

These pleasing similarities come to an end when we consider the non-radiogenic noble gases. Taking Earth as our standard, we find that Venus has far more neon and argon than Earth possesses, more even than exists in type I carbonaceous chondrites, per gram of rock. Furthermore, the relative abundances of krypton and argon on Venus are dramatically different from those on Earth, Mars, or in the meteorites. The ratio of Ar/Kr on Venus is closer to the solar value than the ratio we find in these other sources (Figure 2).

In contrast, Mars shows nearly the same relative abundance pattern as the Earth, but the absolute abundances of noble gases per gram of rock are over two orders of magnitude *lower* than those in Earth's atmosphere or the ordinary chondrites (Figure 2). Finally, on both Mars and Earth, Kr/Xe ~ 10, whereas in the

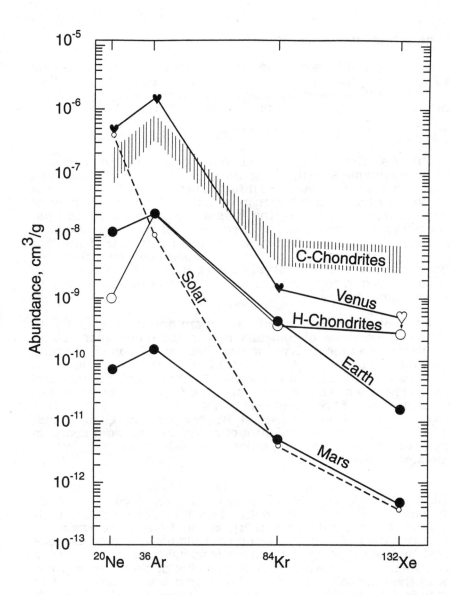

Fig. 2.   Abundances of the noble gases per gram of rock for the atmospheres of Venus, Earth, Mars and the ordinary and carbonaceous chondrites. Solar relative abundances are shown for reference.

meteorites, this ratio is near unity.

On Earth, this last deficiency is often referred to as the "missing xenon problem." For many years, people assumed that this missing xenon must be trapped in shales, ice, or clathrate hydrates. It now seems clear that it simply isn't there [19,20].

4. A COMETARY SOLUTION

This was the background for our attempt to see if comets might be a suitable substitute for the meteorites. While this idea has certainly come up before [21], it has suffered from the absence of data. We now have much better information about abundances of elements and compounds in the interstellar medium, in comets, and in planetary atmospheres. There are still no observations of noble gases in comets, which is not surprising in view of the fact that the ground-state transitions of these elements produce emission lines only in the far ultraviolet region of the spectrum. The mass spectrometers on the Giotto spacecraft also failed to detect any noble gases [22].

We have therefore relied on laboratory investigations to simulate the formation of cometary ices in the outer solar nebula [23,24]. These experiments have shown that amorphous ice forming at temperatures in the range 16 — 100K will trap gases in amounts that are a strong function of the temperature at which the trapping occurs. Hydrogen and helium are only trapped at very low temperatures; above 25K even neon is not retained by the ice. (This is why we do not expect Ne to be enriched in giant planet atmospheres [§2].) A solar mixture of argon, krypton and xenon is trapped in its original proportions at 30K, but is severely fractionated at 50K.

This result led us to suspect that comets might indeed serve as the volatile carriers we were seeking. We tried the simple hypothesis of assuming that both Mars and Earth obtain their heavy noble gases (argon, krypton and xenon) from two reservoirs — one internal: the rocks that make up the bulk of the planet, and one external: impacting icy planetesimals (comets). We constructed a three-element plot on which the proportions of these three noble gases in the atmospheres of Mars and Earth would correspond to two points (Figure 3). The ratios of Ar/Xe and Kr/Xe found in various chondritic meteorites and the sun are also illustrated, as is the domain of possible values for the atmosphere of Venus, where only an upper limit for xenon exists at the present time.

In our simple model, a straight line connecting Mars and Earth in such a plot will extend into the domain of the external reservoir at one end and into the internal reservoir at the other. In the logarithmic plot illustrated in Figure 3, this "mixing line" appears curved, since it is a simple linear function. Here we see that the point from the laboratory experiments that corresponds to

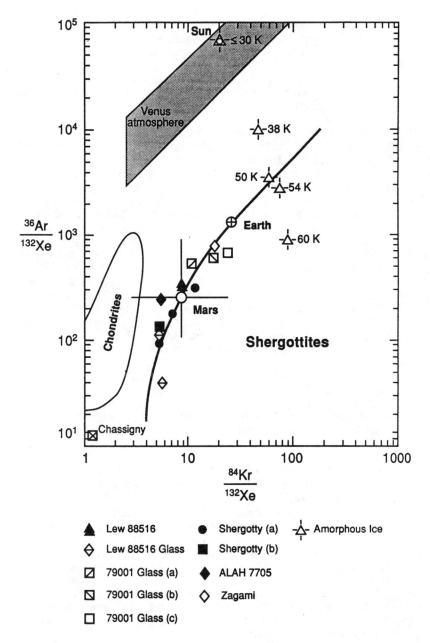

Fig. 3.   Ratios of the heavy noble gases in planetary atmospheres, chondritic meteorites and the SNC meteorites. The mixing line described in the text is the heavy line through the points for the atmospheres of Earth and Mars. Only an upper limit exists for Xe on Venus, hence the trapezoidal uncertainty.

the mixture of noble gases trapped by ice at 50K falls right on an extrapolation of this mixing line. Icy planetesimals that formed at about 50K thus constitute a plausible external reservoir for the heavy noble gases on these two planets [25,26]. This is a significant result, because 50K is the canonical temperature for the Uranus-Neptune region of the solar nebula where most of the Oort cloud comets are thought to have formed [27,28].

The rocky reservoir is then located at the other end of the mixing line, somewhere below the point for Mars. Why are Mars and Earth so far apart on this diagram? Evidently Mars is missing much of its cometary complement of gases. This conclusion is consistent with the present thinness of the Martian atmosphere and the low abundances of the noble gases on Mars that we see in Figure 2. Melosh and Vickery [29] have shown that Mars must have lost an amount of atmosphere equivalent to at least 100 times the atmosphere we see today as a result of impact erosion during the early heavy bombardment. Impact erosion also appears to be the only process that can account for the high values of $^{129}Xe/^{130}Xe$ and $^{40}Ar/^{36}Ar$ that we find on Mars, respectively 2.5 and 10 times the terrestrial ratios [18,30]. Evidently many of the cometary impacts that on Earth would have delivered volatiles, failed to do so on Mars. Both impacting comets and asteroids would contribute to an erosion of the Martian atmosphere that would have been more severe than that which occurred on the more massive Earth.

It seems that the early bombardment, with its delivery and removal of gases, must have played a dominant role in the early history of these atmospheres. Hence attempts to explain planetary surface temperatures that were evidently more clement than expected from the insolation of the dim young sun must take these impact processes into account.

With this model, we can explain the unusual noble gas pattern found in the atmosphere of Venus by invoking an impact by one or more comets from the Kuiper Disk [31,32,33]. These objects will have formed at temperatures near 30K, thereby trapping the noble gases in nearly solar proportions. A single Kuiper-Belt comet with a radius of 60 km could bring in all the heavy noble gases on Venus, while a 40 km Uranus-Neptune comet could supply the Earth [25]. In reality, the accumulation of smaller impacts may have been responsible for what we find today. Obviously comets from both the Kuiper Disk and from the Uranus-Neptune region must have struck all these inner planets. What we are therefore suggesting is that Venus received a dominating share of its heavy noble gases from the colder source.

Because neon is only trapped in ice at temperatures below 25K, we suggest that the neon in these atmospheres is a relic of gas trapped almost exclusively in the rocks that make up most of the mass of these planets. The severe fractionation of $^{20}Ne/^{22}Ne$ that we observe on Earth and Mars today may then reflect early escape processes whose effects on nitrogen and the other noble gases have been masked by subsequent cometary delivery of these volatile. On Venus, the higher abundance of neon per gram of rock (Figure 2) is consistent with the higher value of $^{20}Ne/^{22}Ne$. On Mars, the low value of $^{14}N/^{15}N$ is attributed to non-thermal escape processes during the last 3.7 billion years [1,25].

The similarity of C/N in the volatile inventories of Venus, Earth and Mars is interpreted in this model as a reflection of the contributions of carbonaceous chondrites and icy planetesimals from the Jupiter-Saturn region [1].

5. MESSAGES FROM THE SNC METEORITES

Figure 3 also contains points for noble gases measured in a number of Shergottite meteorites [25,26]. These meteorites are now generally acknowledged to have originated on Mars, as demonstrated in part by the gases they contain [34,35]. What we see in Figure 3 is that samples of the noble gases extracted from these meteorites (sometimes different samples of the same stone or even different temperature releases from the same sample) exhibit proportions of argon, krypton and xenon that lie along the mixing line established by Mars and Earth. Some samples fall above the Mars atmosphere point (from the Viking Lander measurements [18]) and some fall below. Our interpretation of these data is that the meteorites represented by the lowest points contain a mixture of Martian atmosphere with samples of the internal reservoir. The recent measurements [36] of noble gases in a glass sample from LEW 88516 tend to support the mixing line formulation. As we move up the mixing line toward the Earth, it appears that the meteorites are adding cometary gas to the mixture they contain. This suggests that the impactor that blew them off the surface of Mars was a comet.

Alternatively, one could argue that the highest points on this line, those corresponding to Zagami and EETA 79001 glass, actually represent the Martian atmosphere [35,37]. These points still fall within the error bars of the Viking measurement. All the lower points would then simply represent an increasing contribution from the internal reservoir. In either of these interpretations, the Chassigny point falls far from the mixing line, although it lies exactly on a straight line connecting Mars and Earth on this log-log plot [38].

It is also instructive to plot these data as $^{129}Xe/^{132}Xe$ vs $^{84}Kr/^{132}Xe$, which allows us to include *all* the SNCs (Figure 4). In Figure 3, we could not display any data points for the Nakhlites, as it is not possible to isolate the trapped argon in these meteorites [39]. In Figure 4, we use only Kr and Xe, so we can include points for the Nakhlites. The result is that we find two distinct mixing lines instead of one.

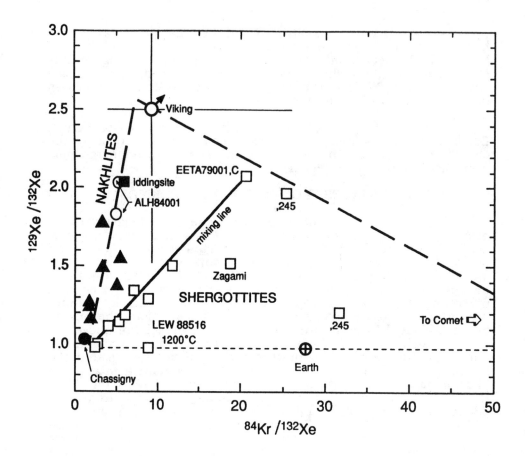

Fig. 4.   Values of $^{129}Xe/^{132}Xe$ and $^{84}Kr/^{132}Xe$ for SNC meteorites. There appear to be two separate mixing lines for nakhlites and shergottites, with some additional scatter for the shergottites.

The increase in $^{129}Xe/^{132}Xe$ along both lines in Figure 4 must represent an increase in the amount of Martian atmospheric gas trapped in the meteorites, since the Viking measurement found an atmospheric value of $^{129}Xe/^{132}Xe = 2.5$ (+2/-1). It is important to note that the progression of points along the Shergottite mixing line in Figure 4 matches that in Figure 3. In other words, the isotopic composition of xenon changes as one proceeds up the line in Figure 3, which would not happen if this were a straight line through Chassigny that simply represented equilibrium partitioning between gas and melt as suggested by Ozima and Wada [41,26].

Drake *et al.* [42] have endorsed the widely held view that the Shergottite line in Figure 4 represents mixing between a mantle source, represented by Chassigny, and the Martian atmosphere, represented by EETA 79001 glass. Drake *et al.* then suggest that the Nakhlite line is basically the same mixing line, shifted to the left as a result of the emplacement of the heavy noble gases in the Nakhlites by water rather than by shock, which was responsible for implanting the gases in the shergottites. (Shergottites are shocked, Nakhlites are not). This idea is consistent with the large fractionations of Kr/Xe found in terrestrial water-lain sediments which are much greater than the factor of ≤ 2 fractionation found in pure water [43].

We suggest as an alternative working hypothesis that these two mixing lines may support the idea that some cometary gas has been captured by some of the Shergottites, but not by the Nakhlites. The idea is that as the shocked rocks are expelled upward through the impact plume with escape velocity, gas from the impactor would be incorporated in melt and microfractures produced by the shock wave. The closest terrestrial analogy is the presence of atmospheric noble gases (in relatively high abundance) in glassy splatter from Kilauea [44].

The krypton and xenon in the Nakhlites may indeed have been emplaced by water solution (their presence in the iddingsite from Nakhla certainly supports this idea [42]), but now the amount of fractionation involved could be very much less, perhaps even lower than the value for pure water. As Drake *et al.* [42] have pointed out, the krypton and xenon abundances from the orthopyroxenite ALH 84001 [45] lie along the Nakhlite line in this diagram. Yet iddingsite has not been identified in this rock; the effects of water alteration appear to be limited to the presence of some carbonate [46] which may be more consistent with the lower degree of fractionation of Kr and Xe that we are suggesting. Note that an extrapolation of the Nakhlite line passes very close to the measurement of atmospheric noble gases made by the Viking Lander GCMS [18]. If this is not simply coincidence, it implies that the Viking measurement is close to the true atmospheric value, and the amount of fractionation experienced by the krypton and xenon implanted in the Nakhlites was indeed less than a factor of 2. (The Viking error bars were deliberately generous to allow for possible unknown systematic effects.) Although we do not yet know the "true" atmospheric value, a recent measurement of $^{129}Xe/^{132}Xe = 2.40 \pm 0.08$ by Marti et al. [47] in a piece of Zagami glass coupled with the values 2.43 ± 0.01 and 2.41 ± 0.01 measured by Swindle et al. [37] in EETA 79001 glass suggests that 2.5 may be an appropriate number.

We think the Shergottite data points that do not fall on the mixing line add some weight to this interpretation. The Zagami point was determined by Ott et al. [48], the LEW 88516 point by Ott and Löhr [49] and the ,245 points are from crushed samples of EETA 79001 glass analyzed by Wiens [50]. The lower ,245 point corresponds to gas simply released by crushing, the upper point characterizes the total gas from crushing and combustion. Ott and Löhr [49] concluded that "Clearly noble gases in the shergottites are more than a simple two-component mixture of EETA 79001 —(Mars atmosphere)—type and Chassigny-type gases."

We agree and we suggest that cometary gas may be the third component. We obviously don't know the isotopic composition of cometary noble gases, but we can assume that $^{129}Xe/^{132}Xe$ is close to 1.0, and the laboratory experiments of Bar-Nun et al. [24] demonstrate that $^{84}Kr/^{132}Xe$ in comets formed near 50 K should be approximately 60. Thus we can plot a point for the cometary gas, which lies off the right hand side of Figure 4, as indicated by the dashed lines.

The key to what is happening here could lie in the details of the xenon and krypton isotope abundances. If the noble gases in the Nakhlites are simply water-fractionated samples of the noble gases in the Shergottites, then the isotope ratios should map from the Shergottite line directly into the Nakhlite line. Solution in water will not change isotopic ratios. If instead the Shergottite line represents the addition of some cometary gas, then the isotopic abundances may differ. Unfortunately, the greatest change between the two lines is presumably in the abundance of krypton, which may exhibit very similar isotope abundances in both the Martian atmosphere and the comet. There is also a serious difficulty in making this comparison because of the effects of spallation, which are different for the different SNCs. We are pursuing this analysis in order to see which of the two hypotheses — sedimentary fractionation or the admixture of cometary gas — is the correct explanation for the two mixing lines. The answer will aid in the determination of which sample from the SNC meteorites provides the gas mixture that is closest to the true Martian atmosphere.

6. CONCLUSIONS, IMPLICATIONS AND TESTS

The laboratory experiments on the trapping of argon, krypton and xenon in ice at temperatures characteristic of the Uranus-Neptune region of the solar nebula demonstrate a fractionation pattern of noble gas abundances that fits a simple mixing model for the atmospheres of Mars and Earth. This suggests that these gases were delivered to the inner planets by icy planetesimals, a.k.a. comets. If that is true, other volatiles would be brought in as well, most notably water, carbon and nitrogen [1]. The hydrogen that is bound up in carbon and nitrogen compounds in the comet nuclei would become available on the planets to make the same or other compounds, offering the potential for early reducing atmospheres. On Mars, because of its small size, impacts apparently carried off more atmosphere than they produced, leaving the thin envelope we find today.

This model can be tested in a number of ways [1]. The first test is simply to see if argon rather than neon is present in dynamically new comets. This will require either a dedicated rocket launch for observations of the UV spectrum of a bright comet, or a comet mission that includes the option of examining volatiles trapped in the interior of a comet nucleus. Feldman et al. [51] have reported upper limits of 6 times the solar value for Ne/O and 10 times the solar value for Ar/O in the far UV spectrum of Comet Levy (1990c). Stern et al. [52] set an upper limit of 30 times the solar value for Ar/O in Comet Austin (1989c1). Both groups employed rocket-borne spectrometers. Using this same technique on a brighter comet should supply the critical data. The Rosetta mission planned by ESA will go to an "old" short period comet. Nevertheless, if the nucleus of this comet could be broken apart near the end of the mission, it may be possible for an on-board mass spectrometer to detect noble gases that have not been baked out.

Another test of the general model for delivery of volatiles by icy planetesimals could come from the mass spectrometer experiment[53] on the Galileo probe that enters Jupiter's atmosphere in December 1995. The model predicts that ice forming in the Jupiter region of the nebula will not trap significant amounts of N_2, which is expected to be the dominant form of nitrogen in the solar nebula. Carbon, on the other hand, should have been mostly in the form of compounds that would have been incorporated in those ices. Hence Jupiter's atmosphere should show a ratio of C/N that is greater than the solar value, although N/H should be close to solar, as nebular N_2 was captured gravitationally by the growing core along with the H_2. Similarly, we expect C/Ne > solar, with Ne/H = solar. However, in this case there is an ambiguity because of the likely dissolution of Ne in Jupiter's liquid hydrogen, which will also reduce the amount of Ne in the gaseous envelope (D. J. Stevenson, private communication). Hence Ne/H may be less than solar. The nitrogen isotopes may also differ from telluric values, because the nitrogen in Jupiter's atmosphere should come predominantly from N_2, while our own nitrogen should have been delivered originally as nitrogen compounds.

ACKNOWLEDGMENTS

The authors thank K. Marti for permission to quote results prior to publication and T. Swindle and S. Epstein for helpful reviews of the manuscript. This research was supported in part by NASA grants NAGW 2631 and NAGW 2650.

REFERENCES

1. T. Owen, and A. Bar-Nun, *Icarus* (in press)(1995).
2. R. O. Pepin, *Icarus* **92**, 2-79 (1991); *Icarus* **111**, 289-304 (1994).
3. B. M. Jakosky, R. O. Pepin, R. E. Johnson, and J. L. Fox, J. *Icarus* **111**, 271 (1994).
4. H. Mizuno, *Prog. Theor. Phys.* **64**, 544 (1980).
4. D. Gautier, and T. Owen, In *Origin and Evolution of Planetary and Satellite Atmospheres,* eds. S. K. Atreya, J. B. Pollack, and M. S. Matthews (Tucson: U. of Ariz. Press), pp 487-512 (1989).
6. T. Owen, In *Planetary Systems: Formation, Evolution, and Detection,* eds. B. F. Burke, J. H. Rahe, and E. E. Roettger. (Dordrecht, Kluwer), pp 1-12 (1984).
7   J. Geiss and H. Reeves, *Astron. and Astrophys.* **93**, 189-199 (1981).
8. T. Owen, B. L. Lutz, and C. de Bergh, *Nature* **320**, 244 (1986).
9. C. de Bergh, B. L. Lutz, T. Owen, and J. P. Maillard, *Astrophys. J.* **355**, 661 (1990).
10. J. Linsky, *et al.*, *Astrophys. J.* **402**, 694 (1993).
11. T. Jacq, C. M. Walmsley, C. Henkel, A. Baudry, R. Mauersberger, and P. R. Jewell, *Astron. Astrophys.* **228**, 447 (1990).
12. P. Eberhardt, M. Reber, D. Krankowsky and R. R. Hodges, abstract submitted to Toulouse Conference on Ice (March 1995).
13. A. M. Lagrange-Henri, A. Vidal-Madjar, and R. Ferlet, *Astron. Astrophys.* **190**, 275 (1988).
14. K. K. Turekian, and S. P. Clark, Jr., *J. Atmos. Sci.* **32**, 1257 (1975).
15. E. Anders, and T. Owen, *Science* **198**, 453 (1977).
16. G. Dreibus, and H. Wänke, *Icarus* **71**, 225 (1987). G. Dreibus, and H. Wänke, In *Origin and Evolution of Planetary and Satellite Atmospheres*, ed. S. K. Atreya, J. B. Pollack, and M. S. Matthews, U. of Ariz. Press, Tucson, pp 268-288 (1989).
17. T. Donahue, and J. B. Pollack, In *Venus*, ed. D. M. Hunten, L. Colin, T. M. Donahue, and V. I. Moroz, U. of Ariz. Press, Tucson, pp 1003-1036 (1983).
18. T. Owen, K. Biemann, D. R. Rushneck, J. E. Biller, D. W. Howarth, and A. L. Lafleur, *J. Geophys. Res.* **82**, 4635 (1977).
19. J. F. Wacker, and E. Anders, *Geochim. et Cosmochim Acta* **48**, 2372 (1984).
20. T. J. Bernatowicz, B. M. Kennedy, and F. A. Podosek, *Geochim Cosmochim Acta* **49**, 2561-2564 (1985).
21. G. T. Sill, and L. L. Wilkening, *Icarus* **33**, 1327 (1978).
22. J. Geiss, *Rev. Mod. Astron.* **1**, 1 (1988).
23. A. Bar-Nun, G. Herman, D. Laufer, and M. L. Rappaport, *Icarus* **63**, 317 (1985); A. Bar-Nun, I. Kleinfeld, and F.

Kochavi, *Phys. Rev. B* **38**, 7749 (1988).

24. T. Owen, A. Bar-Nun, and I. Kleinfeld, In *Comets in the Post-Halley Era*, ed. R. L. Newburn, Jr., M. Neugebauer, J. Rahe, Kluwer, Dordrecht, pp 429-438. (1991).

25. T. Owen, A. Bar-Nun, and I. Kleinfeld, *Nature* **358**, 44-46 (1992).

26. T. Owen and A. Bar-Nun, *Nature* **361**, 693-694.

27. A. P. Boss, G. E. Morfill, and W. M. Tscharmutter, In *Origin and Evolution of Planetary and Satellite Atmospheres*, ed. S. K. Atreya, J. B. Pollack, and M. S. Matthews, U. of Ariz. Press, Tucson, pp 487-512 (1989).

28. J. H. Oort, In *Physics and Chemistry of Comets*, ed. W. F. Huebner, Springer-Verlag, Berlin, pp 235-245 (1990).

29. J. Melosh, and A. M. Vickery, *Nature* **338**, 487 (1989).

30. T. Owen, In *Mars*, ed. H. H. Kieffer, B. M. Jakosky, C. W. Snyder, and M. S. Matthews, Univ. of Ariz. Press, Tucson, pp 818-834 (1992).

31. G. P. Kuiper, In *Astrophysics*, ed. J. A. Hynek, McGraw-Hill, New York, pp 357-424 (1951).

32. M. Duncan, T. Quinn, and S. Tremaine, *Astrophys. J.* **328**, L69 (1988).

33. D. C. Jewitt, and J. Luu, *Nature* **362**, 730 (1993).

34. D. D. Bogard, and P. Johnson, *Science* **221**, 651 (1983).

35. R. H. Becker, and R. O. Pepin, *Earth and Planet. Sci. Lett.* **69**, 225 (1984).

36. R. H. Becker, and R. O. Pepin, *Lunar Planet. Sci.* **24**, 7778 (1993).

37. T. D. Swindle, M. W. Caffee, C. M. Hohenberg, *Geochim. Cosmochim. Acta*, **50**, 1001 (1986).

38. U. Ott, and F. Begemann, *Nature* **317**, 509 (1985).

39. U. Ott, *Geochim. et Cosmochim. Acta*, **52**, 1937 (1988).

40. T. Owen, K. Biemann, D. R. Rushneck, J. E. Biller, D. W. Howarth, and A. L. La Fleur, *Science* **194**, 1293-1295 (1976).

41. M. Ozima, and N. Wada, *Nature* **361**, 693 (1993).

42. Drake et al (1994).

43. M. Ozima, and F. A. Podosek, *Noble Gas Geochemistry*. Cambridge Univ. Press. 367 pp (1983).

44. I. Kaneoka, N. Takaoka, and K. Aoki, In *Terrestrial Rare Gases*, ed. E. C. Alexander, Jr., and M. Ozima (Japan Sci. Soc. Press, Tokyo), p. 71 (1978).

45. T. D. Swindle, J. A. Grier, and M. K. Burkland, *Meteoritics* (in press) (1995).

46. A. Treiman. Private communication (1994).

47. K. Marti, J. S. Kim, A. N. Thakur, T. J. McCoy, and K. Keil, *Science* (in press) (1995); D. W. Mittlefehldt, *Meteoritics*, **29**, 214 (1994).

48. U. Ott, H. P. Löhr, and F. Begemann, 1988, *Meteoritics*, **23**, 295 (1988).

49. U. Ott, and H. P. Löhr, *Meteoritics*, **27**, 271 (1992).

50. R. C. Wiens, *Earth Planet Sci Lett.*, **91**, 55 (1988).

51. P. D. Feldman *et al.*, *Astrophys. J.* **379**, L37-L40 (1991).

52. S. A. Stern, J. C. Green, W. Cash, and T. A. Cook, *Icarus* **95**,

157-161 (1992).

53.   H. B. Niemann, D. N. Harpold, S. K. Atreya, G. R. Carignan, D. Hunten, and T. Owen, *Space Sci. Rev.* **6**, 111 (1992).

54.   J. Geiss, In *Origin and Evolution of the Elements*, eds. N. Prantzos, E. Vagioni-Flan and M. Casse (Cambridge: Cambridge University Press), pp 89-106 (1993).

55.   D. Gautier, B. J. Conrath, T. Owen, I. de Pater, and S. K. Atreya, In *Neptune*, eds. D. P. Cruikshank and M. S. Matthews (Tucson, University of Arizona Press), in press (1995).

DYNAMICS OF VOLATILE DELIVERY
FROM OUTER TO INNER SOLAR SYSTEM

William M. Kaula
University of California, Los Angeles, CA 90024

ABSTRACT

Owen et al[1] propose that the argon excess of Venus compared to Earth was caused by impact of a large icy planetesimal from the outer solar system, where temperatures were low enough for argon to adhere to ice, about 30 K. A body of solar Ar/Si and C/H similar to Pluto and Triton less than 100 km diameter would suffice. However, direct delivery from the Uranus-Neptune zone to Venus would result in a very high approach velocity, causing erosion rather than accretion of volatiles. It would also be an extremely improbable event. Virtually all icy bodies scattered from the Uranus-Neptune zone to the terrestrial zone were strongly perturbed by Jupiter, but even then arrive at Venus at high velocities. Low approach velocities to Venus require Earth perturbations of bodies scattered inward from Jupiter.

Such extreme perturbations are dominated by close approaches; hence an Opik algorithm is an appropriate reconnaissance tool. A multi-stage computation obtains that on the order of 10^{-8} of bodies from the Uranus-Neptune zone are scattered so that their aphelia are within Jupiter's orbit. But extended runs of this population did not obtain approach velocities to Venus less than 12 km/sec. Hence for this mechanism to be effective, the vaporization and ejection of large high-velocity impactors must be less complete than assumed in models to date.

INTRODUCTION

It has long been conjectured that the Earth's volatiles must be supplemented by a "late veneer" delivered from further out in the solar system[2]. However, this is a relatively modest matter of scattering small bodies with compositions like carbonaceous chondrites, which plausibly resulted from condensation closer than Jupiter's orbit, and which can be decelerated by atmospheres.

Much more problematical is the delivery of volatiles by from much further out: the Uranus-Neptune zone and beyond, as has been hypothesized for the provision of organic molecules to the early Earth[3] and for the excess of primordial argon in Venus[1]. The provision of organic molecules can be explained by delivery in bodies small enough to be decelerated by the Earth's atmosphere, but the argon excess depends on delivery by a single large body. The primordial argon excess of Venus (relative to Earth) amounts to about 10^{16} kg. A body carrying 10^{16} kg of argon and having a 2.0 Mg/m^3 density and solar Ar/Si would have a mass less than 10^{-6} Earth masses-- a diameter less than 100 km, of which there plausibly were millions in the Uranus-Neptune zone. The problem is the dynamics of delivery. It is unlikely to be direct: typical approach velocity to Venus are on the order of 30 km/sec, making impact very improbable, and, if occurring, quite erosive, instead of accretive. Such erosive effects appear necessary to explain the paucity of volatiles on Mars relative to Earth and Venus[4,5], and to remove excess carbon dioxide from the Earth's atmosphere[6] so that the water can rainout to form oceans when planetesimal bombardment declines.

But for volatile accretion, rather than erosion, perturbations by Uranus &/or Neptune must drop its perihelion to the vicinity of Jupiter, which in turn must remove enough energy to flip its perihelion into an aphelion. Then, most difficult, Earth &/or Venus must quickly remove enough energy to bring its aphelion comfortably within Jupiter's orbit. The problems are determining how much the approach velocities can be reduced, and the probability of getting one low enough to be accretive.

It is desirable that this problem be examined with an integrator that allows for close approaches, such as been applied to the provenance of periodic comets[7]. However, it would be a long and expensive computation compared to explaining periodic comets to accomplish the third "deboost" phase. But since such distant scattering is required, an Opik algorithm-- one which assumes orbit changes are dominated by approaches within the spheres of influence of the planets-- seems the appropriate reconnaissance tool before a full integration of a sufficient number of test particles.

OPIK ALGORITHM

The algorithm applied was one already developed to explore the plausible range of outcomes in the formation of the terrestrial planets[8]. Probabilities of encounter between a prescribed sets of planets and planetesimals are calculated taking into account the eccentricities and inclinations of both orbits, and each encounter and its location is selected randomly in accord with these probabilities. Encounters between planetesimals are not taken into account. In the original application, the planetesimals have mass, and the effects on planets and their orbits are calculated; in the present calculation, the planetesimals are calculated as massless test particles.

To keep the computation from becoming excessively long, it must be staged, with only a fraction of the output from each stage being retained for the next. These fractions were defined by maximum perihelia and aphelia. The effective definition of a stage was retaining a statistically significant population: i.e., a few hundred bodies. A further computational economy was making several runs with the same starting population, but different Monte Carlo selections. Most arbitrary was probably the ratio of close approach events to number of bodies n_b, which varied from six for the initial phase to almost twenty for the final. The primary stages applied were: I. from the Uranus: Neptune zone to perihelia within Jupiter's orbit, aphelia within Uranus's orbit; II from the Jupiter: Saturn zone to perihelia within Mars's orbit, aphelia within Saturn's orbit; III, from this broad Jupiter-centered zone to perihelia just beyond Earth's orbit, aphelia just beyond Jupiter's orbit; and IV, from the narrower Jupiter-centered zone to perihelia within Earth's orbit, aphelia within Jupiter's orbit. Originally, one deboost stage, combining III and IV, was envisioned. However, the dominance of Jupiter compelled the separation of III and IV, and the weakness of the Earth led to dividing stage IV into three sub-stages, with gradually decreasing maxima in perihelion and aphelion. The results of each stage are summarized in Table I. The starting maximum perihelion was starting number n_b was 1000 for stage I, with eccentricities up to 0.30, inclinations up to 20 degrees, and semi-major axes 18 to 40 AU. For subsequent stages, the starting number was the number retained n_k from the previous stage. The number of runs n_r were 40 for stage I and 80 for the remainder. Hence the portion retained, $n_k/n_b/n_r$, stayed in the range 0.0112-0.0124.

Table I   Results of Monte Carlo Scattering Stages

Stage	Bodies Kept n_k	Per. (AU)	Aph. (AU)	Duration (My)	Venus Collisions No.	Min (km/sec.)
I	465	3.0	14.0	252	0	--
II	420	1.5	9.5	1.5	0	--
III	389	1.3	5.6	3.6	0	--
IVa	347	1.2	5.0	2.2	0	--
IVb	316	1.1	4.8	225.	42	19.
IVc	311	1.0	4.6	28.	47	13.

The most arbitrary parameter, the number of events per body, was checked with a run that tripled it, using output from stage IVc. The change in results was slight-- lowering of the minimum approach velocity to 11 km/sec.

A check on Stage II is how well the distribution compares to that observed for short period comets, whose recent perturbation history should be very similar. This is shown in Figure 1. The agreement is excellent in the range that is best observed, perihelia 0.9-1.7 AU. The deficiency of observed perihelia within 0.8 AU apparently is due to the effects of solar radiation. Rigorous integrations also get more close to the sun than observed[7].

Observed Comets: Perihelion vs. Aphelion Distribution

AU	2.7	3.2	3.7	4.1	4.6	5.1	5.5	6.0	6.5	6.9	7.4	7.9	8.3	8.8	9.3	AU
2.9				1	1									1		2.9
2.7						2										2.7
2.5					2	4	4	1			1	1	1		1	2.5
2.3					2	6	4		1						1	2.3
2.1					5	6	2		1		2					2.1
1.9					2	5	10	3	2	2	2	1	1		2	1.9
1.7					1	8	13	7	2	1	2					1.7
1.5					5	6	20	3	2	2						1.5
1.3					6	7	9	6	1	1	1					1.3
1.1					1	9	6	6	2	2			4			1.1
0.9						6	4	10	1							0.9
0.7						2	3				2					0.7
0.5							5									0.5
0.3			1			1										0.3
0.1								1								0.1
AU	2.7	3.2	3.7	4.1	4.6	5.1	5.5	6.0	6.5	6.9	7.4	7.9	8.3	8.8	9.3	AU

Computed Survivors: Perihelion vs. Aphelion Distribution

AU	2.7	3.2	3.6	4.1	4.6	5.1	5.5	6.0	6.5	7.0	7.4	7.9	8.4	8.9	9.3	AU
2.9						1	3	4	2	1	2	4		1	1	2.9
2.7						1	4	2		2	1	3		1	1	2.7
2.5						1	6	2	5	4			3	3	1	2.5
2.3						6	9	1	4		3	3	2	4	2	2.3
2.1						4	10	7	4	2	4	1	4	3	10	2.1
1.9						12	18	8	8	3	3	3		4	2	1.9
1.7						8	8	6	4	2		1	3	1	1	1.7
1.5					2	9	12	4	3	4		2	3		1	1.5
1.3					1	3	11	5	2	2	2	1	1	5	2	1.3
1.1					1	4	8	5	5	1						1.1
0.9			1		1	4	9	4	1		2					0.9
0.7		1		1	2	3	3	1	1							0.7
0.5	1			1	6	6	5	1						1	1	0.5
0.3			1	5	8	4	1	1	1	1		1				0.3
0.1				1	9	7	1	2	1							0.1
AU	2.7	3.2	3.6	4.1	4.6	5.1	5.5	6.0	6.5	7.0	7.4	7.9	8.4	8.9	9.3	AU

Fig. 1 Comparison of scattered body distributions in the inner solar system.

IMPLICATIONS

For small body impacts, only 10 percent or so of the impact energy goes into kinetic energy of ejection[9]. If this applied to an impact into Venus with approach velocity of 12 km/sec, it would be accretive. However, for a large body, the energy going into internal energy leads to vaporization of material and the hydrodynamic expansion thereof[4,5]. It is assumed in modeling hypervelocity impacts of 100 km, or 10^{18} kg, bodies, that the body is entirely vaporized, and forms the core of the plume of gas which is lost by the planets. However, an actual impact may not be so neat, and factors such as material heterogeneity and obliqueness of the impact may lead to some impactor material being retained: perhaps even inert gases adsorbed deep within the body. Hence the hypothesis remains open, but undemonstrated until more detailed impact calculations are made. Meanwhile, there remains the problem of accounting for the excess neon in Venus, and well as finding adequate alternative processes to account for the inert gas abundances of the terrestrial planets[10].

REFERENCES

1.   Owen, T., Bar-Nun, A., & Kleinfeld, I., Nature, **358,** 43 (1992).
2.   Anders, E. & T. Owen, Science, **198**, 453 (1977).
3.   Chyba, C. F., P. J. Thomas, L. Brookshaw, & C. Sagan, Science, **249**, 366 (1990).
4..   Ahrens, T. J. , Ann. Rev. Earth Plan. Sci., **21**, 525 (1993).
5.   Melosh, H. J., Vickery, A. M., & Tonks, W. B., In *Protostars and Planets III* (E. H. Levy & J. I. Lunine, eds., Arizona, Tucson, 1993), p. 1339.
6.   Abe, Y., Lithos, **30**, 223 (1993).
7.   Levison, H. F. & M. J. Duncan, Icarus, **108**, 18 (1994).
8..   Kaula, W. M., In *Origin of the Earth* (H. E. Newsom & J. H. Jones, eds., Oxford, New York, 1990) p. 45.
9.   O'Keefe, J. D. & T.J. Ahrens, Proc. Lun. Sci. Conf., **8**, 3357 (1977).
10.   Pepin, R. O., Icarus, **92**, 2 (1991).

SURFACE ICES IN THE OUTER SOLAR SYSTEM

Ted L. Roush
San Francisco State University
San Francisco, CA 94132
c/o NASA Ames Research Center
Moffett Field, CA 94035-1000

Dale P. Cruikshank
NASA Ames Research Center
Moffett Field, CA 94035-1000

Tobias C. Owen
University of Hawaii
Honolulu, HI 96822

ABSTRACT

We review relatively recent spectroscopic remote sensing observations, from Earth-based telescopes, that have been used to identify condensed volatile ices on the surfaces of satellites in the Jovian, Saturnian, and Uranian systems, Neptune's moon Triton, Pluto, and Pluto's satellite Charon. Water ice appears ubiquitous on the surfaces of the larger satellites in the Jovian, Saturnian, and Uranian systems and on Charon. More volatile species of ices appear to dominate the surfaces of Triton (N_2, CO, CH_4, and CO_2) and Pluto (N_2, CO, and CH_4).

INTRODUCTION

Information regarding the current compositions and relative abundances of volatile species present on planetary surfaces provides insight into both the internal and surface-atmosphere evolutionary history of a given body. This is because planetary volatile inventories are products of several factors: (1) condensation-accretion of pre-planetary material which determines the initial bulk volatile inventory; (2) energy history of a planet, including timing, causes, and mechanisms of degassing; (3) the volatile sinks, including temporary, long-term, and permanent; and (4) external processes operating on the volatile inventory. Hence, models of planetary evolution must remain consistent with currently observed volatile compositions and abundances. There are several reviews of the surface composition of bodies in the outer solar system[1-4]. Here we focus solely upon bodies located from Jupiter outward and rely heavily upon the previous reviews, providing an update that includes more recent observations and interpretations.

In the context of formation and evolution of solar system bodies, the interesting ices typically considered are simple molecules formed from elements having high cosmic abundances. These mainly include ices of H_2O, NH_3, SO_2, H_2S, CH_4, CO, CO_2, and N_2. In the solid state, these molecules have vibrational spectral features that lie in the near- and mid-infrared and are analogous to their gaseous counterparts with the exception that rotational transitions are quenched. The overtone and combination modes, occurring in the visible and near-ir regions, are of particular importance as standard observational techniques used to identify these ices rely upon reflected solar energy, as discussed in more detail below.

OBSERVATIONAL AND INTERPRETATIONAL TECHNIQUES

Remote sensing observations of objects in the outer solar system have historically been obtained at visible and near-ir wavelengths because the intensity of the solar spectrum is strongly peaked in the visible and rapidly decreases at longer wavelengths, see top left inset of Figure 1. This inset also illustrates how the intensity of the incident solar energy (F) is inversely correlated with the square of the heliocentric distance of a body. As a result, the amount of energy available to interact with the surface components decreases by about 4 orders of magnitude from the distance of Jupiter (~5 Astronomical Units, AU) to Pluto (~30 AU). The energy that penetrates the surface, traverses the material(s), and is ultimately reflected to the detector (I, see upper right inset of Figure 1) provides direct information regarding the composition(s) of the surface material(s). The ratio I/F will exhibit absorption features due to the surface material(s), see right inset of Figure 1. By interfacing spectrometers with telescopes, passive remote sensing of the outer solar system has been performed from Earth-based observatories. The techniques are quite amenable to spacecraft and both the Galileo and Cassini missions to Jupiter and Saturn, respectively, are carrying near-ir mapping spectrometers.

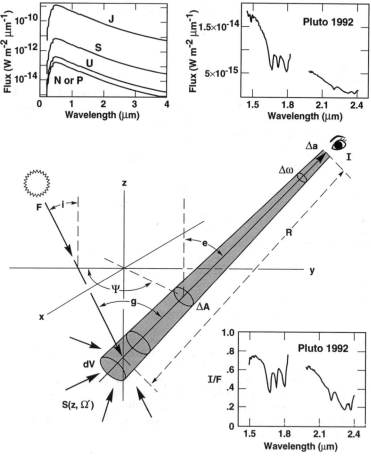

Figure 1. Viewing geometry of sunlight reflected from a planetary surface. Upper left, incident solar flux at Jupiter (J), Saturn (S), Uranus (U), and Neptune or Pluto (N or P). Upper right, measured flux from Pluto. Lower right, ratio of the reflected to incident flux for Pluto.

Interpretation of the observations relies upon both laboratory measurements of analog materials and theoretical calculations. Figure 2 shows the measured reflectance of several laboratory ices[5]. All of these ices exhibit spectral features in the near-ir region of the spectrum, and direct comparison of such laboratory data to the observational data can be used to identify specific ices on other surfaces. However, quantitative interpretation using this comparative technique would require a prohibitively large number of laboratory measurements.

Quantitative analysis can be efficiently accomplished by computation of synthetic spectra of scattering surfaces. Such modeling provides the capability to investigate grain size variations and mixtures of components that have not been directly measured in the laboratory. Hapke has developed a series of equations describing the interaction of energy with particulate surfaces[6,7]. The general equation describing the bi-directional reflectance of a surface is

$$r(i,e,g,\overline{w}) = \frac{\overline{w}}{4\pi} \frac{\mu_o}{\mu_o+\mu} \left\{ [1+B(g)]P(g) + H(\mu_o)H(\mu)-1 \right\}, \tag{1}$$

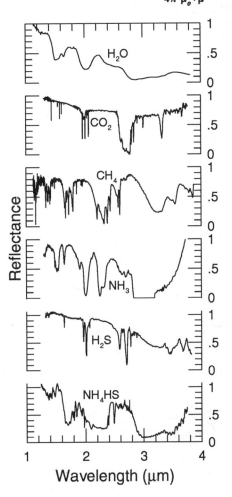

Figure 2. Measured reflectance of several laboratory ices, after Fink and Sill[5].

where, as shown in Figure 1, i is the angle of the incident light ($\mu_o=\cos i$), e is the angle of the emergent light ($\mu=\cos e$), g is the phase angle between i and e, and $\overline{w}$ is the average single scattering albedo of the surface (see Hapke[6,7] for a more detailed discussion of these parameters). Compositional information is contained in the parameter $\overline{w}$ that is related to the absorption coefficients of individual components in a mixture. Given the optical constants (real and imaginary indices of refraction) of potential materials, the reflectance can be calculated via equation 1 and directly compared to the observational data. Such calculations have been shown to agree reasonably well with laboratory measurements[7,8].

DISTRIBUTION OF VOLATILES IN THE OUTER SOLAR SYSTEM

Table I summarizes the ices found on various bodies in the outer solar system. H_2O is the most abundant surface material in the inner and middle regions while more volatile species appear to dominate surfaces in the outermost edge of the outer solar system.

Satellites of Jupiter. The geometric albedos of the Galilean satellites of Jupiter are shown in Figure 3. Spectra of Io, Jupiter's innermost large satellite, exhibit features due to SO_2 ice and frosts[9-12]. It has been suggested that trace amounts of H_2O and H_2S ices also occur on Io[13-17], but these suggestions are somewhat controversial. As seen in Figure 3, the other three Galilean satellites exhibit varying strengths of H_2O ice absorption features near 1.04-, 1.25-, 1.5-, 2.0-, and 3.0 μm. Interpretation of these features has led to the suggestion that the proportion of H_2O ice decreases relative to a non-ice component on these surfaces as a function of increasing distance from Jupiter[18]. Modeling of the reflectance spectra of Callisto suggests that a significant non-ice component is present whose spectral behavior is similar to terrestrial serpentines[19-21]. Recent observations of Ganymede have been interpreted to indicate the presence of condensed O_2 trapped in a H_2O ice matrix[22], yet O_2 is not present on the other satellites. The presence of SO_2 ice on Europa is indicated by recent UV observations[23]. The smaller satellites of Jupiter have low albedos (~6%) and their spectral shapes are similar to C- and D-type asteroids (*i.e.*, neutral or slightly reddish colors). Photometric observations reveal Jupiters rings are composed of low albedo particles but no spectroscopic information is available, hence their composition remains unknown.

Table I. Surface Ices in the Outer Solar System

Jovian Satellites	Io: SO_2, H_2S?, H_2O? Europa: H_2O, SO_2 Ganymede: H_2O, O_2 Callisto: H_2O
Saturnian Satellites	Mimas: H_2O Enceladus: H_2O Tethys: H_2O Dione: H_2O Rhea: H_2O Hyperion: H_2O Iapetus: H_2O Rings: H_2O
Uranian Satellites	Miranda: H_2O Ariel: H_2O Umbriel: H_2O Titania: H_2O Oberon: H_2O
Neptunian Satellites	Triton: N_2, CH_4, CO, CO_2
Pluto	N_2, CH_4, CO Charon: H_2O, CO_2? CH_4?

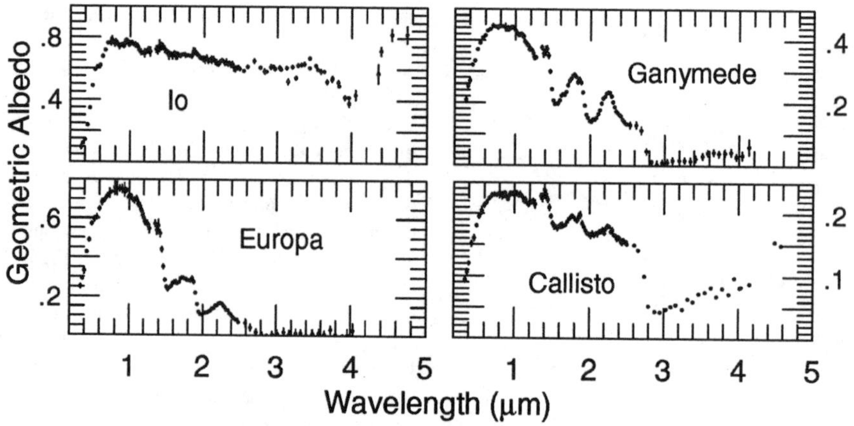

Figure 3. Spectral geometric albedos of the Galilean satellites, after Clark *et al.*[2]

Satellites of Saturn. The presence of H_2O ice on the surfaces of a number of satellites in the Saturnian system has been known for about twenty years[2-4]. The geometric albedos of six of the eight largest satellites of Saturn are shown in Figure 4. The spectra of all these satellites exhibit absorption features due to the presence of H_2O ice at their surfaces. Some of these data are of sufficient precision to show the diagnostic 1.04 μm H_2O ice absorption band. The strength of this feature suggests that relatively pure H_2O ice surfaces exist on Rhea, Tethys, Dione, and the trailing hemisphere of Iapetus[2-4,24]. One significant attribute of the inner Saturnian satellites is their high albedo relative to the Jovian or Uranian satellites. This is likely due to a smaller abundance of the non-H_2O ice component[4] and/or compositional differences between these satellite systems. Overall, the Saturnian satellites are less dense than the Jovian and Uranian satellites. Therefore the Saturnian satellites may owe their higher albedos to a bulk composition that contains fewer other materials, such as silicates, although the connection between surface composition and bulk density is somewhat tenuous. Theoretical arguments suggest that clathrate hydrates of CH_4, NH_3, CO, and CO_2, ices might be expected on the Saturnian satellites[25-27]. To date no evidence has been obtained suggesting any of these other ices are present[4]. Near-infrared spectra of the rings of Saturn have been interpreted to indicate they are composed of nearly pure water ice having an effective grain diameter of 30 μm^2. Recent analysis and modeling of the color properties of Saturn's rings, as measured by Voyager in the visible (0.4-0.8 μm), suggest that the rings are quite red, reflectance increasing with increasing wavelength, and that some organic material, mixed at the molecular scale with the H_2O ice, may be responsible for the reddening[28].

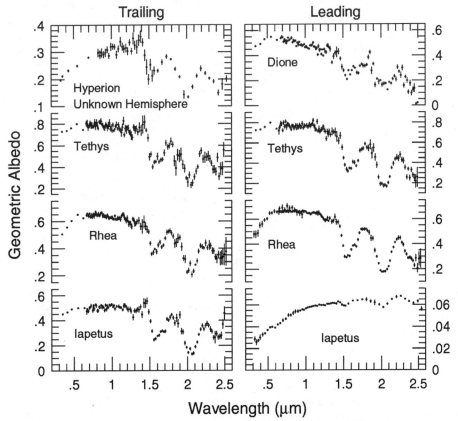

Figure 4. Spectral geometric albedos of the Saturnian satellites.

Satellites of Uranus. Due to their small size and distance from the Sun and Earth (see inset in Fig. 1) observations of the large Uranian satellites (Miranda, Ariel, Umbriel, Titania, and Oberon) have been possible for only about the last 15 years. The albedos of these satellites are shown in Figure 5. All spectra exhibit absorption features due to the presence of H_2O ice at their surfaces. Compared to the Saturnian satellites, the visual geometric albedos of the Uranian satellites are quite low, more similar in value to Ganymede and Callisto[3,4,29,30], suggesting that the surfaces of the Uranian satellites contain an abundance of dark non-ice contaminants similar to that of Ganymede and Callisto[3,4]. Modeling of the spectra by representing the surfaces as discrete regions of water ice and dark spectrally neutral material, such as charcoal or carbonaceous chondritic material, suggested varying degrees of areal coverage by the dark material[3,4]. Voyager 2's imaging of these surfaces[31] revealed geologically complex terrains consisting of discrete regions of bright and dark materials. The five color filters of the narrow angle camera confirmed the spectrally neutral behavior of the dark regions[32]. Theoretical formation scenarios of bodies in the outer solar system suggest the incorporation of other volatiles such as NH_3, CH_4, and CO[26]. At best only a tantalizing hint of any of these other volatiles may be contained in the existing spectra of the Uranian satellites, and more definitive interpretation awaits analysis of newer data at higher spectral resolution[33]. Voyager 2

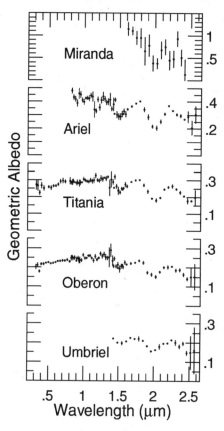

Figure 5. Spectral geometric albedos of the Uranian Satellites, after Cruikshank and Brown[3]. Note, Miranda data are scaled to 1.0 near 1.7 μm.

imaging of the Uranian rings revealed that they are spectrally neutral and have a low albedo (0.032 ± 0.003)[34]. For the rings, a 2.2 μm absorption band has been suggested to indicate NH_3 frost, OH in silicates, or XCN[35-37].

Satellites of Neptune. There is no diagnostic spectral information for Neptune's smaller satellites but during the Voyager encounter the camera system determined the geometric albedo of several; Nereid has an albedo near 0.14, and the other satellites have albedos near 0.04[38]. Nereid and Proteus appear to have neutral color properties in the visual[38]. Diagnostic spectral information is also lacking for Neptune's rings although Voyager 2 imaging observations suggest that the ring particles have a low albedo (0.05)[38].

Early infrared spectroscopic observations of Triton initially revealed several absorptions due to CH_4 located near 0.85-, 1.05-, 1.15-, 1.45-, 1.65-, and 2.3 μm[39-41]. About a decade ago a weak feature, at 2.16 μm, was attributed to N_2[42], and it's strength suggested relatively long path lengths in condensed N_2. This led to debate as to whether N_2 existed as a liquid or solid given the uncertainties in plausible surface temperatures of Triton[42,43]. This question was directly resolved using instruments aboard Voyager[44] that determined the surface temperature was 38^{+3}_{-4}K, well below the melting point of N_2. Voyager also revealed a thin, ~ 16 μbar[45,46], atmosphere on Triton, leading to the conclusion that the CH_4 and N_2 spectral features must arise from surface ices. Figure 6 shows the albedo determined from ground-based infrared spectral observations of Triton obtained in 1992 at resolving powers of ~350. These data revealed the presence of absorption features due to solid CO and CO_2, as well as several previously unresolved CH_4 bands[47], (see Figure 6). Based upon a comparison of the positions of the observed CH_4 features to laboratory spectra of pure CH_4 it appears that much or all of the CH_4 on Triton is dissolved in N_2 ice[47]. Theoretical modeling of these data show that solid N_2 is the dominant surface component ($\geq 99\%$ by weight) contributing to the spectrum of Triton and that CH_4 and CO are minor components[47]. The shape of the 2.16 μm N_2 band is temperature-sensitive and has been used to derive an independent estimate of $38 \pm 1K$ for Triton's surface temperature[48-50]. More recent ground-based observations at resolving powers of ~700-1000 confirm the presence of the previous features and have extended the total wavelength coverage[51].

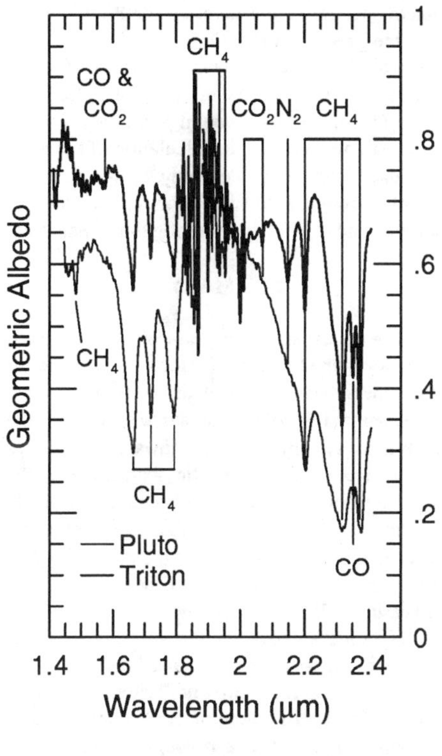

Figure 6. Spectral geometric albedos of Triton (heavy line) and Pluto (thin line). Absorption features due to various ices are indicated. The 1.85-2.0 μm spectral region is strongly influenced by telluric absorptions.

The visual color of Triton is weakly yellowish-red, uncharacteristic of CH_4 or N_2 but similar to colors produced when CH_4 ice or gas is exposed to high energy sources, such as UV light[3]. There are two time scales associated with visual color changes observed for Triton. A long-term secular "blueing" of Triton's visual color has occurred from the 1950s to the present[52]. These changes have been attributed to volatile transport or the α-β phase transition of N_2[52]. There are also short-term or transient color changes on Triton. For example, observations in the late 1970s[53,54] recorded evidence of a sudden reddening of Triton's visual color that has been attributed to some form of geologic activity, such as volcanism or the geyser-like plumes observed by Voyager 2[52].

Pluto and Charon. It has been nearly two decades since Earth based telescopic observations were used initially to identify the presence of CH_4 ice on Pluto[55]. Subsequent observations have confirmed this identification[40,56-62] and recent high spectral resolution data have revealed the presence of absorptions due to N_2 and CO[62], as illustrated in Figure 6. Based upon spectral modeling, it appears that N_2 ice comprises ≥ 98 wt.% of the surface ices, and that CH_4 and CO are more abundant than on Triton[62], causing the N_2 2.16 μm absorption to appear as a weak feature on the short wavelength edge of the CH_4 band. Like Triton, the positions of the CH_4 ice absorptions indicate that some of the CH_4 on Pluto occurs as a solid solution dissolved in the N_2 matrix[62]. Newer observations at higher spectral resolution[63], and additional spectral modeling suggests that up to ~15% of the surface could be pure CH_4 deposits and the positions of the CH_4 bands would still remain consistent with the observations[64]. Spectroscopic modeling of the N_2 band shape indicates the surface temperature of Pluto is $40 \pm 2K$[50]. Residual features seen on the short wavelength side of the 2.35 μm CH_4 feature have been attributed to the presence of saturated hydrocarbons[65].

Pluto has a reddish visual color[3] that is inconsistent with pure CH_4 or N_2 ices but similar to colors produced when CH_4 ice or gas is exposed to high energy sources, such as UV light[3]. That Pluto is redder than Triton is likely due to the greater abundance of CH_4 on Pluto. Comparisons of observations in the 1950s to those obtained in the late 1970s show no indication of color changes on Pluto[3]. As seen from the Earth, the unique orbital geometry of Pluto and the revolution of its moon Charon provide a unique opportunity of observing a series of eclipses and mutual occultations twice during each of Pluto's 248 year orbits[66]. A series of these events were observed during the mid- to late-1980s and have provided a unique opportunity to map the albedo distributions of Pluto and Charon[64,65]. Several independent studies[59,61,67-71] indicate: (1) high albedos (~ 0.7) associated with high southern latitudes; (2) darker albedos (0.4-0.5) within the mid-southern latitude regions and equatorial zones; (3) high albedos (0.6-0.65) at mid-northern latitudes; and (4) dark albedos (~ 0.3) at high northern latitudes. Near-ir albedo maps indicate that the bright south polar region is a candidate region of high methane concentration[72].

During these mutual events, it was possible to obtain spectral data separately for Pluto's moon Charon[60,73,74]. The near-ir data revealed evidence for the presence of H_2O ice on Charon[73,74] as illustrated in Figure 7. Recent modeling of the existing highest spectral resolution data of Charon included considerations of both two and three component spatial and intimate mixtures of H_2O, CO_2, and CH_4 ice[75]. The results suggest that in addition to H_2O ice, up to 40 weight percentage (wt %) of other icy components may be present on Charon, yet remain undetected[75]. Calculation of the near-ir geometric albedo of Charon[76], based on the older data, can eliminate many of the potential candidate mixtures reducing the non-H_2O ice component to roughly 30 wt % but further analyses await future observations involving higher spectral resolution and separately resolving the signals from Pluto and Charon.

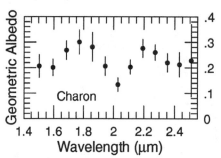

Figure 7. Spectral geometric albedo of Charon[76], adapted from Buie et al.[74].

SUMMARY

The results of many years of observational, laboratory, and theoretical efforts by a number of researchers have provided direct compositional information regarding the surface ices in the outer solar system. The near-infrared region has proven ripe with diagnostic features and the spectral resolutions currently becoming available continue to reveal new ices that were previously unrecognized. Table I shows a general trend of increasing volatility of surface ices with increasing heliocentric distance, as expected.

H_2O ice appears on the surfaces of small bodies throughout the outer solar system, with the notable exception of comet nuclei, the dark D-type asteroids, Io, Pluto and Triton. Current models for the formation of the solar system predict that H_2O ice will be the most abundant solid at and beyond Jupiter's orbit. Thus its nearly ubiquitous appearance is not surprising. In the case of the comet nuclei, we know the ice is there; it is simply hidden by a coating of dark, presumably carbon-rich material. The same may be true for the D-asteroids, we presently lack observations of outgassing that would betray the presence of sub-surface ice. The absence of H_2O ice on Io is easily explained by the intense, tidally-driven volcanism this satellite experiences. It is obviously important to establish whether the present indications that there may be some H_2O on this exotic object can be confirmed. On Pluto and Triton, the more volatile CO_2, CH_4, CO and N_2 are presumably covering thick mantles of solid H_2O.

The views of satellite surfaces returned by the Voyager spacecraft indicate that wherever an "old" surface is present, as indicated by the higher density of impact craters, there is often geomorphic evidence that the surfaces have undergone endogenic geological evolution. This has led to speculation that NH_3-hydrate may be present on these bodies, allowing the ice to be more plastic or even to melt at lower temperatures than would be possible for pure H_2O ice. To date there is no firm spectroscopic evidence identifying either this compound or pure NH_3 ice on any surface in the solar system. However, the Giotto mass spectrometer did identify NH_3 in Halley's Comet[77]. Perhaps the ease with which UV photons can dissociate NH_3 has converted any exposed form of this compound to hydrogen and N_2.

The accretion of these small bodies must have had a major effect on the surface compositions we see today. High-velocity impacts will degas the impactors, and these volatiles will be lost from the growing planetesimals. Low-velocity impacts, becoming more common with increasing distance from the sun, will allow retention of volatiles with relatively little chemical alteration. Because of the tidal energy dissipated during its capture by Neptune, Triton may have had an especially complex history, with some post-formation internal processing of its constituents[78]. The geological evolution of some of the other satellite surfaces, referred to above, also suggests some type of internal activity on these bodies. It is against this background that we must struggle to understand which of the ices we see today represent primordial materials, and which are the result of chemistry driven by the energy released during accretion or by subsequent events.

ACKNOWLEDGEMENTS

This research is supported by NASA's Planetary Geology and Geophysics Program via RTOP 151-01-60-01 and NASA's Planetary Astronomy Program via RTOP 196-41-67-03. We thank reviewers Marc Buie and Wendy Calvin for helping to improve the original manuscript.

REFERENCES

1. G. T. Sill and R. N. Clark, in *Satellites of Jupiter*, D. Morrison, Ed., Univ. Arizona Press, Tucson, Arizona, p. 174 (1982).

2. R. N. Clark, *et al.*, in *Satellites*, J. Burns and M. S. Matthews, Eds., Univ. Arizona Press, Tucson, Arizona, p. 437 (1986).

3. D. P. Cruikshank and R. H. Brown, in *Satellites*, J. Burns and M. S. Matthews, Eds., Univ. Arizona Press, Tucson, Arizona, p. 836 (1986).

4. D.P. Cruikshank and R.H. Brown, *Remote Sensing of Ices and Ice-Mineral Mixtures in the Outer Solar System*, in *Remote Geochemical Analysis: Elemental and Mineralogical Composition*, C.M. Pieters and P.A.J. Englert, Eds., Cambridge Univ. Press, NY, p. 455 (1993).

5. U. Fink and G. T. Sill, *The Infrared Spectral Properties of Frozen Volatiles*, in *Comets*, L. L. Wilkening, Ed., Univ. Arizona Press, Tucson, Arizona, p. 164 (1983).

6. B. W. Hapke, *Combined Theory of Reflectance and Emittance Spectroscopy*, in *Remote Geochemical Analysis: Elemental and Mineralogical Composition*, C.M. Pieters and P.A.J. Englert, Eds., Cambridge Univ. Press, New York, p. 31 (1993).

7. B. W. Hapke, *Reflectance and Emittance Spectroscopy*, Cambridge Univ. Press, New York, 455 pp. (1993).

8. R. N. Clark, *et al.*, unpublished manuscript, (1986).

9. F. P. Fanale, *et al.*, *Nature* **280**, 760 (1979).

10. W. D. Smythe, *et al.*, *Nature* **280**, 766 (1979).

11. R. R. Howell, *et al.*, *Icarus* **78**, 27 (1989).

12. B. Schmitt, *et al.*, *Icarus* **111**, 79 (1994).

13. D. B. Nash and R. R. Howell, *Science* **244**, 454 (1989).

14. F. Salama, *et al.*, *Icarus* **83**, 66 (1990).

15. F. Salama, *et al.*, *Icarus* **107**, 413 (1994).

16. D. B. Nash, *Icarus* **107**, 418 (1994).

17. S. A. Sandford, *et al.*, *Icarus* **110**, 292 (1994).

18. J. B. Pollack, *et al.*, *Icarus* **36**, 271 (1978).

19. T. L. Roush, *et al.*, *Icarus* **86**, 355 (1990).

20. W. M. Calvin and R. N. Clark, *Icarus* **89**, 305 (1991).

21. W. M. Calvin, *et al.*, *J. Geophys. Res. - Planets*, in press (1995).

22. J. R. Spencer, *et al.*, *J. Geophys. Res. - Planets*, in press (1995).

23. K. S. Noll, *et al.*, *J. Geophys. Res. - Planets*, in press (1995).
24. R. N. Clark, *et al.*, *Icarus* **58**, 265 (1984).
25. J. I. Lunine and D. J. Stevenson, *Astrophys J. Suppl. Ser.* **58**, 493 (1985).
26. D. J. Stevenson, *et al.*, *Origins of Satellites*, in *Satellites*, J. Burns and M. S. Matthews, Eds., Univ. Arizona Press, 39 (1986).
27. J. I. Lunine, *et al.*, *Icarus* **94**, 333 (1991).
28. P. R. Estrada and J. N. Cuzzi, *Icarus*, in press (1995).
29. R. H. Brown, *et al.*, *Icarus* **52**, 188 (1982).
30. R. H. Brown and R. N. Clark, *Icarus* **58**, 288 (1984).
31. B. A. Smith, *et al.*, *Science* **233**, 43 (1986).
32. J. Veverka, *et al.*, *J. Geophys. Res.* **92**, 14895 (1984)
33. R. H. Brown and D. P. Cruikshank, in preparation, (1995).
34. M. E. Ockert, *et al.*, *J. Geophys. Res.* **92**, 14969 (1987).
35. P. D. Nicholson and T. J. Jones, *Icarus* **42**, 54 (1980).
36. B. T. Soifer *et al.*, *Icarus* **45**, 612 (1981).
37. D. P. Cruikshank *et al.*, *Icarus* **94**, 345 (1991).
38. P. Thomas and J. Veverka, *J. Geophys. Res.* **96**, 19261 (1991).
39. D. P. Cruikshank and P. Silvaggio, *Astrophys. J.* **233**, 1016 (1979)
40. J. Apt, *et al.*, *Astrophys. J.* **270**, 342 (1983).
41. D. P. Cruikshank and J. Apt, *Icarus* **58**, 306 (1984).
42. D. P. Cruikshank, *et al.*, *Icarus* **58**, 293 (1984).
43. J. I. Lunine and D. J. Stevenson, *Nature* **317**, 238 (1985)
44. B. Conrath, *et al.*, *Science* **246**, 1454 (1989).
45. A. L. Broadfoot et al., *Science* **246**, 1459 (1989).
46. G. L. Tyler et al., *Science* **246**, 1466 (1989).
47. D. P. Cruikshank, *et al.*, *Science*, **261**, 742 (1993).
48. K. A. Tryka, *et al.*, *Science* **261**, 751 (1993).
49. W. M. Grundy, *et al.*, *Icarus* **105**, (1993).
50. K. A. Tryka, *et al.*, *Icarus*, **112**, 513 (1994).
51. D. P. Cruikshank *et al.*, in preparation, (1995).
52. B. J. Buratti, *et al.*, *Icarus* **110**, 303 (1994).
53. J. F. Bell, *et al.*, *Bull. Am. Astron. Soc.* **11**, 570 (1979).
54. D. P. Cruikshank, *et al.*, *Icarus* **40**, 104 (1979).
55. D. P. Cruikshank, *et al.*, *Science* **194**, 835 (1976).
56. L. A. Lebofsky, *et al.*, *Icarus* **37**, 554 (1979).
57. B. T. Soifer, *et al.*, *Astron. J.* **85**, 166 (1980).
58. D. P. Cruikshank and P. Silvaggio, *Icarus* **41**, 96 (1980).
59. M. W. Buie and U. Fink, *Icarus* **70**, 483 (1987).
60. U. Fink and M. A. DiSanti, *Astron. J.* **95**, 229 (1988).
61. R. L. Marcialis and L. A. Lebofsky, *Icarus* **89**, 255 (1991).
62. T. C. Owen, *et al.*, *Science* **261**, 745 (1993).
63. T. C. Owen *et al.*, in preparation (1995).
64. D. P. Cruikshank, *et al.*, *Surface Properties, Composition, and Geological Context*, in *Pluto and Charon*, D. J. Tholen and A. Stern, Eds., Univ. Arizona Press, Tucson, submitted (1995).
65. R. B. Bohn, *et al.*, *Icarus* **111**, 151 (1994).
66. R. P. Binzel, *et al.*, *Science* **228**, 1193 (1985).
67. D. J. Tholen and M. W. Buie, *Astron. J.* **96**, 1977 (1988).
68. M. W. Buie, *et al.*, *Icarus* **97**, 211 (1992).
69. E. F. Young and R. P. Binzel, *Icarus* **102**, 134 (1993).
70. K. Reinsch, *et al.*, *Icarus* **108**, 209 (1994).
71. W. F. Drish Jr., *et al.*, *Icarus* in press, (1995).
72. E. F. Young and M. W. Buie, *Bull. Am. Astron. Soc.* **25**, 1130 (1994).

73. R. L. Marcialis, *et al.*, *Science* **237**, 1349 (1987).
74. M. W. Buie, *et al.*, *Nature* **329**, 522 (1987).
75. T. L. Roush, *Icarus* **104**, 243 (1994).
76. T. L. Roush, *et al.*, *Icarus* submitted (1995).
77. R. Meier, *et al.*, *Astron. and Astrophys.* **287**, 268 (1994).
78. E. L. Shock, and W. B. McKinnon, *Icarus* **106**, 464 (1993).

WATER ON MARS AND VENUS

T.M. Donahue
Department of Atmospheric, Oceanic and Space Sciences
University of Michigan, Ann Arbor, MI 48109

ABSTRACT

This paper reviews evidence relating to the abundance of water on early Mars and Venus from measurements of the present abundance of hydrogen compounds, deuterium to hydrogen (D/H) ratios and escape fluxes. For Mars, recent measurements of D/H ratios in SNC hydrous minerals provide data on the ratios at earlier times to augment present atmospheric values. Interpretation of these data shows that they are consistent with the presence of scores to hundreds of meters of liquid water on early Mars, as well as less, but still abundant, water in today's crust. They also require concentrations of hydrogen compounds in the early atmosphere orders of magnitude higher than is present today to support large scale hydrogen escape. For Venus, a very large D/H enhancement (160 fold) implies at least 3 to 4 meters liquid equivalent of early water (depending on how much hydrogen is in the atmosphere today). It is consistent with much more, even the equivalent of a full terrestrial ocean. The low escape flux and high fractionation factor place severe constraints on volcanic or cometary sources. Some of the present water can have been injected by volcanism but comets as important sources appear to be excluded.

INTRODUCTION

Among the terrestrial planets only Earth now can boast of an ocean of water. Mercury and the Moon are bone-dry. So is the atmosphere of Venus, where the water vapor mixing ratio below the clouds is only 30 ppm. How much water is below its surface is unknown. On Mars too the water content of the atmosphere is variable, but low; the average value is 10 precipitable μm. Again, as on Venus, how much water is in the crust and mantle is not known but is the subject of much speculation because of evidence for vigorous fluvial activity in the recent as well as the distant past. The polar caps clearly provide a reservoir for water under present climatic conditions, water that could be mobilized under certain circumstances, such as higher obliquity or more favorable orbital characteristics.

The purpose of this paper is to examine evidence that both Venus and Mars once had much more water in their atmospheres, hydrospheres (perhaps) and crusts than they do today. Even the early equivalent of a full terrestrial ocean on Venus cannot be excluded. The evidence in question is the highly fractionated state of hydrogen and deuterium on both planets, especially Venus, compared to Earth and other solar system objects. On Mars R, the ratio of deuterium to hydrogen in atmospheric water vapor is $(8.1 \pm 0.3) \times 10^{-4}$.[1,2] Here

$$2R = [HDO]/[H_2O], \qquad (1)$$

where $[x]$ is the atmospheric columnar abundance of species x. R is greater on Mars than in terrestrial ocean water (SMOW) by a factor of 5.2 ± 0.2. (There is evidence that Martian mantle water is not more than 50% richer in deuterium than SMOW.[3]) On Venus R is huge -- 2.5×10^{-2}, 160 times that of SMOW.[4,5,6,7,8,9] If we assume -- as is reasonable -- that all three planets started with about the same value of R in their water, we can show that these results most probably mean that both Mars and Venus have lost a lot of the hydrogen associated with water, or other hydrogen compounds that resided in reservoirs containing much more hydrogen in the past than they do today. The challenge is to use the measurements of R, of the present abundance of atmospheric hydrogen compounds and of hydrogen and deuterium escape fluxes to infer what the abundances were in the past. Recent measurement of the

D/H ratio of water in the Martian crust as it was hundreds of millions or perhaps billions of years ago[3] will also be extremely useful in this exercise.[10]

   Suppose atomic hydrogen is escaping from the atmosphere of a planet with a flux ϕ_1 and deuterium with a flux ϕ_2, where the fluxes are related by the fractionation factor f,

$$f = \phi_2/R\phi_1 .$$ (2)

Here

$$R = [D]/[H].$$ (3)

f, which also can be expressed in the form

$$f = \frac{d[D]}{[D]} \Big/ \frac{d[H]}{[H]} ,$$ (4)

measures the relative ease with which deuterium escapes. The abundance of hydrogen and deuterium are then governed by the differential equations

$$\frac{d[H]}{dt} = P - \phi_1 = P - K[H],$$ (5)

$$\frac{d[D]}{dt} = R_S P - Rf\,KH,$$ (6)

where it is assumed that there is a source of hydrogen of strength P and of deuterium $R_S P$. K is a proportionality constant. In case P is small or absent, Eq. (4) can be integrated at once to give the ratio of the size of the reservoir at time t_2 to a later time t_1

$$r(t_1, t_2) = \frac{[H(t_2)]}{[H(t_1)]} = \left(\frac{R(t_1)}{R(t_2)}\right)^{\frac{1}{1-f}} .$$ (7)

This is independent of ϕ_1 and depends only on the change in the D to H ratio and the fractionation factor. Otherwise the solution of Eqs. (5) and (6) for $R^{11,12,13}$ is the time dependent expression

$$R = R_S/f - (R_S/f - R_o)\exp(-\int \phi_1\,fdt'/[H_{SS}]),$$ (8)

where R_o is the initial D/H ratio and $[H_{SS}]$ is the steady state hydrogen abundance that prevails when escape balances production. The steady state D to H ratio is given by

$$R_{SS} = R_S/f$$ (9)

(In integrating Eqs. (4), (5) and (6), we assume that f is time independent. This is a very shaky assumption. Both f and ϕ_1 have surely varied considerably during the lifetime of Mars and Venus. There has been only one attempt to model this variation.[14] It needs to be repeated for Venus and extended to Mars in the light of today's knowledge.) For Jeans escape ϕ_2/ϕ_1 and R covary unless the exospheric temperature changes. In terms of the characteristic hydrogen lifetime,

$$\tau_1 = [H_{SS}]/\phi_1 ,$$ (10)

the approach to a steady state in Eq. (8) is described by a lifetime

$$\tau_{SS} = \tau_1/f.$$ (11)

WATER ON MARS

In the case of Mars, the present rate of loss of hydrogen in the form of H and H_2 by thermal or so-called "Jeans" escape can be measured and modeled. A recent calculation of ϕ_1 averaged over the solar cycle, based on observed atomic hydrogen densities in the upper atmosphere, gives[15]

$$\phi_1 = 2.4 \times 10^8 \, cm^{-2}s^{-1},$$ (12)

where

$$\phi_1 = \phi(H) + 2\phi(H_2) + \phi(HD),$$ (13)

$$\phi_2 = \phi(D) + \phi(HD).$$ (14)

The fractionation factor for Jeans escape of D and HD from Mars under present conditions has been calculated to be 0.32.[16] Other escape mechanisms are known to be important for Earth[17] and Venus[18,19] and may also be so for Mars. To allow for this eventuality, f will be treated as a parameter in this paper. It should be kept in mind also that ϕ_1 for Mars may be larger than Eq. (12) today, and that it may have varied even during the recent few hundred million years and, thus, that the contents of Martian water reservoirs may in fact have been considerably different (probably higher) than those determined in this paper. On the other hand, because Jeans escape flux from Mars is so large compared to the Jeans flux from either of the other planets, it probably dominates other escape modes on that planet. Measurements that enable an evaluation of all escape processes certainly should be included in any future mission to Mars, so crucial it is to understand the evolution of water. Relationship (9) shows that Martian water can be in a steady state with a source and escape if R_S is about twice that of SMOW and the strength of the source at least $2.5 \times 10^8 \, cm^{-2}s^{-1}$. The first is possible, the second unlikely, as we shall eventually discuss in greater detail.

We shall assume that water in the crust exchanges freely with atmospheric water vapor at present and in the past, as indicated by the evidence for interaction of highly deuterated crustal water with minerals found in SNC meteorites. Thus, the D to H ratio, R, will covary in the atmosphere and in crustal water. If some of the crustal water is stored as ice in equilibrium with liquid water, the D to H ratio in the ice will be about 1.27 times the value in the vapor and liquid phase.[16] To the extent that there is exchangeable water in the solid phase, the estimates of reservoir sizes in this paper are lower limits. Following a treatment by Yung et al,[16] we assume that there is very much more water in the crust than in the atmosphere. The total amount of hydrogen present in the two reservoirs is

$$h(t) = c(t) + a(t).$$ (15)

Integration of Eq. (4), where

$$\phi_i = -\frac{dh_i}{dt},$$ (16)

gives Eq. (7) for the ratio of the water reservoir at t_2 to the reservoir at a later time t_1 in the form

$$r(t_1, t_2) = \frac{a(t_2) + c(t_2)}{a(t_1) + c(t_1)} = \left(\frac{R(t_1)}{R(t_2)}\right)^{\frac{1}{1-f}},$$ (17)

which, as we have already seen, is independent of ϕ_1. The actual size of the present reservoir does depend on the escape rate, however. The smaller the change in R over a given period of time the larger the reservoir must be compared to the amount of hydrogen (and deuterium) lost. In fact, since

$$c(t_1) = c(t_2) - \int_{t_1}^{t_2} \phi_1 \, dt, \tag{18}$$

and

$$c(t) >> a(t), \tag{19}$$

$$c(t_1) \cong \int_{t_1}^{t_2} \phi_1 \, dt/(r-1). \tag{20}$$

$c(t_1)$ increases rapidly as r approaches unity.

Table I. Early and late crustal water reservoirs (constant escape flux)

f	0	0.1	0.32	0.6
r	5.2	6.2	11.3	62
t_1		$c(t_1)$ meters		
Present	1.3	0.98	0.5	0.084
t_2		$c(t_2)$ meters		
4.5 Ga	6.7	6.1	5.7	5.2

Here time is measured backward from the present so that $t_1 < t_2$. Table I gives r and the depth of the water reservoir expressed as meters of liquid water at $t_1 = 0$ and $t_2 = 4.5$ Ga if ϕ_1 has a constant value given by Eq. (12). To allow for the possibility of an important contribution to the escape flux by mechanisms in which escape of deuterium is inefficient compared to Jeans escape, r and $c(o)$ and $c(t_2)$ are shown for several values of f. No matter how effectively deuterium was retained, these crustal reservoirs are small. As others have pointed out,[16,20] increasing ϕ_1 causes $c(t_2)$ and $c(o)$ to increase correspondingly. And if Mars was much warmer and wetter during part of its life than it is today, escape rates then could have been much higher than today.

In all cases, the early reservoirs $c(t_2)$ are very small compared with the amount of water that was almost surely mobilized when valley networks and catastrophic flood channels were produced on the Martian surface. Production of the flood channels apparently requires that there exist a planet-wide reservoir of crustal water hundreds of meters deep.[21] In early times, after the late heavy bombardment, the climate may have been warm enough for precipitation to produce such fluvial features as the valley networks.[22-24] Erosion of early surface features, possibly by fluvial activity, was extensive.[25,26] Some have argued that there is geological evidence that the planet has undergone frequent episodes of moderate climate with abundant surface water. These may have been triggered by release of water and CO_2 during periods of volcanism as recently as 500 million years ago.[27] Furthermore, the inclination of the Martian spin axis varies secularly. During periods of high obliquity, water now in polar deposits could have been mobilized.[21] Thus, ϕ_1 was almost surely much larger at times in the past than it is today. But, until recently, there has been no guidepost to determine how much.

Recent measurements of the D to H ratio in hydrous minerals contained in three of the ten SNC meteorites[3] may give R in crustal water and thus in the atmosphere at one time in the past for comparison with the present and original values. In the case of Zagami, one of the Shergottites examined, the time of crystallization is controversial. Early whole rock chronometry set the time at 1.3 Ga,[28] whereas some favor a date as late as 180 Ma.[29] Very surprisingly, the SNC D to H ratio measurements showed that the ratio for some apatites in Zagami was as large as today's atmospheric value. These elevated values apparently require that crustal water, which interacted with the igneous SNC rocks shortly after they crystallized and exchanged with deuterium-poor mantle water, had a D to H ratio, R, 180 Ma to 1.3 Ga, indistinguishable from R today. There must have been an active hydrothermal system exchanging atmospheric and crustal water as recently as the time at which Zagami crystallized. The authors of the paper reporting these measurements[3] suggest that the lower values of R found for most hydrogen samples of hydrous minerals resulted from incomplete exchange of crustal water with pristine, low R mantle water. Presumably magmatic water would have intruded into the crust along with the rest of the magma. They also argued[3] that hydrothermal alteration would not have significantly post-dated primary crystallization of the magmas. Donahue[10] has examined the possibility that some of the water forming the hydrous phase was crustal water with the same R as atmospheric water vapor, at the time, t_z, when Zagami apatites crystallized. The largest apatite D to H ratio would then be the value of R for atmospheric and crustal water at that time or at the time the water was last exposed to the atmosphere. Hence, R was about as high then as it is today. From the published data and the precision claimed for the measurements, it is difficult to see how R could have been more than five percent lower than the present R. In fact, any value of $R(t_z)$ up to $R(o)$ seems to be possible. At a constant loss rate equal to today's value of ϕ_1, R would have fallen to 80% of today's 180 Ma ago and 40% 1.3 Ga ago. Such low values are certainly excluded. If, indeed, $R(t_z)$ was, say, five percent lower than its present value when these minerals crystallized, the amount of hydrogen and deuterium that have since escaped must be small compared to the amount in the reservoir 0.18 or 1.3 billion years ago. But in 0.18 Gyr hydrogen associated with 0.2m of water in liquid form and in 1.3 Gyr that associated with 1.4m of water would have escaped at present escape rates. These amounts of water are much larger or comparable to all values of $c(t)$ in Table I.

Table II. Crustal water (meters), escape fluxes ($cm^{-2}s^{-1}$), Zagami 0.18 Gyr old, $R(o) = 1.04R(t_z)$

f	0	0.32	0.6	0.8
$c(t_1)$ (present)	5.5	3.7	2.1	1.0
$c(t_z)$ (180 Ma)	5.7	3.9	2.3	1.2
$c(t_2)$ (4.5 Ga)	28	42	132	3800
$\phi_1(t_z, t_2)$	1.1 (9)	1.8 (9)	6.3(9)	1.9 (10)
ϕ_1 (3.5 − 4.5 Ga)	4.8 (9)	8 (9)	2.7 (10)	8 (11)

In fact, examination of the published SNC data[3] indicates that an extremely conservative estimate of the lower limit to the ratio of $R(o)$ to $R(t_z)$ is 1.04. r in this case would be given by

$$r = (1.04)^{\frac{1}{1-f}}, \qquad (21)$$

and

$$c(o) = \frac{\phi_1 t_z}{r-1}. \qquad (22)$$

$c(o)$ is given in Table II for several values of f. Finally, the crustal water content when fractionation began, $c(t_2)$, is calculated, where

$$c(t) = r_o\, c(o) \qquad (23)$$

with

$$r_o = r(o,t_z) = 5.2^{\frac{1}{1-f}}. \qquad (24)$$

$c(t_2)$ is also tabulated in Table II. All of these quantities are, of course, considerably larger than the corresponding entries in Table I. They are lower limits for the crustal inventories.

Also entered in Table II, as are the average fluxes

$$\phi_1(t_z, t_2) = \frac{c(t_2) - c(t_z)}{t_2 - t_z}, \qquad (25)$$

where t_z is 180 Ma, t_2 is 4.5 Ga and c is expressed in terms of hydrogen atoms per cm^2 column. Shown also as fluxes 4.32 times as large as these, such as would have been required if the hydrogen was lost mainly during the first billion years as would have been the case if the warm, wet[26] time on Mars was as early as geological evidence indicates. In fact, this period may have lasted only for the first 500 Myr. In this case the fluxes listed must be doubled.

Table III. Crustal water (meters) and escape fluxes (cm^{-2}s^{-1}), Zagami 1.3 Ga, $R_1(o) = 1.04\, R(t_z)$

f	0	0.32	0.6	0.8
$r(t_z)$	1.038	1.057	1.1	1.21
$c(t_1)$ (present)	37	25	14	6.7
$c(t_z)$ (1.3 Ga)	38.4	26.4	15.4	8.1
$c(t_2)$ (4.5 Ga)	191	278	880	2.5 (4)
$\phi_1(t_z, t_2)$	7.5(9)	1.2 (9)	4.2 (10)	1.2 (12)
ϕ_1 (3.5-4.5 Ga)	3.2 (10)	5.3 (10)	1.8 (11)	5.3 (12)

$c(t_2)$ still falls very much short of the 400 to 500 meters of water called for in some channel and basin creating scenarios, if f is fixed at 0.32. Another scenario is provided if Zagami apatites are assumed to be 1.3 Gyr old. During those 1.3 Gyr, 1.4m of water would have escaped at the present loss rate. All reservoirs and fluxes in Table II would be multiplied by a factor of 1.4/0.2. The various values of c and ϕ_1 are listed in Table III. Here much more respectable early water inventories are called for, particularly if ϕ_1 was larger than $2.4 \times 10^8 \text{cm}^{-2}\text{s}^{-1}$ for much of the past 1.3 Gyr. Early escape fluxes orders of magnitude larger than today's would be called for.

But the Zagami results do not constrain the D to H ratio in crustal water at t_z to be as small as 96% of the present R. They allow an arbitrarily close match, so that $r - 1$, in principle, can be as small as we choose. It is legitimate, for example, to ask how large R must have been when the apatites crystallized, if the original water reservoir was arbitrarily large, say 500m. To accommodate any desired reservoir depth $c(t)$, 4.5 Gyr ago, the ratio of R at t_z to R(o) must be

$$R(t_z)/R(o) = \left[(c(t)/11.3)/(\phi_1 t_z + c(t)/11.3) \right]^{0.68}. \tag{26}$$

for Jeans escape. For $c(t_2)$ 500 km, $c(o)$ would be 44 meters and $R(t_z)$ 0.997 today's R if t_z is 1.3 Ga. The average escape fluxes would then have been about 400 times those of today if enhanced escape was restricted to the first billion years. Table IV shows the ratios of $R(t_z)$ to $R(o)$ for a range of fractionation factors and for $t_z = 180$ Ma and 1.3 Ga if $c(t_2)$ was 500 meters 4.5 Ga ago.

Table IV. Crustal reservoirs in meters now and at t_z if the original Reservoir $c(t)$ was 500m Deep. Ratio of $R(t_z)$ to $R(o)$ required if $c(t_2)$ was 500m.

f	0		0.32		0.6	
r_o	5.2		11.3		62	
time	c	$R(t_z)/R(t_1)$	c	$R(t_z)/R(t_1)$	c	$R(t_z)/R(t_1)$
present	96.1		44.3		8.1	
$t_z = 180$ Ma	96.3	0.998	44.5	0.997	8.3	0.990
$t_z = 1.3$ Ga	97.4	0.987	45.5	0.980	9.5	0.940
$t_2 = 4.5$ Ga	500		500		500	

Escape fluxes orders of magnitude larger than those of today are possible if the hydrogen concentration in the atmosphere was correspondingly larger. A fundamental limitation to the escape flux is the so-called limiting flux determined by the mixing ratio of hydrogen in all of its forms in the lower atmosphere below the homopause.[30] In the case of the Martian atmosphere, the dominant hydrogen species there (at least under today's conditions) is H_2,[14,31] so

$$\phi_l = 2 \text{ bf}_{H_2}/H_a = 2.25 \times 10^{13} f_{H_2} \text{cm}^{-2}\text{s}^{-1}, \tag{27}$$

where f_{H_2} is the mixing ratio of H_2 near the homopause, H_a is the mean atmospheric scale height there, and

$$b = D_{H_2}/n_a. \tag{28}$$

Here D_{H_2} is the H_2 molecular diffusion coefficient and n_a the atmospheric density. In today's atmosphere, with $f_{H_2} = 2 \times 10^{-5}$, ϕ_l is larger than the escape flux by a factor of about 2. This is because much of the escaping hydrogen is in the form of H_2, which escapes slowly compared to H. To attain an escape flux of $10^{11} \text{cm}^{-2}\text{s}^{-1}$, f_{H_2} would need to be equal to or larger than 0.40%. If it is warranted, scaling from today's atmosphere, where a homopause mixing ratio f_{H_2} of 2×10^{-5} corresponds to a water vapor abundance of 10 precipitable μm in the lower Martian atmosphere, would require that the average Martian atmosphere for the first 3.2 billion years contain an average of 2 precipitable centimeters of water vapor. However, the whole issue of how the hydrogen mixing ratio in the upper atmosphere relates to the amount of water vapor in the troposphere on an early warmer, wetter Mars given the likely presence of a cold trap,[32] is one of the major uncertainties besetting this entire problem.

Enhanced escape of hydrogen can be effected by increasing the exospheric temperature of the upper atmosphere until the flow bottleneck is transferred from the exosphere to the region near the homopause and the flux becomes diffusion limited. This is something that almost surely would have happened during the period of high euv solar luminosity when the solar system was young. The fractionation factor for Jeans escape would increase by a significant amount if the exospheric temperature were to increase appreciably. This is because the escape flux of the more massive species D, HD and H_2 increase much more rapidly with temperature than that of H does. Taking account of the change in densities at the exobase n_{ci} as well as the change in effusion velocity w_{ci} where

$$\phi_i = n_{ci} w_{ci},$$

$\phi(H)$ would change by a factor of 4.4, $\phi(H_2)$ and $\phi(HD)$ by a factor of 10.5 and $\phi(HD)$ by a factor of 34 if the exospheric temperature should change from 365K to 500K.[16] The ratio ϕ_2/ϕ_1, and consequently the fractionation factor, would increase by a factor of 1.53 making $r_o(o,t_2)$ grow from 11.3 to 27. Secondly, the escape rate would increase if the supply of hydrogen compounds (H_2 in today's atmosphere) below the diffusion barrier should be increased. In fact, when limiting flux applies, the escape flux depends only on the total hydrogen mixing ratio in this region, as we have seen. Ultimately, however, increasing the mixing ratio of the dominant hydrogen constituent in the upper atmosphere and the escape rate by orders of magnitude, as this discussion demands, would appear somehow to require a corresponding increase in the dominant hydrogen constituent in the lower atmosphere, water vapor.

Something has to be done to dispose of the oxygen counterpart of the escaping hydrogen. McElroy and Donahue[33] in fact argued that energetic oxygen atoms resulting from dissociative recombination of O_2^+, escape from the upper atmosphere at a rate half the hydrogen escape rate, and control hydrogen escape as the result of a feedback process that maintains the stability of the CO_2 atmosphere. However, it now appears that neither dissociative recombination of O_2^+ [34] nor sputtering of oxygen atoms[35] can remove oxygen nearly fast enough to prevent a significant change in the redox state of the atmosphere in a time of the order of 10^5 years. If this is true, it is difficult to see how the redox state of the atmosphere can be maintained without another oxygen sink -- such as surface oxidation.[36] Presumably, this would be true even in the very different circumstances of the early atmosphere. However, the necessary investigation of the aeronomy of that atmosphere remains to be carried out. There are essential elements missing in our understanding of an early Mars atmosphere which must be warm, wet and support a hydrogen escape flux hundreds of times as large as the one today. As Kasting[32] has pointed out, a simple $CO_2 - H_2O$ greenhouse cannot be invoked to provide us with a warm wet Mars because of the effects of condensation and cloud formation. Alternative processes capable of producing early valley networks without precipitation[26] and thus without requiring a very humid atmosphere will *not* result in the loss of massive amounts of hydrogen which the D to H measurements seem to require.

Another limit on the maximum possible escape flux, no matter how wet the atmosphere, is set by the solar insolation available below 185nm to photolyze H_2O at 1.5 AU. After allowance for enhanced uv from the young sun,[37] solar cycle effects, absorption by other atmospheric species and solar cycle variation, a generous estimate of this limit is $5 \times 10^{11} cm^{-2} s^{-1}$. If escape occurred mainly during the first billion years this means that $c(t_2)$ could not have exceeded 2200m and $c(o)$ consequently, 190m. Thus it is difficult to allow more than 190m of water in the modern crust to drive hydrothermal systems or account for catastrophic outflow channels if the early warm wet period of enhanced escape lasted only 1 Gyr. Extending the period buys only a little relief because the ultraviolet luminosity of the sun should have been reduced to its present level 3.5 Ga. This would reduce the maximum photolysis rate significantly, by a factor of about 2.

There may be the planet-wide equivalent of 30m of water stored today in the polar caps.[38] This water may be mobilized at intervals of 10^5 to 10^6 years because of changes in obliquity and eccentricity. Because these interludes are short on planetary time scales or the crystallization age of Zagami, this polar cap water can be taken to exchange effectively with atmospheric and crustal water. The treatment presented here would still be valid except that $c(o)$ would have to be at least as large as the polar cap reservoir. If the polar caps sequester a large amount of water which has only infrequently been mobilized, the planet's exchangeable water would have been occasionally diluted with water having a low D to H ratio. The present R would be lower than it would have been if straightforward fractionation had occurred and the enhancement factor 11.3 would be an *underestimate*. Numerical modelling would be required to explore this contingency further, but the basic conclusions reached here, that an early large crustal reservoir is allowed by deuterium to hydrogen ratio measurements, would not change. This is true also, of course, for any sizable reservoir that might ordinarily be sequestered but is mobilized occasionally. Whenever that might occur, large amounts of deuterium-poor water would mingle with the rest of the water in the planet's interchangeable reservoirs.

This analysis assumes that there is no important exogenous source of hydrogen and deuterium, such as would result from deposition of water on Mars by incoming comets. If the D to H ratio in the source is low and the rate of hydrogen input is comparable to the escape flux, the effect of such a source would be to produce a low net escape rate with a large fractionation factor. Consequently, R would change more slowly with time than if the source were absent, and smaller crustal reservoirs would be implied by any given ratio of $R(t_z)$ to $R(o)$. If, for example, the production term in Eq. (6) is assumed to be $\alpha\phi_1$ and the D to H ratio in the source assumed to be βR, the differential equations (5) and (6) that control the evolution of [H] and [D] during the time from 0 to t_z, when ϕ_1 is constant, become

$$\frac{d[\text{H}]}{dt} = (\alpha - 1)\phi_1 \tag{29}$$

$$\frac{d[\text{D}]}{dt} = (\beta\alpha - \text{f})R\phi_1. \tag{30}$$

Thus the effective fractionation factor f′ defined in (4) becomes

$$\text{f}' = (0.32 - \beta\alpha)/(1 - \alpha). \tag{31}$$

If $\beta < 0.32$, that is if the D to H ratio in cometary water is less than 2.6×10^{-4}, f′ increases slowly with α until α approaches 1. At $\alpha = 1$, where the net hydrogen flux vanishes, f′ is singular. Above $\alpha = 1$, f′ increases from large negative values to a node at $\alpha = 0.32/\beta$ after which f′ approaches β as α goes to infinity. On the other hand, if the D to H cometary ratio is greater than 2.6×10^{-4}, so that $\beta > 0.32$, f′ goes to 0 at $\alpha = 0.32/\beta$. Beyond the singularity at $\alpha = 1$, f′ decreases rapidly at first and then slowly approaches β as α gets large. The interesting D to H ratios for water in comets bracket 2.6×10^{-4}. One case is that of SMOW, 1.56×10^{-4}, the other is that of comet Halley, which has recently been set[39] at $3.08(+0.38, -0.68) \times 10^{-4}$. In either case the effect of injection of water from comets on the estimated size of the crustal reservoir is small until α is about 0.8. There, if $\beta = 0.37$, f′ becomes 0.024, r becomes 1.039 and $c(o)$ 36m instead of 25m, as it was without comets. If $\beta = 0.19$ these quantities change to f′ = 0.83, $r = 1.245$ and $c(o) = 5.7$m (t_z is 1.3 Ga). A flux of 1.9×10^8 hydrogen atoms cm s^{-1} would be effected by one 10^{18} gm comet impacting on average every 3.4 million years if 5% of the comets mass is assumed to be hydrogen. This impact rate is higher by orders of magnitude than the generally accepted rate for Mars.[40]

WATER ON VENUS

Space missions have provided a wealth of information about hydrogen compounds and hydrogen escape for Venus. Particularly productive were the Pioneer Venus Multiprobe and Orbiter (PVMP and PVO) missions and remote spectrographic observations from Earth and Galileo. Direct measurement of the ratio of HDO to H_2O densities by the Pioneer Venus Large Probe Neutral Mass Spectrometer (LNMS),[5] inferences from ionospheric observations of 3 amu and 2 amu ions,[4,6,7] two near infrared spectral measurements of HDO and H_2O[8,9] mixing ratios all agree that R on Venus is 120 to 160 times that of SMOW. PV ion and neutral density measurements revealed that there are only two significant channels of hydrogen and deuterium escape. Almost all of the escaping hydrogen originates from the post-midnight hydrogen bulge in the upper atmosphere of the planet. Calculations indicate that charge exchange of hot H^+ ions with thermal H atoms causes a planet-wide average hydrogen escape flux of $9 \times 10^6 \, cm^{-2} s^{-1}$ during solar maximum.[19] The fractionation factor is 0.02. Ion density profiles[18] show that H^+ and D^+ ions are accelerated out of this region by a charge separation electric field. The planet-wide average flux of H^+ produced by this mechanism is $1.4 \times 10^7 \, cm^{-2} s^{-1}$ during solar maximum. f is 0.17. No other escape mechanism is important. Ejection of hydrogen atoms after collisions with fast oxygen atoms, once thought to be important,[41] has been shown by Hodges[42] to make only a meager contribution to the total escape rate. Most energetic oxygen atoms are created so deep in the atmosphere that they lose most of their energy in collisions with other oxygen atoms well below the base of the exosphere. Thus, during solar maximum in 1978 and 1979, the planet-wide escape flux was $2.3 \times 10^7 \, cm^{-2} s^{-1}$ with an average fractionation factor of 0.11. During the PVO reentry in the latter half of 1992, the solar 10.7cm radio flux was at a level of $125 \times 10^{22} \, Wm^{-2} Hz^{-1}$ compared to $200 - 250 \times 10^{22} \, Wm^{-2} Hz^{-1}$ in 1978-1979. The H^+ densities in the bulge region were only 0.125 times those measured during solar maximum,[19] so the electric field driven hydrogen escape flux was reduced to $1.7 \times 10^6 \, cm^{-2} s^{-1}$. On the other hand, the atomic hydrogen density increased in the bulge as solar activity decreased, because the outflow due to escape was much reduced while the trans-terminator flow into the bulge continued.[43] The charge transfer escape flux, being proportional to the product of the H and H^+ densities, was only slightly reduced to $0.7 \times 10^7 \, cm^{-2} s^{-1}$. This made the escape flux in 1992 only $0.87 \times 10^7 \, cm^{-2} s^{-1}$ and the fractionation factor 0.05. Thus, the solar cycle average escape fluxes are $0.8 \times 10^7 \, cm^{-2}$ for the electric field driven flow and $0.8 \times 10^7 \, cm^{-2}$ for charge transfer, making a total average flux of $1.6 \times 10^7 \, cm^{-2} s^{-1}$ with a fractionation factor of 0.1.

It is, however, questionable that the charge transfer escape flux should be added to the electric field driven flow. Instead, it appears more likely that included in the latter flow are the ions which will undergo charge transfer collision at higher altitude. In this case, the escape flux varies from $1.4 \times 10^7 \, cm^{-2} s^{-1}$ to $0.17 \times 10^7 \, cm^{-2} s^{-1}$ with solar activity. The average flux is $0.8 \times 10^7 \, cm^{-2} s^{-1}$ and the fractionation factor 0.17. This interpretation is strongly reinforced by an analysis of the rate at which the hydrogen density in the bulge increased as loss of hydrogen from the bulge decreased with reduced solar activity.[43] The rate at which the column density grows should equal the rate at which it decreases when solar activity is turned on again, which is the escape flux. This rate was $8 \times 10^7 \, cm^{-2} s^{-1}$ for hydrogen and $2.3 \times 10^5 \, cm^{-2} s^{-1}$ for deuterium from the bulge, resulting in a fractionation factor of 0.12. The planet-wide average hydrogen flux would thus be $1.6 \times 10^7 \, cm^{-2} s^{-1}$, which is very close to the measured electric field driven flow during solar maximum without the contribution from charge exchange.

Both the PV LNMS and infra-red studies agree that the water vapor mixing ratio below the clouds is 30 ppm.[44,45] The LNMS also detected 2 amu and 3 amu species. Analysis of these data indicate that, whereas most of the 2 amu species are terrestrial H_2 molecules emanating from mass spectrometer surfaces and some of the 3 amu species are H_3^+ produced by dissociative ionization of internally generated methane, there is a possibility that some of these species are H_2 and HD from the Venus atmosphere. Mixing ratios as small as 0 and as large as 10 ppm below 25 km are possible.[45] We consider two possibilities: the hydrogen budget amounts to 30 ppm of H_2O and that it amounts to 40 ppm of H_2O and H_2. In either case, from Eq. (7), with f = 0.11, we have an r of 260. This means that, when fractionation began, Venus had 260 times as much hydrogen as it has now, if there are no important endogenous or exogenous sources of hydrogen. If the hydrogen was all in the form of H_2O, there would have been 320 or 430 g/cm^2 of water, which is equivalent to 3.2 or 4.3m of liquid water, 0.10% or 0.13% of a full terrestrial ocean. τ_H would be 230 or 300 Myr. If, however, the escape flux is only $0.8 \times 10^7 cm^{-2}s^{-1}$ and the fractionation factor 0.17, these quantities would charge to 5.1 and 6.9m of water and 460 to 600 Myr.

The characteristic time for establishing a steady state τ_{SS} would range from 2.3 to 3.5 Gyr. But R_{SS} would only be 9 times R_S (or 5.9 times R_S if f is 0.17), the D/H ratio in the putative source. Thus, the source of H_2O and HDO would need to have had a D/H ratio 17 (or 25) times terrestrial. This seems to rule out exogenous sources such as comets for the small bit of water in the atmosphere. This is quite surprising in that the escape fluxes are so small. A hydrogen flux $0.8 \times 10^7 cm^{-2}s^{-1}$ would be matched by the input of one 10^{18} gm comet every 375 million years. It may be possible to develop a scenario for water on Venus that is somewhat like that suggested for Mars, in which fractionated water is cycled from the atmosphere into the interior and back again during periods of widespread vulcanism.[47] If so, the sequestration of fractionated water, with a D/H ratio about 20 times that of SMOW, would have occurred long ago. This mechanism, however, differs only in detail from a simple Rayleigh fractionation of the sort we described for Mars.

There may have been a time when Venus had on its surface or in its atmosphere much more water than that equivalent to a few tenths of a full terrestrial ocean. If so, the hydrogen and deuterium must have blown off, driven by solar euv or impact erosion. Attempts have been made to model the effect of the increasing luminosity of the faint early sun on an early "wet" Venus, and they make a plausible case. Disposal of the oxygen left behind would not be a problem, given the probable presence of much CO, and eons of volcanic production of fresh surface magmas to be oxidized.

DISCUSSION

The most direct and plausible interpretation of measurements of the deuterium-hydrogen ratio in atmospheric and crustal species on Mars and Venus is that each planet had abundant water in both of these regions long ago but that most of it has since been photolyzed and disappeared. The similarity in the deuterium-hydrogen ratio in the present Martian atmosphere and in some hydrous minerals of igneous rocks formed between 180 million and 1.3 billion years ago means that recently there has been some kind of hydrothermal activity that cycles water between the crust and the atmosphere. This exchange may still be occurring and almost surely has occurred frequently or continuously in the past if the D to H ratio of the water in the crust is determined by escape of hydrogen from the atmosphere by Jeans escape during the past 3.5 to 4.5 billion years. The measurements also place a fairly large lower limit on the amount of water in the present and ancient crustal reservoirs which is comparable to the amount inferred from study of outflow channels, especially if the crystallization age of Zagami apatites is large.[10] But they also allow for the (likely) possibility of much more crustal water, since the D to H ratio in some samples was measured to be as large as in the present atmosphere. The atmospheric and Zagami observations alone cannot determine when the bulk of the original crustal water was lost, but allow it to be very early, provided the period of loss was not so short as to cause a violation of maximum escape flux constraints. Clearly needed to resolve issues raised by these observations and this interpretation is a better understanding of the hydrothermal regime on Mars and the physical and chemical processes accompanying massive loss of hydrogen from the ancient atmosphere.

How escape of hydrogen and deuterium evolved over the lifetime of each planet, needs to be modeled. Only when this has been done can we decide if the fractionation observed can be understood in terms of the evolution of the hydrogen isotopes during the past 4.5 Gyr. An effort was made to do this once for Venus,[14] but the calculation needs to be repeated in the light of present day understanding of escape mechanisms. Even the calculation of charge exchange loss from Venus needs to be redone, using neutral, ionic and temperature data provided by the PV Orbiter, and taking into account the effect of the solar cycle on these densities.

Thus the D to H ratio measurements in atmospheric water vapor and Zagami apatites are compatible with the presence of larger crustal water reservoirs freely exchanging with the atmosphere, now and long ago, with fractionation a consequence of loss of H and D to space since planetary accretion was completed. The problem of incompatibility of D/H measurements and geological requirements discussed by Carr[38] is thereby removed and an alternative provided to the scenarios discussed in his paper. If D enriched crustal water interacts with the atmosphere, now and in the past, in amounts as large as proposed here, most of Carr's strictures should not apply. There are problems posed by such an evolutionary history, particularly in coupling large changes in tropospheric water to corresponding changes in hydrogen mixing ratios in the upper atmosphere, in order to account for the large loss of hydrogen required early on and in disposing of the oxygen counterpart of the escaping hydrogen. Other explanations of the D to H ratio measurements exist. These measurements are, for example, clearly consistent with the presence of a crustal reservoir in contact with the atmospheric water vapor that is extensive compared to the 5m or so of water that would have been lost if the present rate of hydrogen escape had been essentially maintained throughout the lifetime of Mars. In that case another explanation of the D to H ratio would need to be found. One unlikely possibility is that Mars accreted crustal (but not mantle) volatiles rich in deuterium compared to Earth or Halley. Another possibility is that fractionation occurred during the dying phase of hydrogen blowoff driven by enhanced solar euv or impact erosion. This mechanism would need to be tested for consistency in accounting as well for the myriad other elemental and isotopic fractionation patterns in Martian volatiles[48,49] that also must be explained.

Acknowledgments. This work was supported in part by a grant from NASA. Valuable criticism, comments and suggestions to improve the presentation were made by Sushil Atreya, Bruce Jakosky, James Kasting and Laurie Leshin Watson.

REFERENCES

1.   T. Owen, et al, Science 240, 1767 (1988).
2.   G.L. Bjoraker, M.J. Mumma and H.P. Larson, Proc. 4th Int. Conf. Mars, Tucson, AZ, Jan. 10-13, 69 (1989).
3.   L.L. Watson, et al, Science 265, 86 (1994).
4.   M.B. McElroy, M.J. Prather and J.M. Rodriguez, Science 215, 1614 (1982).
5.   T.M. Donahue, J.H. Hoffman, R.R. Hodges, Jr. and A.J. Watson, Science 216, 630 (1982).
6.   R.E. Hartle and H.A. Taylor, Jr., Geophys. Res. Lett. 10, 965 (1983).
7.   S. Kumar and H.A. Taylor, Jr., Icarus 62, 494 (1985).
8.   B. deBergh, B. Bézard, T. Owen, D. Crisp, J.-P. Maillard and B.L. Lutz, Science 251, 547 (1991).
9.   G.L. Bjoraker, M.J. Mumma and H.P. Larson, Bulletin of the Am. Astron. Soc. 24, 1067 (1992).
10. T.M. Donahue, Nature, to be published (1995).
11. M.B. McElroy, T.Y. Kong and Y.L. Yung, J. Geophys. Res. 82, 4379 (1977).
12. V.A. Krasnopolsky, Icarus 62, 221 (1985).
13. D.H. Grinspoon and J.S. Lewis, Icarus 74, 21 (1988).
14. S. Kumar, D.M. Hunten and B. Pollack, Icarus 55, 369 (1985).
15. V.A. Krasnopolsky, Icarus 101, 33 (1993).
16. Y. Yung, et al, Icarus 76, 146 (1988).
17. T.M. Donahue and D.M. Hunten, Ann. Rev. Earth and Planet Sci. 4, 265 (1976).
18. R.E. Hartle and J.M. Grebowsky, Adv. Space Res. (1992).
19. T.M. Donahue and R.E. Hartle, Geophys. Res. Lett. 19, 2449 (1992).
20. B.M. Jakosky, J. Geophys. Res. 95, 1475 (1990).
21. M.H. Carr, J. Geophys. Res. 84, 2995 (1979); Nature 326, 30 (1987).
22. D.C. Pieri, Icarus 27, 25 (1976).
23. M.H. Carr and G.D. Clow, Icarus 48, 91 (1981).
24. D.C. Pieri, Science 210, 895 (1980).

25. R.A. Craddock and T.A. Maxwell, J. Geophys. Res. 95, 14265 (1990).
26. S.W. Squyres and J.F. Kasting, Science 265, 747 (1994).
27. V.R. Baker, et al, Nature 352, 589 (1991).
28. C.Y. Shih, et al, Geochim. Cosmochim. Acta 46, 2323 (1982).
29. J.H. Chen and G.J. Wasserberg, Geochim. Cosmochim. Acta 50, 955 (1986).
30. D.M. Hunten, J. Atmos. Sci. 30, 736 (1973); 1481 (1973).
31. S.C. Liu and T.M. Donahue, Icarus 28, 231 (1976).
32. J.F. Kasting, Icarus 94, 1 (1991).
33. M.B. McElroy and T.M. Donahue, Science 177, 986 (1972).
34. J.L. Fox, Geophys. Res. Lett. 20, 1747 (1993).
35. J.G. Luhmann and J.U. Kozyra, J. Geophys. Res. 96, 5457 (1991).
36. H. Nair, et al., Icarus 111, 124 (1994).
37. K.J. Zahnle and J.C.G. Walker, Rev. Geophys. and Space Phys. 20, 280 (1982).
38. M.H. Carr, Icarus 87, 210 (1990).
39. H. Balsinger, K. Altwegg and J.J. Weiss, Geophys. Res., in press (1995).
40. W.-H. Ip and J.A. Fernandez, Icarus 74, 47 (1988).
41. M.A. Gurwell and Y.L. Yung, Planet. Space Sci. 41, 410 (1993).
42. R.R. Hodges, Jr., J. Geophys. Res. 98, 10833(1993).
43. R.E. Hartle, J.M. Grebowsky, W.T. Kasprzak and T.M. Donahue, J. Geophys. Res., to be published (1995).
44. T.M. Donahue and R.R. Hodges, Jr., Geophys. Res. Lett. 20, 591 (1993).
45. J. Pollack, et al, Icarus 103, 1 (1993).
46. T.M. Donahue and R.R. Hodges, Jr., J. Geophys. Res., to be published, (1995).
47. D.H. Grinspoon, Nature 363, 428 (1993).
48. B.M. Jakosky, et al, Icarus 111, 271 (1994).
49. R.O. Pepin, Icarus 111, 289 (1994).

NITROGEN AND ITS ISOTOPES IN THE EARLY SOLAR SYSTEM

John F. Kerridge
Department of Chemistry & California Space Institute
UCSD, La Jolla, California 92093

ABSTRACT

Evidence from meteorites and from solar atoms trapped in lunar regolith samples over the past few billion years suggests that nitrogen in the early solar system was characterised by extreme isotopic inhomogeneity. However, the number of distinct isotopic components, their isotopic compositions, and their physical and chemical nature are all unknown at this time. The forms in which nitrogen was incorporated into planetesimals, and ultimately planets, in the inner solar system, are also unknown, though organic matter of at least partially interstellar origin seems likely.

INTRODUCTION

Although nitrogen is the sixth most abundant element in the solar system, and a key element in the terrestrial biosphere and atmosphere, we know surprisingly little about its cosmochemical behavior during formation and early evolution of the solar system. In this article, I shall review what is currently known about nitrogen in the early solar system, concluding that improved understanding of the isotopic distribution and chemical forms of nitrogen in the solar nebula would greatly facilitate use of nitrogen as a tracer for volatile elements in the early solar system and would aid in reconstruction of the reaction pathways that led to production of organic matter, and hence to the origin of life, in the solar system.

The first-order questions to be asked are: Was nitrogen isotopically homogeneous in the solar nebula, i.e., was there a unique solar-system-wide value for the $^{15}N/^{14}N$ ratio? If so, what was that value? If not, how many isotopic components were present in the nebula and what were their compositions? And, in what chemical forms was nitrogen present while planets were accreting from the solar nebula, and how and where did those chemical compounds form? Unfortunately, only partial answers to those questions are presently available.

Ratios of the two stable nitrogen isotopes, ^{15}N and ^{14}N, are customarily measured by means of either dynamic or static gas-source mass spectrometry, though secondary-ion mass spectrometry (ion microprobe) is also sometimes used for suitable samples. The ratio in the terrestrial atmosphere, $^{15}N/^{14}N= 0.00366$, is used as the international standard. Here we shall express measured values as deviations in parts per thousand (per mil), either positive or negative, from that value. Experimental uncertainties vary widely among different measurements, ranging from +/-0.1‰ for laboratory analysis of a large gas sample to about +/-160‰ for the Viking analysis of the martian atmosphere.

It is worth noting that the two nitrogen isotopes are believed to have been synthesised in quite different nucleosynthetic sites, i.e., in quite different stellar environments: ^{14}N predominantly in hydrostatic helium burning in medium-sized stars, and ^{15}N predominantly in Type II supernovae. Consequently, it is reasonable to expect that the distribution of these isotopes in the interstellar medium would vary widely with location and that the interstellar $^{15}N/^{14}N$ ratio would show similar variations. We now consider how this ratio varies among objects within the present-day solar system.

TERRESTRIAL PLANETS

Although the $^{15}N/^{14}N$ ratio in the terrestrial atmosphere is known with great precision, the value characteristic of the terrestrial mantle is controversial. Nitrogen associated with apparently "primitive" helium exhaled above mantle hot-spots is about 14‰ enriched in ^{15}N relative to the atmospheric value [1], whereas apparently old, mantle-derived diamonds yield a value that is about

11‰ depleted in ^{15}N [2], Fig. 1. Both lines of reasoning seem sound and perhaps mantle heterogeneity cannot be ruled out. This author's preference is for the heavier value, largely because of the good agreement with lunar indigenous nitrogen (see below), though it must be admitted that it is then difficult to understand how common fractionation processes could account for the lighter value. Because of the uncertainties in both abundance and isotopic composition of mantle nitrogen, it is not possible to define the whole-Earth ^{15}N/^{14}N ratio better than "close to the atmospheric value".

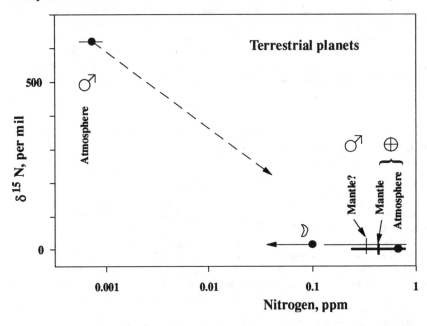

Fig. 1. Abundance and isotopic composition of nitrogen in the mantles and atmospheres of the Earth and Mars, and in the moon. The dashed line from the martian atmospheric value towards the mantle value corresponds to the isotopic fractionation believed to have been experienced by nitrogen during dissociative recombination in the martian exosphere. The error bars represent the range of present estimates. Abundances expressed as μgN per g of planet. Isotopic compositions expressed as deviations in parts per thousand from the terrestrial atmospheric value.

The value measured in the atmosphere of Mars [3] corresponds to a 620‰ enrichment in ^{15}N relative to the terrestrial standard, Fig. 1. This enrichment is generally understood to result from an isotopically competitive loss mechanism operating at the top of the martian exosphere [4], which leads to preferential loss of ^{14}N. This process, dissociative recombination acting upon a gravitationally stratified exosphere, has been quantitatively modelled and is commonly believed to be consistent with a starting ^{15}N/^{14}N ratio equal to the terrestrial value [4], but the uncertainty associated with this estimate is hard to evaluate. In principle, the question of the initial ^{15}N/^{14}N ratio of the solid planet could be addressed using the Mars-derived SNC meteorites. However, analyses of those meteorites have revealed a range of possible values, from a depletion in ^{15}N of about 18‰ to an enrichment of about +29‰ [5-7], Fig. 1, curiously reminiscent of the case for the terrestrial mantle. Obviously, the difference between these estimates is not sufficiently large to call into question the identification of the dissociative-recombination fractionation mechanism.

The Moon is so depleted in volatiles that only an upper limit of about 0.1ppm by weight can be obtained for nitrogen in lunar crystalline rocks [8] and no isotopic measurement is possible on such samples. However, a volatile-rich deposit on the surfaces of fumarolic glass beads collected by the

Apollo 17 mission contains enough nitrogen, of presumably indigenous origin, to permit a precise isotopic analysis [9]. This yielded a value corresponding to a 14‰ enrichment in ^{15}N, Fig. 1.

LUNAR REGOLITH/SOLAR WIND

In contrast to the nitrogen-free rocks, the lunar regolith contains an appreciable quantity of nitrogen, exceeding 100ppm by weight in some cases. The abundance of this regolith nitrogen correlates strongly with measures of surface exposure, including abundances of elements known to be implanted in the lunar surface by the solar corpuscular radiation (solar wind and higher-energy particles). Consequently, much if not all of the regolith nitrogen is believed to consist of implanted solar atoms [10], though a non-solar source is sometimes invoked for a fraction of the nitrogen [11-14]. Examples of non-solar sources that have been proposed include nitrogen from meteorites [13], the terrestrial atmosphere [14], and the lunar interior [15]. However, it is difficult to understand how a significant contribution from such a source could avoid disturbing the excellent correlation between abundances of nitrogen and those of other solar elements, such as the noble gases that are substantially depleted in such sources [10].

The principal motivation for appealing to a significant non-solar nitrogen component on the lunar surface stems from the large variability observed in $^{15}N/^{14}N$ ratio for regolith nitrogen. Measured values for bulk samples range from a 180‰ depletion in ^{15}N to 100‰ enrichment [16,17], with yet more extreme values measured during stepwise thermal release of nitrogen from such a sample [18,19]. The isotopic variability observed during stepwise nitrogen release is not a result of diffusive fractionation during pyrolysis but reflects the presence of at least two isotopically distinct components apparently residing at different depths within individual regolith grains [20]. It is tempting to identify these components with solar nitrogen implanted with different energies into the grains [20], as appears to be the case for solar neon in such samples [21].

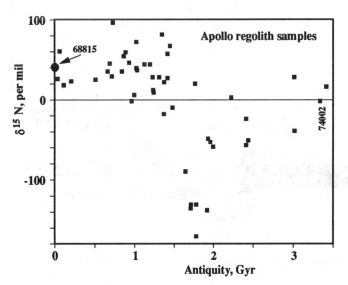

Fig. 2. Isotopic composition of nitrogen in lunar regolith samples expressed as a function of the epoch in which they acquired their nitrogen by implantation from the solar corpuscular radiation. Dating of that epoch (antiquity) is performed using abundances of parentless radiogenic ^{40}Ar from decay of ^{40}K in the lunar interior [50]. Sample 68815 is a rock whose surface was exposed on the lunar surface during the past 2Myr [51]. Antiquity of sample 74002 was also derived [50] using Xe isotopes from the neutron-induced fission of ^{235}U.

A striking feature of the bulk-sample $^{15}N/^{14}N$ ratios is that they vary systematically with the epoch when surface exposure took place. Fig. 2. The same trend is exhibited by the minimum values observed during stepwise release and, to a less extent, by the maximum values also [10]. The cause of this variability is presently unknown. Numerous explanations have been advanced, some of the most popular being: (a) long-term changes in $^{15}N/^{14}N$ in the solar convective zone [22]; (b) long-term changes in fractionation of the solar corpuscular radiation [23]; and (c) mixing of solar nitrogen, taken to be isotopically invariant, with variable proportions of another isotopically invariant component of putatively non-solar origin [11,13,14]. All of these explanations conflict with either observational data or established models of solar structure, evolution or radiation [10,11,14]. Obviously, this issue is of considerable interest in its own right, to solar physics as well as to cosmochemistry, but for present purposes the significance of the lunar regolith data is that they constitute *prima facie* evidence for at least two nitrogen components in the solar system of dramatically different isotopic composition. The extent to which those components might have been available for incorporation into accreting planetary bodies is, of course, unknown at this time.

METEORITES

Meteorites are commonly regarded as our best source of information about the earliest history of the solar system [24]. However, nitrogen analyses of meteorites reveal a bewildering variety of isotopic compositions [25-34], spanning an even wider range of $^{15}N/^{14}N$ values than the lunar regolith, from a 150‰ depletion in ^{15}N to a 1600‰ enrichment, Fig. 3. No systematic pattern is

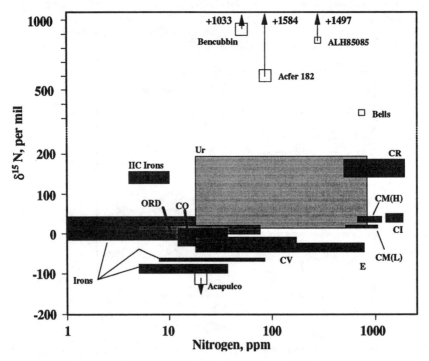

Fig. 3. Abundance and isotopic composition of nitrogen in different types of meteorite [25-34]. A few anomalous meteorites are named. The great majority of meteorites fall into various chemically and petrographically defined groups: CI, CM, CO, CV and CR are carbonaceous chondrites; E and ORD are enstatite and ordinary chondrites, respectively; Ur stands for ureilites.

apparent in the data, except that most meteorite groups, defined by chemical or petrographic criteria, exhibit fairly similar $^{15}N/^{14}N$ ratios. It is noticeable that several meteorite groups contain quite high concentrations of nitrogen, which is one reason why the volatile inventories of the terrestrial planets have often been attributed to late accretion of a population of such meteorites [35]. By comparing Figs. 1 and 3, it is clear that several meteorite groups have $^{15}N/^{14}N$ ratios close to that of the terrestrial mantle, whichever value turns out to be the true one, but when other isotopic data and the effects of post-accretional processing are factored in, the identification of such a putative volatile-rich veneer becomes less than straightforward.

The most nitrogen-rich groups of meteorites are the carbonaceous chondrites, Fig. 3. The bulk of the nitrogen in those meteorites is present in organic matter, either in the form of characterisable organic molecules, such as amino acids, or present as a poorly characterised, macromolecular complex resembling terrestrial kerogen. The amino acids are strikingly enriched in ^{15}N, by 90‰ relative to the terrestrial standard [36], but the kerogen-like material contains nitrogen exhibiting a significant range of $^{15}N/^{14}N$ ratios, from a 10‰ depletion to a 90‰ enrichment in ^{15}N [37]. From other isotopic studies, there is good reason to believe that much, if not all, of the meteoritic organic matter was derived from molecules that were synthesised in interstellar clouds prior to formation of the solar system [38,39]. The extent to which interstellar-cloud processes, such as ion-molecule reactions, might have been responsible for the $^{15}N/^{14}N$ ratios observed in meteoritic organic matter is unknown, though at interstellar-cloud temperatures, ion-molecule reactions could in principle fractionate the nitrogen isotopes by a factor of a few [40]. Note that the organic molecules now present in the meteorites were probably themselves synthesised in the early solar system from interstellar precursor molecules which apparently no longer exist in meteorites though they may be present in cometary nuclei. Meteoritic evidence points only to organic synthesis in planetesimals formed early in solar-system history; there is no evidence in meteorites for organic synthesis from the gas phase in the nebula, though that does not rule out the possibility of such a synthesis having produced a population of organics not sampled by the carbonaceous meteorites.

The issue of organic synthesis in the early solar system takes on an additional significance in connection with the questions: In what form was nitrogen present in the solar nebula? and, How was nitrogen accreted by planets and planetesimals in the inner solar system? Thermochemical calculations [41] have shown that condensation processes from the gas phase would have led to nitrogen concentrations in condensed solids no higher than a few ppm by weight. The much higher concentrations observed in meteorites, Fig. 3, therefore demand a non-condensation, or at least a non-equilibrium, origin. For the case of the carbonaceous meteorites, it is possible to finesse the question by appealing to accretion of nitrogen-rich organic matter or their (probably interstellar) precursors, which could have been in the form of either ices or condensed organic molecules. However, there are many non-carbonaceous meteorites, most notably the enstatite chondrites but also including the abundant ordinary chondrites and several iron-meteorite groups, with concentrations well above 10ppm, Fig. 3. For these meteorites, also, it seems most plausible to attribute their nitrogen contents to accretion of presumably interstellar organic molecules or ices into their parent planetesimals with subsequent secondary processing serving to transfer nitrogen into inorganic phases such as nitrides or solid solutions in either metal or silicates [41].

Organic synthesis in the early solar system is also of importance in connection with the question of how life originated on Earth. The presence on the early Earth of organic matter, which would eventually evolve prebiotically into the first organisms, requires either an extraterrestrial source of such organics or a mechanism operating on the early Earth that was capable of synthesising such material. Although theories of the origin of life have tended to favor the latter scenario [42], recent findings [43] that the atmosphere of the early Earth was apparently less favorable to the synthesis of organic matter than previously believed, have tended to favor the idea that extraterrestrial organics may have played an important role. The evidence from nitrogen systematics certainly suggests that the early Earth probably accreted a significant proportion of organic matter, of which a less-degraded record remains in the meteorites. However, it would be premature to discount the importance of local production of organics on the early Earth, particularly in light of recent research into organic synthesis in hydrothermal systems [44].

Before leaving the subject of meteoritic nitrogen, we should mention the recent discovery of dramatically anomalous nitrogen isotopes in interstellar graphite, diamond, and silicon carbide grains isolated from certain primitive meteorites [45]. The range of $^{15}N/^{14}N$ ratios observed in these grains, which contain nitrogen as a major impurity, is remarkable, Fig. 4, but it is important to note that this material constitutes a trivial fraction of the nitrogen in such a meteorite. Furthermore, it is probably unrelated to the bulk of the meteoritic nitrogen; whereas the "interstellar" *ices* and *organics* are believed to originate in dense molecular clouds in the interstellar medium [38,39], the "interstellar" *grains* are thought to be formed in outflows from a variety of stellar objects, such as AGB stars or supernovae [45]. Furthermore, interstellar oxide and silicate grains can become enriched in ^{15}N by spallation reactions induced by galactic cosmic rays [33].

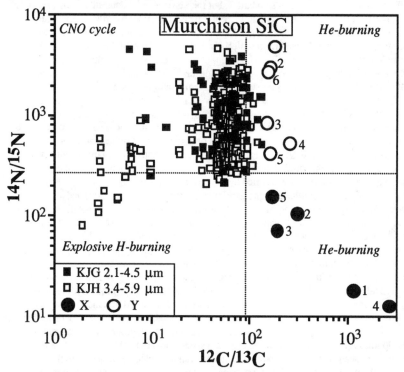

Fig. 4. Isotopic compositions of nitrogen and carbon in SiC grains extracted from the Murchison carbonaceous chondrite. Note enormous range of values observed, and the different nucleosynthetic processes responsible for production of these isotopes. From Anders and Zinner [45].

OUTER SOLAR SYSTEM

Before concluding our survey of nitrogen in solar-system objects, we should note that attempts have been made to measure $^{15}N/^{14}N$ in comets [46] and in Jupiter [47], but in both cases observational uncertainties encompass most of the values found in the inner solar system. Fortunately, there is a reasonable likelihood that the neutral mass spectrometer on the Galileo entry probe [48] will yield a value of $^{15}N/^{14}N$ for jovian ammonia, and ground-based observational techniques are now in place that should yield reasonably precise $^{15}N/^{14}N$ ratios for future bright comets[49].

CONCLUSIONS

The meteoritic data, Fig. 3, point strongly towards an inhomogeneous distribution of the nitrogen isotopes in the early solar system, i.e., towards the proposition that several quantitatively major, isotopically distinct components of nitrogen were present in the primordial solar nebula. The chemical and isotopic compositions of those components, and whether they were solid or gaseous, or a combination of both, are unknown at this time, though several lines of reasoning implicate either organic molecules or their precursors. How those components were spatially, and/or temporally, distributed in the early solar system is also unknown, as are their sources. However, an origin as interstellar molecules derived from dense molecular clouds seems likely for at least some of them. The record of solar corpuscular radiation trapped in the lunar regolith, Fig. 2, is also consistent with the presence of isotopically distinct components with dramatically different compositions in the solar system, though it is much less straightforward to relate those data to conditions in the solar nebula.

It follows that these uncertainties currently prevent us from applying the nitrogen isotopes to the question of the distribution of volatile-rich material in the earliest solar system. However, if we become successful at identifying the important components of nitrogen in the solar nebula, they may well serve as powerful tracers of preplanetary volatiles.

ACKNOWLEDGEMENTS

I thank Kurt Marti and Thomas Graf for advice and assistance, and David DesMarais and an anonymous reviewer for helpful comments. I also thank Ernst Zinner for providing Fig. 4. This study is supported by the Planetary Materials and Geochemistry, Exobiology, and Origins of Solar Systems Programs of NASA.

REFERENCES

1. R.A.Exley, S.R.Boyd, D.P. Mattey and C.T.Pillinger, Earth Planet.Sci.Lett. 81, 163 (1987)
2. M.Javoy, F.Pineau and D.Demaiffe, Earth Planet.Sci.Lett. 68, 399 (1984)
3. A.O.Nier and M.B.McElroy, J.Geophys.Res. 82, 4341 (1977)
4. M.B.McElroy, T.Y.Kong and Y.L.Yung, J.Geophys.Res. 82, 4379 (1977)
5. A.E.Fallick, R.W.Hinton, D.P.Mattey, S.J.Norris, C.T.Pillinger, P.K.Swart and I.P.Wright, Lunar.Planet.Sci.XIV, 183 (1983)
6. I.P.Wright, R.H.Carr and C.T.Pillinger, Geochim.Cosmochim.Acta 50, 983 (1986)
7. I.P.Wright, M.M.Grady and C.T.Pillinger, Geochim.Cosmochim.Acta 56, 817 (1992)
8. D.J.DesMarais, Geochim.Cosmochim.Acta 47, 1769 (1983)
9. J.F.Kerridge, O.Eugster, J.S.Kim and K.Marti, Proc.Lunar Planet.Sci.Conf. 21, 291 (1991)
10. J.F.Kerridge, Rev.Geophys. 31, 423 (1993)
11. J.Geiss and P.Bochsler, Geochim.Cosmochim.Acta 46, 529 (1982)
12. P.Signer, H.Baur, P.Etique and R.Weiler, in: Workshop on Past and Present Solar Radiation (Lunar and Planetary Inst., Houston, 1986), p.36
13. D.R.Brilliant, I.A.Franchi, J.W.Arden and C.T.Pillinger, Meteoritics 27, 206 (1992)
14. J.Geiss and P.Bochsler, in:The Sun in Time (Univ.Arizona, Tucson, 1991), p.98
15. R.H.Becker and R.N.Clayton, Proc.Lunar Sci.Conf. 6, 2131 (1975)
16. R.N.Clayton and M.H.Thiemens, in: The Ancient Sun (Pergamon, NY, 1980) p.463
17. J.F.Kerridge, I.R.Kaplan, C.Petrowski and S.Chang, Geochim.Cosmochim.Acta 39, 137 (1975)
18. S.J.Norris, P.K.Swart, I.P.Wright, M.M.Grady and C.T.Pillinger, Proc.Lunar Planet.Sci.Conf. 14, B200 (1983)
19. L.P.Carr, I.P.Wright and C.T.Pillinger, Meteoritics 20, 622 (1985)

20.   J.F.Kerridge, J.S.Kim, Y.Kim and K.Marti, Proc.Lunar Planet.Sci.Conf. 22, 215 (1992)
21.   R.Wieler, H.Baur and P.Signer, Geochim.Cosmochim.Acta 50, 1997 (1986)
22.   J.F.Kerridge, Science 188, 162 (1975)
23.   P.Bochsler and R.Kallenbach, Meteoritics 29, 653 (1994)
24.   J.F.Kerridge and M.S.Matthews, Meteorites and the Early Solar System (Univ.Arizona, Tucson, 1988)
25.   I.A.Franchi, I.P.Wright and C.T.Pillinger, Nature 323, 138 (1986)
26.   I.A.Franchi, I.P.Wright and C.T.Pillinger, Meteoritics 22, 379 (1987)
27.   M.M.Grady and C.T.Pillinger, Nature 331, 321 (1988)
28.   M.M.Grady and C.T.Pillinger, Earth Planet.Sci.Lett. 97, 29 (1990)
29.   M.M.Grady, C.T.Pillinger and J.W.Arden, Meteoritics 27, 226 (1992)
30.   J.F.Kerridge, Geochim.Cosmochim.Acta 49, 1707 (1985)
31.   C.C.Kung and R.N.Clayton, Earth Planet.Sci.Lett. 38, 421 (1978)
32.   C.A.Prombo and R.N.Clayton, Science 230, 935 (1985)
33.   C.A.Prombo and R.N.Clayton, Geochim.Cosmochim.Acta 57, 3749 (1993)
34.   Y.Kim and K.Marti, Lunar Planet.Sci.XXV, 703 (1994)
35.   E.Anders and T.Owen, Science 198, 453 (1977)
36.   S.Epstein, R.V.Krishnamurthy, J.R.Cronin, S.Pizzarello and G.U.Yuen, Nature 326, 477 (1987)
37.   J.F.Kerridge, Earth Planet.Sci.Lett. 64, 186 (1983)
38.   J.Geiss and H.Reeves, Astron.Astrophys. 93, 189 (1981)
39.   J.F.Kerridge, S.Chang and R.Shipp, Geochim.Cosmochim.Acta 51, 2527 (1987)
40.   N.G.Adams and D.Smith, Astrophys.J.Lett. 247, L123 (1981)
41.   B.Fegley, Proc.Lunar Planet.Sci.Conf. 13, A853 (1983)
42.   S.L.Miller and L.E.Orgel, The Origins of Life on Earth (Prentice-Hall, Englewood Cliffs, 1974)
43.   J.C.G.Walker, Origins of Life 16, 117 (1986)
44.   E.L.Shock, Origins of Life 22, 67 (1992)
45.   E.Anders and E.Zinner, Meteoritics 28, 490 (1993)
46.   S.Wyckoff and E.Lindholm, Adv.Space Res. 9, #3, 151 (1989)
47.   A.T.Tokunaga, R.F.Knacke, S.T.Ridgeway and L.Wallace, Astrophys.J. 232, 603 (1979)
48.   H.B.Niemann, D.N.Harpold, S.K.Atreya, G.R.Carignan, D.M.Hunten and T.C.Owen, Space Sci.Rev. 60, 111 (1992)
49.   S.Wyckoff, Pers.Comm.
50.   O.Eugster, J.Geiss and N.Grögler, Lunar Planet.Sci.XIV, 177 (1983)
51.   J.S.Kim, Y.Kim, K.Marti and J.F.Kerridge, Nature, submitted (1994)

HOW MANY MARTIAN NOBLE GAS RESERVOIRS HAVE WE SAMPLED?

Timothy D. Swindle
Lunar and Planetary Laboratory, University of Arizona, Tucson AZ 85721

ABSTRACT

The most precise estimates of the elemental and isotopic composition of the martian atmosphere come from martian meteorites, in particular EETA79001. These meteorites also provide substantial evidence for at least one additional component, presumably associated with the martian interior. This evidence includes noble gases within Chassigny that are isotopically and elementally distinct from either the terrestrial or martian atmospheres, and low-δ^{15}N and low ^{40}Ar/^{36}Ar components within EETA79001. Much of the martian meteorite data can be explained by simple mixing of these two components, with some elemental fractionation generated by adsorption processes. There is no compelling evidence for additional components, although there are several hints within the data.

INTRODUCTION

Understanding the origin and evolution of volatiles in the interior of a planet is crucial to understanding the planet itself. The volatiles provide the material for the atmosphere, and they often provide the gases that produce the pressure to drive volcanic activity at the surface. However, it is hard to get a handle on the abundance and composition of volatiles within the Earth's interior, simply because we can only use remote means to sample the interior. The problem is even worse on other planets, where we lack the detailed gravity and seismic data that we have for Earth, and generally lack even the rock samples that we have terrestrially.

The one planet, other than the Earth and Moon, for which we have enough rocks to develop and test models for its volatile evolution is Mars. The rocks are the 10 martian meteorites (Table I), which represent a variety of types of surface and near-surface rocks[1]. Identification of these meteorites as martian remains somewhat controversial. However, if these meteorites do not come from Mars, it is hard to imagine a reasonable alternative parent body[2]. Probably the best-studied volatiles are the noble gases.

As discussed below, one of the martian meteorites, EETA79001, seems to contain a nearly-pristine sample of the martian atmosphere[3,4]. Another, Chassigny, contains noble gases that are distinctly different[5], and must represent a separate component. At this stage, we can begin to ask what the relationship is between the two components identified, and how the history of martian volatiles (particularly noble gases) compares to that on Earth. But there are some more fundamental (if not quite as exciting) questions that remain, such as the exact composition of those two components, and the question of the existence of additional components remaining to be discovered. Although I'll briefly review some of the progress that has been made on the first type of question, this paper will focus on the second type of question, in particular, the number of reservoirs. After all, no model is any better than its input parameters, and a model that either leaves out a component that exists or calls on a component that doesn't exist has an obvious flaw.

There have been two fundamentally different reasons for suggestions for an additional martian component. Some workers[5-10] have measured compositions in various martian meteorites that can't be easily modeled as mixtures of the components seen in Chassigny and EETA79001. A different approach was taken by Owen[11], who noted that experiments on trapping noble gases in ices yielded impressive elemental fractionation, and suggested that much of the relationship between the noble gas inventories of Earth, Mars and Venus could be explained by mixing varying proportions of cometary noble gases. Further, he suggested

Table I: Martian meteorites

Meteorite	Fall Date	Location	Mineralogy
(S)hergottites			
Shergotty	1865	India	Basalt
Zagami	1962	Nigeria	Basalt
EETA79001	Find	Antarctica	Basalt
ALHA77005	Find	Antarctica	Feldspathic harzburgite
LEW88516	Find	Antarctica	Feldspathic harzburgite
(N)akhlites			
Nakhla	1911	Egypt	Pyroxenite
Lafayette	Find	Indiana	Pyroxenite
Governador Valadares	Find	Brazil	Pyroxenite
(C)hassignite			
Chassigny	1815	France	Dunite
Other			
ALH84001	Find	Antarctica	Orthopyroxenite

that mixing of cometary gases might be important not only for interplanet variations, but also for intraplanet variations on Mars.

In the spirit of Occam's Razor, this paper will adopt the attitude that the least complicated models are best. In particular, can the martian meteorite data be explained by only two fundamental components? There will certainly have to be some processes occurring to alter some features of these components, but can we come up with plausible processes or are we forced to have additional components, either within the planet or delivered to it?

MARTIAN ATMOSPHERE

Identification

The first measurements of the noble gases in the martian atmosphere came from the mass spectrometers on the Viking landers[12]. They determined the abundance of all of the noble gases except He, and made crude determinations of their isotopic ratios. Within the rather large error bars (10% for most ratios, up to 50% for Xe), most ratios were consistent with the terrestrial atmosphere, but the ratios $^{40}Ar/^{36}Ar$, $^{129}Xe/^{132}Xe$ and $^{15}N/^{14}N$ were all considerably higher than in the terrestrial atmosphere (Table II).

We now believe we have much more precise measurements, thanks to martian meteorites. The key is EETA79001, a meteorite collected in Elephant Morraine, Antarctica in

Table II. Key isotopic ratios in the martian atmosphere

	$^{40}Ar/^{36}Ar$	$^{132}Xe/^{129}Xe$	$\delta^{15}N$
Terrestrial atmosphere	295.5	0.983	$\equiv 0\%o$
Martian atmosphere			
Viking lander	3000±1100	$2.5^{+1.5-1.0}$	620±160‰
EETA79001 glass	2300±700	2.39±0.03	(620±160‰)

Martian data from compilation by Pepin and Carr[24]. Nitrogen composition of atmosphere in EETA 79001 glass is only constrained to be >300‰, but N/Ar systematics, as in Fig. 4, are consistent with the Viking measurement[13]

the 1979-80 field season. Analyses of the noble gases in this meteorite revealed the presence of tremendous amounts of trapped noble gases, particularly in some glassy inclusions[3,4]. More detailed studies of the gas in the glass produced more precise isotopic ratios[13,14] (Table II).

Intriguingly, the abundances of the noble gases and N_2 in the EETA79001 glass match those measured by Viking in the martian atmosphere, not just in relative concentration, but even in the number of atoms per cm^3 (Fig. 1). The fact that the relative abundances match the martian atmosphere is not surprising -- the glass is produced by shock melting, and shock can implant noble gases and nitrogen without fractionation[15]. The agreement with the absolute abundances is a little more surprising, since the experimental shock studies did not achieve 100% retention. However, the retention was as high as 50%, and it is possible that the shock event produced a locally enhanced atmospheric pressure[15].

Comparison to Earth

The relative elemental abundances of the noble gases in the martian atmosphere are remarkably similar to those in the terrestrial atmosphere (Fig. 2). Like the gas trapped in most meteorites, the terrestrial and martian atmospheres appear to be fractionated relative to the solar wind. The fractionation is typically attributed to adsorption[16]. For many years, the terrestrial pattern was assumed to be identical to the meteoritic pattern, except for some "missing Xe" that remained sequestered somewhere. However, the "missing Xe" has consistently failed to show up either in studies of samples from potential hiding places[17] (e.g., shales, polar ice caps) or experiments designed to simulate deep Earth conditions[18]. The fact that Mars shows exactly the same pattern suggests that the process causing the "missing Xe" is not unique to the Earth. One promising proposal is that of Zahnle[19], who suggests that because Xe is the least soluble noble gas in silicate melts, it outgassed first, and most of it was lost in processes (such as hydrodynamic escape) that ceased before significant amounts of the other noble gases had been outgassed. It is not clear whether that model is quantitatively able to account for all the properties it explains qualitatively, but it at least provides a plausible way to preferentially lose Xe.

The overall isotopic pattern of martian atmospheric Xe is also very similar to its terrestrial counterpart. Compared to the solar wind, or to the Xe trapped in most meteorites, there is a mass-dependent fractionation of about 2-3% per amu (Fig. 3), although the comparison is not as clear for the isotopes that can be produced in nuclear decays ($^{129,131,132,134,136}Xe$). Like the "missing Xe", the fractionation appears to be a common property of the Xe in the atmospheres of the terrestrial planets. Again, with few exceptions[20], explanations typically centered on Earth-specific mechanisms for this fractionation until the determination of the isotopic composition of martian Xe. Many recent models address both planets[16,19,21].

There are some important differences in the two atmospheres, though. Even Viking was able to detect the high ratios of $^{129}Xe/^{132}Xe$ and $^{40}Ar/^{36}Ar$ in the martian atmosphere. These ratios reflect the initial volatile inventory and the degassing history of the planet. The higher values on Mars suggest a large loss of things like ^{132}Xe and ^{36}Ar from the atmosphere very early in martian history, before ^{129}Xe and ^{40}Ar were built up by the decay of ^{129}I ($T_{1/2}$ = 16 Ma) and ^{40}K ($T_{1/2}$ = 1.28Ga), or else I/Xe and K/Ar ratios very much higher than in the Earth. That in itself is a fascinating topic, and the high

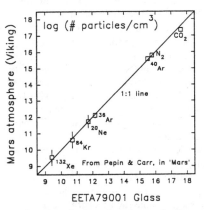

Fig. 1. Abundances of various volatiles as measured by Viking[12], compared with abundances inferred from EETA79001 glass[24].

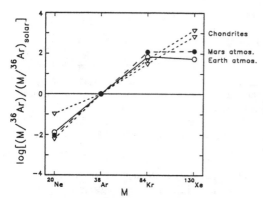

Fig. 2. Elemental abundances of noble gases in the martian and terrestrial atmospheres, normalized to ^{36}Ar and the solar wind[24,35]. Two typical chondrites are included for comparison.

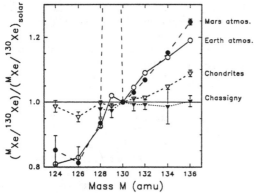

Fig. 3. Isotopic composition of martian atmospheric Xe as measured in EETA79001[14] and Chassigny[5], normalized to ^{130}Xe and the solar wind[35]. The terrestrial atmosphere and a value typical of chondrites are included for comparison[35].

^{129}Xe/^{132}Xe ratio is particularly hard to explain[16,22], but will not concern us further, except as a known property of the atmospheric reservoir.

Similarly, the ^{38}Ar/^{36}Ar and ^{15}N/^{14}N ratios are higher in the martian atmosphere than in the terrestrial atmosphere. These are usually explained as the result of some mass-dependent atmospheric loss process. Recently, attention has been focussing on sputtering of the top of the atmosphere, associated with the solar wind[23]. Again, since we are most interested in ratios as tracers of the martian atmosphere, we will not worry about exactly how they came to be what that are. The best estimates of the isotopic and elemental composition of the martian atmosphere has been summarized elsewhere[24].

A SIMPLE TWO-COMPONENT MODEL AND ITS IMPLICATIONS

Evidence for a second component

The presence of martian atmospheric gases within at least EETA 79001 seems well established, as does the elemental composition of the atmosphere and the isotopic composition of several gases in the atmosphere. There are several pieces of evidence that there is at least one other noble gas component present in martian meteorites. However, it is much less clear exactly how many components there are or what the elemental and isotopic compositions are. This section will review the evidence for additional components, and present the simplest two-component model.

The meteorite Chassigny is a dunite (an olivine-rich cumulate) that contains trapped noble gases that are distinctly different from the martian atmosphere[5]. Chassigny contains no signature of radiogenic ^{129}Xe, the overall structure of the Xe isotopes is much more similar to the solar wind than it is to the martian or terrestrial atmospheres or the gas trapped in ordinary chondrites (Fig. 3), and the relative elemental abundances are also distinct. In fact, the differences are so striking that it is tempting to suggest that if EETA79001 is martian, perhaps Chassigny is not. However, Chassigny has the same cosmic-ray exposure age and crystallization age as the nakhlites, and its oxygen isotopic composition matches all other martian meteorites. Furthermore, as will be discussed in more detail below, several martian meteorites require the presence of some Xe component other than martian and terrestrial atmosphere.

In EETA 79001, the atmospheric signature in the glassy inclusions is spectacular, but some extractions from the glass yield a $^{40}Ar/^{36}Ar$ ratio that is significantly less than either the martian or terrestrial atmospheric values[3,4,14]. Since this effect has been measured in three different laboratories, it is unlikely to represent an analytical artifact. Although it could represent gas from a less-radiogenic projectile[3], it seems more plausible that it represents gas that was in EETA 79001 before it was shocked.

Finally, there is evidence for a second component of nitrogen. Since the $^{15}N/^{14}N$ ratio correlates with the presence of martian atmospheric ^{40}Ar and ^{36}Ar (Fig. 4a,b), the combined nitrogen-argon data in EETA79001 glass could be explained by a two-component mixture of martian atmosphere and one other component[13]. Although the composition of the other component is close to that of the terrestrial atmosphere in both Fig. 4a and Fig. 4b, the correlation lines misses by more than 1σ in both cases, again suggesting at least one more martian component.

Are the second components observed for Ar, Xe and N really part of a single component? The low $^{40}Ar/^{36}Ar$ signal seen in EETA79001 has not been seen in any other martian meteorites, but it would tend to be swamped by radiogenic and/or cosmogenic Ar everywhere but in EETA79001. The Xe signal found in Chassigny does seem to be present in other martian meteorites, as will be discussed below. The most interesting constraints come from the N-Ar data. Fig. 4c displays the allowed $^{40}Ar/^{36}Ar$ values for the low δ^{15}N component in the EETA79001 glass, obtained at any δ^{15}N value by taking the $^{40}Ar/^{14}N$ value on the correlation line in Fig. 4a and dividing it by the $^{36}Ar/^{14}N$ value from Fig. 4b, including 1σ uncertainties on the Ar/N ratios. If the low δ^{15}N component has δ$^{15}N>0$, then the $^{40}Ar/^{36}Ar$ ratio associated with it must be greater than the terrestrial value. On the other hand, if the $^{40}Ar/^{36}Ar$ value is lower than the terrestrial value, δ^{15}N must be less than -5‰ (at the 1σ level). Interestingly enough, δ^{15}N values of less than -30‰ in Chassigny and Shergotty have been reported[25]. Fig. 4c can not be considered strong evidence for a low-δ^{15}N, low $^{40}Ar/^{36}Ar$ component, but it is consistent with such a story. The link between the low-δ^{15}N component

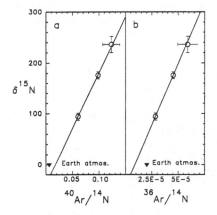
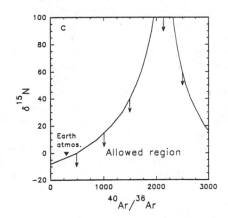

Fig. 4. Combined N and Ar systematics of samples of EETA 79001 glass[4,13]. In a) and b), data appears to be consistent with a two-component mixture. Values in a) and b) have been corrected for spallation. In c), the allowed $^{40}Ar/^{36}Ar$ ratios in the low δ^{15}N component has been calculated from the $^{40}Ar/^{14}N$ and $^{36}Ar/^{14}N$ values at particular values of δ^{15}N. The δ^{15}N value is constrained to lie between about +100‰ (the lowest value observed in the glass samples) and about -40‰ (the value of the ordinate intercepts in a) and b)).

and Chassigny Xe is similarly tenuous, hanging only on the fact that the lowest $\delta^{15}N$ measurement does come from Chassigny.

Comparison with Earth

On Earth, there is a well-established second noble gas component, most easily identified in mantle samples such as ocean island basalts (OIB) and mid-ocean ridge basalts (MORB). However, the relationship of these components to the terrestrial atmosphere is fundamentally different than the relationship of the suggested martian samples to their atmosphere.

On Earth, the mantle is characterized by higher ratios of $^{40}Ar/^{36}Ar$ and $^{129}Xe/^{132}Xe$ (and $^{20}Ne/^{22}Ne$, a ratio that can not be adequately measured in martian samples because of cosmic-ray-produced Ne). This is suggested to be the result of the atmosphere having been extracted from the mantle, or a reservoir like the mantle, long enough ago to have had significant ingrowth of radiogenic ^{40}Ar and ^{129}Xe since then[26]. Although there are mantle-derived samples that seem to indicate various amounts of mixing of radiogenic crustal noble gases with mantle gases[27], the idea that there is more than one fundamental noble gases reservoir in the mantle (e.g., distinct upper- and lower-mantle components[26]) is controversial[28].

On Mars, the $^{40}Ar/^{36}Ar$ and $^{129}Xe/^{132}Xe$ ratios are higher in the atmosphere than in the suggested second component. Several possible explanations have been suggested, although there is no consensus[22]. The composition of the later-accreted source of the atmosphere might have been distinct from the that of the mantle if Mars accreted heterogeneously[29]. Also, the communication between reservoirs must be much different on Mars, where plate tectonics does not recycle the crust and the volatiles within it. Finally, since Mars is smaller, it is likely to have lost a more significant portion of its atmosphere as a result of such processes as ion sputtering[23].

EVIDENCE FOR MORE THAN TWO COMPONENTS?

Even though we have not explained the origin of the second component nor even demonstrated that there is a single well-defined second component, let us for the moment assume that there is. If there is a single second component, then we only have to explain its origin and relationship to the atmosphere to have a fairly complete picture of the evolution of noble gases inside Mars[16]. On the other hand, a demonstration of the need for more than two components would make it necessary to decipher how those components are related to one another, what features of what meteorites are related to which components, and, of course, how many other components there are.

There are only a few places where we can make comparisons to test whether a two-component model is sufficient. As mentioned above, the only element where the isotopic composition of the trapped gas might be expected to be dominant is Xe, so we will rely heavily on Xe isotopes. In addition, it is sometimes possible to measure, or at least set limits on, the relative abundances of different elements with a sufficient precision to make meaningful comparisons, since the differences in elemental ratios are much larger than the differences in isotopic ratios. We will consider three different three-isotope systems. The attitude taken here will be that we wish to minimize the number of noble gas reservoirs floating around inside (or above) Mars. If some reasonable physical process can be identified that could generate the deviations, we will provisionally adopt that process, and consider the implications.

The three-element plot

The first plot we will consider (Fig. 5) involves two elemental ratios ($^{36}Ar/^{132}Xe$ and $^{84}Kr/^{132}Xe$). Since the denominator on both axes is the same, it will obey all the rules of a normal three-isotope plot -- two component mixtures on a straight line between the endpoints, and multi-component mixtures fall inside a polygon whose vertices are the compositions of the individual components.

Much of the martian meteorite data is consistent with a straight line on Fig. 5, but there are significant deviations. Owen et al.[11] noted that at certain temperatures, ices can incorporate large quantities of noble gases with high Ar/Xe and Kr/Xe ratios. Because of this, they suggested that there might be a cometary component in the martian data. In fact, even the noble gases in EETA79001 glass (which are usually assumed to be representative of the martian atmosphere) could be mixtures of a cometary component and Chassigny, or even a cometary component and the real martian atmosphere. Remember, the Viking lander measurements of the martian atmosphere are imprecise enough to allow the atmosphere to be significantly different from EETA79001 glass.

Actually, since the axes on Fig. 5 are on a log-log scale, a two-component mixture between Chassigny and EETA79001 would plot along the solid curve (which would be straight on a linear scale) rather than the dashed line on which they fall. Viewed in that way, most of the data falls below the curve. However, consider some mass-dependent elementally fractionating process in which we preferentially lose the lighter gases or retain the heavier gases. Both ratios will be reduced, but the Ar/Xe ratio will be reduced more, driving the points below the line, where they are observed. Adsorption onto surfaces, which will be discussed in more detail below, is a plausible process which could produce the observed results.

There are two problems with the simple two-component mixing plus adsorptive fractionation. First, there is the variability in the measurements of EETA79001 glass. If this does not represent analytical uncertainties (and the measurements do come from three different laboratories), it may be telling us about real, hard-to-explain, sample heterogeneity. The second problem is the point to the northwest of the mixing curve, which represents an average of several samples of lithology A (non-glassy) from EETA79001[30], which scatter considerably. Again, if not a result of analytical uncertainty, the data could be tough to explain, even with the addition of a cometary component (which would be unlikely to have a high enough $^{40}Ar/^{132}Xe$ ratio[11]). Since most chondrites plot to the northwest of the martian trend, it could indicate gas from an impactor, but why is it seen in these particular samples?

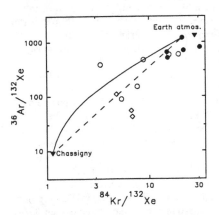

Fig. 5. Elemental abundances of noble gases in martian meteorites[2,5,6,30,31], compared to Chassigny[5] and the martian atmosphere (as measured in EETA79001)[4]. Data has been corrected for spallation. Symbols are: shergottites - circles; ALH84001 - diamonds. Filled circles are EETA79001 glass.

Xenon isotopic evidence

The next plot we will consider involves only Xe isotopes. On Fig. 6, the $^{129}Xe/^{132}Xe$ and $^{136}Xe/^{132}Xe$ ratios of the martian atmosphere and the (interior?) component in Chassigny are distinctly different. The terrestrial atmosphere is only slightly different from Chassigny in $^{129}Xe/^{132}Xe$, but has a $^{136}Xe/^{132}Xe$ ratio that is distinct. The presence of fission Xe from in situ decay of uranium will drive points to higher $^{136}Xe/^{132}Xe$ (spallation Xe should have a smaller effect, and is more easily corrected for). Thus a mixture of only martian atmosphere, in situ fission and terrestrial contamination should plot on or above the upper line. On the other hand, if there is a contribution from a Chassigny-like component, points should plot between the lines. Points below the lower line would suggest another component altogether.

Most martian meteorite data falls on or above the mixing line between the terrestrial and martian atmospheres on the Xe isotope plot (Fig. 6). That data could be explained without any martian components other than the atmosphere

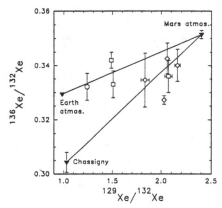

Fig. 6. Three-isotope plot for Xe from martian meteorites[2,5,9,10,30,31]. Data has been corrected for spallation and fission (except for ALH84001, where the low U content and uncertain age makes fission corrections uncertain, but probably negligible). Symbols as in Fig. 5, except that EETA79001 glass is now called "Mars atmos.", and nakhlites are denoted by squares.

(and in situ fission of actinides). However, some of the nakhlite points fall below that line, requiring an additional component, consistent with the Chassigny data.

The data for ALH84001, the most recently recognized martian meteorite, is more interesting. Much of the data falls on the Chassigny-martian atmosphere mixing line, but there is one point[2] in the "forbidden" region. That one point represents a summation of stepwise heating experiment, in which several extractions are consistent with the total, though with larger uncertainties. However, another sample from the same laboratory[2], and two from another[31], all fall on the mixing line. If the one aberrant point is real, then it may represent inhomogeneous mixing of a third component. The crystallization age of ALH84001 is uncertain, but might be close to 4.5 Ga[32], which is old enough for the rock to have once contained live ^{129}I. If other experiments reproduce the data point below the Chassigny-martian atmosphere mixing line, and if it can be shown that it is not the result of in situ decay of ^{129}I, then another component seems inevitable.

Krypton/Xenon systematics

The final plot we will consider (Fig. 7) involves one isotopic ratio ($^{129}Xe/^{132}Xe$) and one elemental ratio ($^{84}Kr/^{132}Xe$). At first, this may seem like an unlikely presentation of the data, but it may be the most valuable of the three. Since the denominator on both axes is the same, it is again a proper three-isotope plot. Furthermore, the differences between the martian atmosphere and Chassigny is large on both axes, and neither parameter is seriously affected by spallation (unlike ^{36}Ar) or in situ fission (unlike ^{136}Xe).

Ott and Begemann[5] were the first to point out that analyses of Shergotty appear to fall on a mixing line between Chassigny and the martian atmosphere on this plot (actually, the choice of EETA79001 glass data used for "Mars atmos." in Fig. 7 makes most shergottite data fall slightly below the line). However, they also noted that Nakhla does not fall on the mixing line. Instead, it appears to be depleted in Kr relative to Xe. The other nakhlites, Lafayette and Governador Valadares, fall in the same region as Nakhla[7,10]. This appears to suggest another martian component. Even if the elemental fractionation can be explained, why is there a strong signature of the martian atmosphere in igneous rocks like the nakhlites? Meanwhile, Zagami[30], LEW88516[8], and gas from the vesicles of the glass in EETA79001[6] appear on fall to the high Kr/Xe side of the mixing line. As we will see, the low Kr/Xe points can probably be explained without resorting to a third martian component, but the high Kr/Xe points pose more of a challenge.

Low Kr/Xe "component": Ott et al.[7] leached Nakhla with HCl, and removed about 15% of the mass of the rock, consistent with the fraction of olivine. The residue fell almost on the mixing line in Fig. 7. They suggested that the third component resided either in olivine or in some minor phase associated with olivine.

Drake et al.[33] took this explanation one step further, suggesting that there is a low-Kr/Xe, high $^{129}Xe/^{132}Xe$ component found in "iddingsite." Iddingsite, a mixture of clays and oxides, forms by aqueous alteration of olivine (and would have been removed by leaching[7]). It

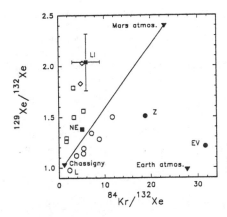

Fig. 7. Plot of $^{129}Xe/^{132}Xe$ ratio vs. $^{84}Kr/^{132}Xe$. Symbols as in Fig. 6. Filled circles and squares are analyses mentioned in the text: LEW88516 - L; Zagami - Z; Vesicles in EETA79001 glass[6] - EV; Iddingsite from Lafayette[34] - LI; Etched sample of Nakhla - NE.

is found in all the nakhlites, but not in Chassigny or any of the shergottites. One of the most attractive aspects of this hypothesis is that the extra component really isn't a separate component at all. Because Xe is more easily adsorbed by rocks than is Kr, formation of sedimentary materials on Earth (of which iddingsite is the martian equivalent) typically involves elemental fractionation of roughly the magnitude required, but not isotopic fractionation. Although it is difficult to separate enough of the iddingsite to analyze it separately, Drake et al.[33] did analyze one 34μg sample from Lafayette, and found that it has the predicted isotopic and elemental composition (Fig. 7). Furthermore, the iddingsite would have to be much richer in noble gas than the bulk rock to explain the strong atmospheric $^{129}Xe/^{132}Xe$ signature. Indeed, the iddingsite separate was rich in noble gases, although the amount fell somewhat short of what is required by mass balance.

Almost simultaneously with the verification that the Lafayette iddingsite matches the predictions, the noble gases in ALH84001 were measured and found to fall in the same region as the nakhlites[2,31]. Unfortunately for the iddingsite hypothesis, this meteorite does not contain iddingsite. It does contain carbonates that apparently formed at low temperatures[34], so they may have played the same role that iddingsite apparently did in nakhlites.

High Kr/Xe component: There are basically three points with unexplained high Kr/Xe ratios. Two of the three, Zagami and LEW 88516, could be explained by the addition of some terrestrial contamination, since they lie within the triangle defined by the terrestrial and martian atmospheres and Chassigny on Fig. 7. Although there is no reason to think that these meteorites would have been more susceptible to terrestrial contamination than the other martian meteorites, the possibility should be kept in mind, particularly for LEW88516, which is gas-poor.

Another way to produce a Kr-rich composition would be through solubility, since Kr is more soluble in a melt than is Xe. Again, there is nothing special about Zagami and LEW88516, so it isn't clear what process could have acted there that didn't act on other shergottites.

The third point ("EV") comes from the experiments of Wiens[6], who crushed a sample of the glass, and analyzed the gas released by crushing (presumably gas from within vesicles). In one sense, this might be expected to be a more pristine sample of the martian atmosphere than that trapped within the glass. However, since Xe is more easily adsorbed than Kr, and the gas amount in the vesicles is lower than in the glass, if some adsorption occurred, either in nature or in the laboratory, the result would be that the released gas would have a higher Kr/Xe ratio than was originally trapped in the vesicle. This is qualitatively consistent with what is observed. Quantitatively, it would require adsorption of >30% of the Xe to make these data consistent with the same mixture of martian components (and terrestrial contamination) on both Fig. 5 and Fig. 7.

The other explanation that has been proffered[11] is that all the high Kr/Xe points might be explained by the addition of some very Kr-rich cometary gas. The $^{129}Xe/^{132}Xe$ ratio of the cometary component doesn't really matter, since it is far off scale to the right. This explanation is particularly attractive for the EETA 79001 vesicles, since that is a likely place to find evidence of an atmosphere temporarily enriched in cometary noble gases. However, it is not clear why the bulk of the EETA79001 glass would not have picked up the same signature, or, if it did, why most of the shergottite analyses would fall on a mixing line between the bulk EETA 79001 glass and Chassigny. Nor is it clear why Zagami and LEW88516 should contain higher relative proportions of the cometary component than other shergottites.

SUMMARY AND CONCLUSIONS

1) There are several pieces of evidence from martian meteorites that we have sampled at least two martian reservoirs of noble gases. One reservoir is almost certainly the martian atmosphere, given the similarities between the noble gases (and N_2) in the glass from EETA79001 and the Viking lander measurements of the martian atmosphere. Evidence for the second component includes:

a) Noble gases in Chassigny that are isotopically (for most Xe isotope ratios) and elementally distinct from the martian atmosphere.

b) $^{40}Ar/^{36}Ar$ ratios in some extractions from EETA79001 that are lower than either the terrestrial or martian atmospheres.

c) Nitrogen/argon systematics in EETA79001 indicating the presence of a low-$\delta^{15}N$ component distinct from either the terrestrial or martian atmospheres.

d) Xe isotopic systematics in ALH84001 and some nakhlites, which are not consistent with a mixture of martian and terrestrial atmosphere, but are (with one exception) consistent with a mixture of Chassigny and martian atmosphere.

It seems to be an inescapable conclusion that Mars contains a component, very probably associated with the interior, that is less radiogenic (for ^{129}Xe and probably, though not certainly, for ^{40}Ar) and has a lower $\delta^{15}N$ than the atmosphere.

2) There are several ways of presenting the data for martian meteorites in which simple mixing of two components is not sufficient. However, most of this data can be explained by either

a) Preferential adsorption of Xe onto rocks, particularly during aqueous alteration, which leads to elemental fractionation unaccompanied by isotopic fractionation, or

b) Terrestrial contamination.

The adsorption story seems reasonably firm, but the idea of terrestrial contamination is merely a fallback position. Zagami would have to have acquired more terrestrial noble gases than other martian meteorites, but there is no reason to see why it would have, since it is a relatively recent fall. There are some other samples in which some, but not all, measurements are inconsistent with simple two-component mixing, but it remains to be seen whether these represent analytical uncertainties or real sample heterogeneity.

3) Is there a third noble gas reservoir represented in the martian meteorites? These data by no means preclude the existence of additional components. In fact, since Mars has undergone less recycling of material, it is more likely that we might find isolated reservoirs with distinct signatures than it is on Earth.

However, there is no compelling reason, based on the noble gases in martian meteorites, to propose the existence of any additional components. Addition of a high Kr/Xe cometary component could explain some features of the data, but the assumptions required (e.g., Zagami contains more of the cometary component) are no more likely a priori than those required for other explanations (e.g., Zagami contains more terrestrial contamination).

ACKNOWLEDGEMENTS
This work was supported by NASA grant NAGW 3611. Constructive reviews by D. Bogard and R. Wiens are gratefully acknowledged.

REFERENCES

1. H. Y. McSween Jr., Meteoritics 29, 757 (1994).
2. T. D. Swindle, J. A. Grier, and M. K. Burkland, Geochim. Cosmochim. Acta 59, in press (1995).
3. D. D. Bogard and P. Johnson, Science 221, 651 (1983).
4. R. H. Becker and R. O. Pepin, Earth Planet. Sci. Lett. 69, 225 (1984).
5. U. Ott, Geochim. Cosmochim. Acta 52, 1937 (1988); U. Ott and F. Begemann, Nature 317, 509 (1985).
6. R. Wiens, Earth Planet. Sci. Lett. 91, 55 (1988).
7. U. Ott, H. P. Löhr, and F. Begemann, Meteoritics 23, 295 (1988).
8. U. Ott and H. P. Löhr, Meteoritics 27, 271 (1992).
9. T. D. Swindle, D. Garrison, C. M. Hohenberg, and C. T. Olinger, Lunar Planet. Sci. XVIII, 984 (1987).
10. T. D. Swindle, R. Nichols, and C. T. Olinger, Lunar Planet. Sci. XX, 1097 (1989).
11. T. Owen, A. Bar-Nun, and I. Kleinfeld, Nature 358, 43 (1992); T. Owen and A. Bar-Nun, this volume.
12. T. Owen, K. Biemann, D. R. Rushneck, J. E. Biller, D. W. Howarth, and A. L. Lafleur, J. Geophys. Res. 82, 4635 (1977).
13. R. C. Wiens, R. H. Becker, and R. O. Pepin, Earth Planet Sci. Lett. 77, 149 (1986).
14. T. D. Swindle, M. W. Caffee, and C. M. Hohenberg, Geochim. Cosmochim. Acta 50, 1001 (1986).
15. D. D. Bogard, F. Hörz, and P. Johnson, J. Geophys. Res. 91, E99 (1986); R. C. Wiens and R. O. Pepin, Geochim. Cosmochim. Acta 52, 295 (1988).
16. R. O. Pepin, Icarus 92, 2 (1991).
17. T. J. Bernatowicz, B. M. Kennedy, and F. A. Podosek, Geochim. Cosmochim. Acta 49, 2561, (1985).
18. J. Matsuda, M. Sudo, M. Ozima, K. Ito, O. Ohtaka, and E. Ito, Science 259, 788 (1993).
19. K. Zahnle, In Conference on Deep Earth and Planetary Volatiles. LPI Contribution No. 845, Lunar and Planetary Institute, Houston. p. 50 (1994).
20. M. Ozima and K. Nakazawa, Nature 284, 313 (1980).
21. K. Zahnle, J. B. Pollack, and J. F. Kasting, Geochim. Cosmochim. Acta 54, 2577 (1990).
22. G. Dreibus and H. Wänke H., Icarus 71, 225 (1987); D. M. Musselwhite, M. J. Drake, and T. D. Swindle, Nature 352, 697 (1991).
23. B. M. Jakosky, R. O. Pepin, R. E. Johnson, and J. L. Fox, Icarus 111, 271 (1994); R. O. Pepin, Icarus 111, 289 (1994).
24. R. O. Pepin and M. H. Carr, In Mars (U. of Arizona, Tucson, 1992), p. 120.
25. A. E. Fallick, R. W. Hinton, D. P. Mattey, S. J. Norris, C. T. Pillinger, P. K. Swart, and I. P. Wright. Lunar Planet. Sci. XIV, 183.
26. C. J. Allègre, T. Staudacher, and P. Sarda, Earth Planet. Sci. Lett. 81, 127 (1987).
27. K. A. Farley, J. H. Natland, and H. Craig, Earth Planet. Sci. Lett. 111, 183 (1992).
28. D. B. Patterson, M. Honda, and I. McDougall, Geophys. Res. Lett. 17, 705 (1990).
29. M. H. Carr and H. Wänke, Icarus 98, 61 (1992).
30. D. D. Bogard, L. E. Nyquist, and P. Johnson, Geochim. Cosmochim. Acta 48, 1723 (1984).
31. Y. N. Miura, N. Sugiura, and K. Nagao, Lunar Planet. Sci. XXV, 919 (1994).
32. E. Jagoutz, A. Sorovka, J. Vogel, and H. Wänke, Meteoritics 29, 478 (1994).
33. M. J. Drake, T. D. Swindle, T. Owen, and D. Musselwhite, Meteoritics 29, 854 (1994).
34. C. S. Romanek, M. M. Grady, I. P. Wright, D. W. Mittlefehldt, R. A. Socki, C. T. Pillinger, and E. K. Gibson Jr., Nature 372, 655 (1994).
35. T. D. Swindle, In Meteorites and the Early Solar System (U. of Arizona, Tucson, 1988), p. 535.

DEGASSING HISTORY AND EVOLUTION OF VOLCANIC ACTIVITIES OF TERRESTRIAL PLANETS BASED ON RADIOGENIC NOBLE GAS DEGASSING MODELS

Sho Sasaki and Eiichi Tajika
Geological Institute, School of Science, University of Tokyo, Tokyo 113, JAPAN

ABSTRACT

The radiogenic noble gas species ^{4}He, ^{40}Ar and ^{129}Xe in the atmospheres of the terrestrial planets provide valuable information on planetary degassing history. The abundance of radiogenic fraction of Martian atmospheric ^{129}Xe relative to the total planetary mass suggests that early degassing, possibly during magma ocean cooling, would be 1/3 as much as that of the Earth. Atmospheric ^{40}Ar amounts relative to the total mass of each planet show that the long-term volcanic degassing of Mars and Venus are 1/20 and 1/4 of that of the Earth, respectively. The duration of plate tectonics is estimated from the total ^{40}Ar amounts: for Mars it is less than 300Ma assuming a plate velocity of 8 cm/s and ridge length of 8000 km and for Venus it is less than 1Ga. Detection of Martian ^{4}He will be evidence of current degassing activity on Mars, whereas the observed abundance of Venusian ^{4}He is the outcome of long-term degassing from the interior.

INTRODUCTION

The current atmospheres of the terrestrial planets were formed by secondary degassing from planetary interiors [1, 2]. The degassing history of terrestrial planets is divided into two categories. One is early degassing during or within a short period after planetary formation. Impact degassing [3] and/or outgassing during magma ocean cooling are plausible processes for early degassing. The other is continuous (or episodic) degassing throughout planetary history, i.e., degassing in the course of magmatic activity at hot spots and during production of oceanic crust at mid oceanic ridges [4].

The principle volatiles that are degassed from the interior are H_2O and CO_2. Because of their high reactivity and retentivity by surface rocks, H_2O and CO_2 can be trapped and incorporated into solids, forming hydrate or carbonate minerals in the crust; their atmospheric abundances do not reflect the total degassed amount. In the Earth, moreover, H_2O and CO_2 may be recycled or "regassed" into the mantle through plate subduction. In the Venusian atmosphere, H_2O has decomposed and hydrogen escaped into space.

On the other hand, atmospheric noble gases provide important information for planetary degassing history in spite of their low concentrations. Because of their low reactivity, noble gases cannot be trapped by solids or incorporated into the interior again. The present noble gas abundances in the atmosphere reflect degassing activities integrated over the planetary history. As for the lightest rare gas He, continuous escape to the space has decreased the atmospheric abundance, which therefore reflects ongoing present degassing activity.

Among noble gas isotopes, radiogenic species such as ^{4}He (produced through decay of ^{235}U, ^{238}U, and ^{232}Th) , ^{40}Ar (a decay product of ^{40}K) and ^{129}Xe (a decay product of ^{129}I) can provide chronological information for degassing since their abundances increase with time. Because of the short half life of ^{129}I (1.57×10^{7}yr), ^{129}Xe can be used in discussing early degassing from the interior [5]. The difference between atmospheric and interior ^{129}Xe/^{130}Xe favors the early ($\ll$ a few 10^{8}yr) catastrophic degassing of the Earth, which was originally proposed on the basis of the difference of atmospheric and mantle ^{40}Ar/^{36}Ar [6]. Comparing radiogenic ^{129}Xe abundances in their atmosphere, Sasaki [7] concluded that early Martian degassing is about one third as much as that of the Earth.

The atmospheric abundance of ^{40}Ar is a useful clue in pursuing the rate of long-term global volcanic activity because of the longer half life of ^{40}K (1.25×10^{9}yr). In discussing the total abundance, the physical process of volatile ^{40}Ar transport from the interior to the surface must be examined precisely. Then we may deduce meaningful information on volcanic activity history. In the Earth, most of degassing is associated with volcanic activity forming oceanic crust at mid oceanic ridges. Tajika and Matsui [8] estimated the long-term history of terrestrial magmatic production, taking into account melt generation through adiabatic decompression of uprising mantle materials, partitioning of ^{40}Ar and ^{40}K between silicate melt and solid, and bubbling and degassing of volatiles during magma eruption. They numerically examined changes in mantle temperature, heat flux, melt generation and degassing rates using a simple convection model. They showed that the average magma generation rate over the Earth's age is about 37 km^3/yr, which is a little larger than the current production rate of oceanic crust of 20 km^3/yr. In Mars, on the basis of absolute ages estimated from surface crater density and volume of hot-spot type volcanics, the magma production rate through time is obtained [9]. Sasaki [7] compared the erupted volcanic volume with the present ^{40}Ar abundance in the Martian atmosphere, considering that partial melting should concentrate volatiles. He concluded that the present ^{40}Ar abundance is compatible with the present apparent volume of volcanics, assuming some amount of intrusive volcanics.

In this study, we present quantitative estimates of Martian and Venusian volcanic activities assuming that atmospheric ^{40}Ar abundances constrain the integrated magmatic production. In the Earth, plate-tectonic activity (degassing at mid oceanic ridges should supply more volatiles as well as magma than hot spot type activity; long-term degassing of ^{40}Ar is mainly controlled by plate-tectonic activity [8]. If there were plate-tectonic activity on Mars or Venus, it would have supplied a significant amount of ^{40}Ar into the atmosphere. From the present ^{40}Ar abundance in the atmosphere, we can obtain the upper limit on the duration of plate-tectonic activity on Mars or Venus. Recently, Sleep [10] proposed that the ancient plate tectonics occurred on the northern hemisphere of Mars. Using some of the parameters in his model, we try to constrain duration of the Martian plate tectonics from ^{40}Ar abundance. A similar estimate is briefly shown for Venus, too. Finally we discuss another radiogenic noble gas, ^{4}He, which constrains current degassing on Mars and possibly long-term degassing on Venus.

Table 1   Normalized abundances of representative noble gas isotopes in CI chondrites, Earth, Venus and Mars: R(X) [kilogram/kilogram-planet] [11].

	^{4}He	^{20}Ne	^{36}Ar	^{84}Kr	^{130}Xe
CI	-	2.89×10^{-10} ±0.77	1.25×10^{-9} ±0.10	3.57×10^{-11} ±0.15	7.0×10^{-12} ±1.9
Earth	6.21×10^{-10} ±0.07	1.00×10^{-11} ±0.01	3.45×10^{-11} ±0.01	1.66×10^{-12} ±0.02	1.40×10^{-14} ±0.02
Venus	1.1×10^{-10} $+2/-0.5$	2.9×10^{-10} ±1.3	2.51×10^{-9} ±0.97	4.7×10^{-12} $+0.6/-3.4$	8.9×10^{-14} $+2.5/-6.8$
Mars	?	4.38×10^{-14} ±0.74	2.16×10^{-13} ±0.55	1.76×10^{-14} ±0.28	2.08×10^{-16} ±0.41

Table 2   Isotopic ratios of Ar and Xe in the atmospheres of the terrestrial planets

	^{40}Ar/^{36}Ar	^{129}Xe/^{130}Xe	^{129}Xe/^{132}Xe
Earth	295.5	6.50	0.983
Venus	1.19 ± 0.07	-	-
Mars Viking	3.01×10^3	-	1.5-4.5
EETA79001	2.26×10^3	16.40 ± 0.8	2.39 ± 0.03

RADIOGENIC ^{129}Xe AND ^{40}Ar ABUNDANCES AND DEGREE OF DEGASSING

Table 1 summarizes noble gas abundances of CI chondrites, Earth, Mars and Venus [11]. Each abundance is expressed as the mass of species X in the atmosphere divided by the total planetary mass (R(X): abundance of X in kilogram/kilogram-planet). Table 2 shows the isotopic ratios involving ^{129}Xe and ^{40}Ar. For Venus, we have no data on Xe isotopes. Although almost all of ^{40}Ar is radiogenic, there is a significant primary non-radiogenic component of ^{129}Xe. According to Pepin and Phinney [12], only 6.7% of ^{129}Xe in the terrestrial atmosphere is the decay product of ^{129}I. Assuming that the initial non-radiogenic component is the same between Mars and Earth (^{129}Xe^n/^{130}Xe = 6.06, where superscript n denotes nonradiogenic species) as a first approximation, we obtain the abundance of radiogenic ^{129}Xe (with

^{129}Xe*/^{130}Xe = 10.34, where superscript * denotes radiogenic species) of the Martian atmosphere from the total abundance of Martian ^{129}Xe (with ^{129}Xe/^{130}Xe = 16.40) [7]. A similar analysis is made also by Pepin [13]. Then 63% of ^{129}Xe in the Martian atmosphere is of radiogenic origin and we have

$$\frac{R\left(^{129}Xe^{*}\right)_{Mars}}{R\left(^{129}Xe^{*}\right)_{Earth}} = \frac{R\left(^{129}Xe\right)_{Mars} \times 0.63}{R\left(^{129}Xe\right)_{Earth} \times 0.067} = 0.353 \tag{1}$$

A similar value was obtained by Zahnle [14]. In contrast to the isotopic ratio, the Martian radiogenic ^{129}Xe abundance is less than the terrestrial value. But the Martian atmosphere still contains one third of radiogenic ^{129}Xe compared with the terrestrial atmosphere in relative abundance (kg/kg-planet).

Because we may consider that all of ^{40}Ar is the radiogenic product of ^{40}K, the relative ratio of ^{40}Ar is more easily obtained both for Mars and Venus: we have

$$\frac{R\left(^{40}Ar^{*}\right)_{Mars}}{R\left(^{40}Ar^{*}\right)_{Earth}} = 0.048 \quad \text{and} \quad \frac{R\left(^{40}Ar^{*}\right)_{Venus}}{R\left(^{40}Ar^{*}\right)_{Earth}} = 0.29 \tag{2}$$

In the above we use data of SNC meteorite EETA79001 for Martian Ar. The Martian atmosphere is deficient not only in non-radiogenic Ar but also in radiogenic Ar; it contains only one twentieth of ^{40}Ar in relative abundance (kg/kg-planet) compared with the terrestrial atmosphere [15-17]. This ^{40}Ar deficiency can be ascribed to either less tectonic activity on Mars or atmospheric escape. Although the Venusian atmosphere contains 70 times as much ^{36}Ar as than the terrestrial atmosphere, it has one fourth of ^{40}Ar compared with the terrestrial atmosphere.

Radiogenic noble gas isotopes are supplied into the atmosphere through degassing from the interior. The above difference between relative Martian ^{129}Xe and ^{40}Ar abundances suggests that atmospheric ^{129}Xe was not supplied into the atmosphere by the inefficient degassing process that has supplied ^{40}Ar. Using a simple two-stage degassing model where sudden degassing occurred at $t = t_1$ and $t = t_2$ ($> t_1$), we try to express the above ratios (1) and (2) by parameters describing the degassing process [7]. Here α is the degassing fraction from the interior at the initial degassing at $t = t_1$, and β is that at the later degassing at $t = t_2$ ($> t_1$), where t_1 is longer than the half life of ^{129}I and much shorter than the half life of ^{40}K and t_2 is at least comparable to the half life of ^{40}K. We assume that the degassing fractions (α and β) are the same for argon and xenon: the fraction α (β) of gas species (radiogenic or non-radiogenic) degassed suddenly at $t = t_1$ ($t = t_2$). So long as the orders of magnitude of α and β are not largely different, we have for the present abundances of radiogenic ^{129}Xe and ^{40}Ar in the Martian atmosphere:

$$\begin{aligned} ^{129}Xe^{*} &= \alpha \,^{129}I_0\left(1 - \exp(-\lambda_{129}t_1)\right) \\ &+ \beta\left[(1-\alpha)^{129}I_0\left(1 - \exp(-\lambda_{129}t_1)\right) + {}^{129}I_0\left(\exp(-\lambda_{129}t_1) - \exp(-\lambda_{129}t_2)\right)\right] \\ &\approx \alpha \,^{129}I_0\left(1 - \exp(-\lambda_{129}t_1)\right) \end{aligned} \tag{3}$$

and

$$
\begin{aligned}
{}^{40}\text{Ar}^* &= \alpha f_{Ar}{}^{40}\text{K}_0\left(1-\exp(-\lambda_{40}t_1)\right) \\
&\quad + \beta\left[(1-\alpha)f_{Ar}{}^{40}\text{K}_0\left(1-\exp(-\lambda_{40}t_1)\right)+f_{Ar}{}^{40}\text{K}_0\left(\exp(-\lambda_{40}t_1)-\exp(-\lambda_{40}t_2)\right)\right] \\
&\approx \beta f_{Ar}{}^{40}\text{K}_0\left(\exp(-\lambda_{40}t_1)-\exp(-\lambda_{40}t_2)\right) \\
&\approx \beta f_{Ar}{}^{40}\text{K}_0\left(1-\exp(-\lambda_{40}t_2)\right)
\end{aligned}
\tag{4}
$$

where λ_{129} and λ_{40} are the decay constants and ${}^{129}\text{I}_0$ and ${}^{40}\text{K}_0$ are the initial abundances of ${}^{129}\text{I}$ and ${}^{40}\text{K}$, respectively, and $f_{Ar} = 0.1048$ is the branching ratio for ${}^{40}\text{Ar}$. Assuming that the abundances of ${}^{129}\text{I}_0$ and ${}^{40}\text{K}_0$ relative to the total planetary mass are the same among the Earth, Mars and Venus, we have

$$
\frac{R\left({}^{129}\text{Xe}^*\right)_{Mars}}{R\left({}^{129}\text{Xe}^*\right)_{Earth}} = 0.353 \approx \frac{\alpha_{Mars}}{\alpha_{Earth}}
$$

$$
\frac{R\left({}^{40}\text{Ar}^*\right)_{Mars}}{R\left({}^{40}\text{Ar}^*\right)_{Earth}} = 0.048 \approx \frac{\beta_{Mars}}{\beta_{Earth}} \quad \text{and} \quad \frac{R\left({}^{40}\text{Ar}^*\right)_{Venus}}{R\left({}^{40}\text{Ar}^*\right)_{Earth}} = 0.29 \approx \frac{\beta_{Venus}}{\beta_{Earth}} \tag{5}
$$

In the above, the timings of degassing events among the planets are assumed to be the same ($t_{M1} \sim t_{E1} \sim t_{V1}$ and $t_{M2} \sim t_{E2} \sim t_{V2}$). So long as $t_{M1}, t_{E1}, t_{V1} \sim 10^8\text{yr}$ and $t_{M2}, t_{E2}, t_{V2} \geq 2\times10^9\text{yr}$, the differences among t_{M1}, t_{E1}, t_{V1} and among t_{M2}, t_{E2}, t_{V2} do not change the above relations. From the above derivations, it is confirmed that differences in the current abundances of radiogenic noble gases among the Earth, Mars, and Venus should correspond to differences in the degassing fraction (α or β) from the interior. And the earlier and later degassing event can be discussed separately using relative abundance ratios of ${}^{129}\text{Xe}$ and ${}^{40}\text{Ar}$.

The ratio of radiogenic ${}^{129}\text{Xe}$ between the Earth and Mars (0.353) suggests that the outgassed fraction at the early degassing ($t = t_1 \sim 10^8\text{yr}$) on Mars is less than that of the Earth but the difference is not so large. Impact degassing during accretion and degassing during magma ocean cooling are plausible mechanisms for the early degassing. Since impacts would not only supply volatiles but remove atmospheric gases of the smaller planet Mars by "atmospheric cratering" [14, 18-20], impacts should produce larger differences in atmospheric volatile inventories between the Earth and Mars. On Mars, atmospheric cratering would have decreased non-radiogenic noble gas abundances greatly and may explain their present small abundances [14]. In this respect, we prefer the magma ocean cooling after the accretion - atmospheric cratering stage as the mechanism of early degassing, which supplied radiogenic ${}^{129}\text{Xe}$ into the atmosphere. Because accretion timescale ($\sim10^7\text{yr}$) is comparable to the half life of ${}^{129}\text{I}$, the atmospheric cratering should have also escaped some radiogenic ${}^{129}\text{Xe}$; this loss might account for a part of the difference of relative radiogenic ${}^{129}\text{Xe}$ abundances between the Earth and Mars (denoted by the relation (5)). Another idea is that I/Xe fractionation occurred on Mars by the difference of water solubility between iodine and xenon [15, 21]: whereas primary xenon was lost from the atmosphere by the atmospheric cratering, iodine should have dissolved into the protoocean, been captured by crustal rocks, and finally supplied radiogenic ^{129}Xe. There is no data on Venusian ^{129}Xe, nor is precise data of Xe elemental abundance. Therefore we cannot discuss about the early degassing on Venus using radiogenic ^{129}Xe abundance.

We assumed that the initial mass abundances (kg/kg-planet) of ^{129}I were the same between the Earth and Mars. From relatively higher iodine abundance of SNC meteorites than terrestrial mantle, Dreibus and Wänke [15] estimated mean abundance of Martian iodine 2.4 times higher than that of the Earth. This might decrease the corresponding ratio of degassing degree $\alpha_{Mars}/\alpha_{Earth}$ in the relation (5) proportionally. Moreover, the initial ^{129}I is affected by the formation age, i.e., the time when iodine is trapped firmly into solids in the cosmochemical stage. We assumed that this age difference would be small in the scenario of the solar system formation. The case where the radiogenic ^{129}Xe difference is ascribed to the different formation ages was discussed previously [7].

Different from radiogenic ^{129}Xe, the above relation (5) suggests that later degassing on Mars deduced from ^{40}Ar abundances is much less than that of the Earth. This corresponds to a smaller volume of melting on Mars after the initial hot epoch during or just after accretion. Indeed, assuming that the present ^{40}K abundance is the same as that of the Earth (5.0×10^{-8} kg/kg), the present Martian atmospheric ^{40}Ar abundance ($R_{Mars}(^{40}Ar) = 5.7\times10^{-10}$ kg/kg) is only about 1% of the total radiogenic ^{40}Ar produced in the interior of 5.0×10^{-8} kg/kg. Later degassing on Venus is also smaller than that of the Earth, and corresponds to a current absence of plate tectonics. In the following sections, using more detailed models, we discuss the history of volcanic activities on Mars and Venus and place upper limits on the duration of their plate-tectonic epoch, if they existed.

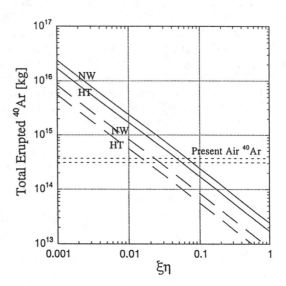

Fig. 1. The total degassed ^{40}Ar for different values of $\xi \times \eta$. Solid lines are based on the volume estimate by Greeley [9]. The dashed lines are based on the volume estimate by Greeley and Schneid [22]. NW denotes the age model by Neukum and Wise [23] and HT denotes the age model by Hartmann et al. [24] and Tanaka [25].

MARTIAN ^{40}Ar AND VOLCANIC ACTIVITY

From surface images of Mars, the erupted volcanic volume at each epoch has been estimated [9, 22]. According to Greeley [9], the total erupted volume of volcanics is 2×10^{17} m^3. Combined with an age model based on crater density, we can estimate the magma generation rate associated with Martian volcanoes, which are apparently hot-spot type. Sasaki [7] compared the total erupted volcanic volume with the present ^{40}Ar abundance in the Martian atmosphere. The outgassed ^{40}Ar mass by volcanic activity is estimated using the product of two parameters, the melt accumulation factor (~ melt fraction) ξ and the contribution of intrusive volcanics (ratio of the extrusive mass to the total erupted mass) η (Fig. 1). When the volcanic eruption rate at the surface is denoted by q, the total magma supply (generation) rate is q/η. The mantle volume, which supplies magma during unit time, is expressed by $q/\eta\xi$, and then ^{40}Ar degassing rate is approximated by $C_{Ar}q/\eta\xi$, where C_{Ar} is ^{40}Ar concentration. Then, the degassing ^{40}Ar amount is inversely proportional to $\xi \times \eta$ (see Fig. 1). In Fig. 1, the initial ^{40}K abundance of Mars is assumed to be the same as that of the Earth. Assuming $\eta \sim 1/4 - 1/2$, a favorable value for melt fraction of $\xi \sim 0.1 - 0.2$ could explain the present Martian ^{40}Ar atmospheric abundance. Scambos and Jakosky [17] compared the erupted volcanics with the present ^{40}Ar abundance in order to constrain non-radiogenic volatiles such as ^{36}Ar and H_2O. Although they obtained a parameter "^{40}Ar release factor" which corresponds to $\xi \times \eta$ in Sasaki [7], they did not pursue discussion on the history of Martian volcanic activity. Pepin [13] also used the erupted volcanic volume of Mars to obtain the degassed ^{40}Ar from the interior.

In the above estimate, we cannot assume an arbitrary small value for η because there is little evidence of intrusive volcanic volumes on Mars. Previously very small value (~ 1/9.5) is used by Greeley and Schneid [22], but it is not possible that Martian large volcanoes should involve nearly large intrusive volumes. A part of discrepancy between the erupted volume and degassed ^{40}Ar might be explained using a larger value for the initial ^{40}K abundance in Mars. However, there is a possibility of other magmatic sources for ^{40}Ar. Recently, Sleep (1994) proposed that a large part of the Martian northern hemisphere had been formed by plate-tectonic activity during the earliest history of Mars; the area is about one third of the total surface of Mars. Like the Earth, ridge magmatism involving crustal formation should supply much volatiles into the atmosphere. Subducted materials may account for the assumed intrusive volcanics (denoted by η) in the previous model.

Here we consider that the present atmospheric ^{40}Ar is supplied both by identified large volcanoes (hot spots) and by ancient plate tectonics. Then we can obtain an upper limit of the duration of the ancient plate-tectonic activity. First, the amount of ^{40}Ar degassed by hot-spot type volcanism is evaluated from the erupted volcanic volume [9], using the cratering age model [24, 25]. It is expressed as

$$M_{Ar-h} = \int_0^{4.6Ga} F_{Ar-h}\, dt$$
$$F_{Ar-h}(t) = fC_{Ar}(t)k_h \approx C_{Ar}(t)q_h(t)\xi_h^{-1} \qquad (6)$$

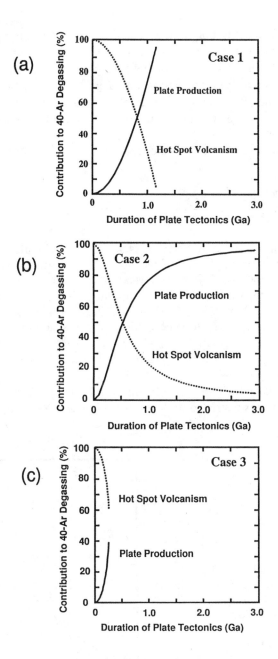

Fig. 2   Contribution of two types of volcanism (hot spot type and plate tectonics type) to ^{40}Ar degassing as a function of the duration of plate tectonics on Mars. (a), (b), and (c) correspond to the results for Cases 1, 2, and 3 respectively.

where subscript h denotes hot spot, $F_{Ar-h}(t)$ is the ^{40}Ar degassing rate, $C_{Ar}(t)$ [kg/m^3] is the ^{40}Ar concentration in the mantle, ξ_h is the accumulation factor (~ melting degree), k_h [m^3/yr] is the volume rate of uprising mantle material that actually contributes to magma production, and then $q_h(t) = \xi_h k_h$ [m^3/yr] is the melt production rate. The efficiency of ^{40}Ar degassing from the uprising volume, f, is close to the unity [8] and is neglected here. The amount of outgassed ^{40}Ar is inversely proportional to ξ_h.

The degassing rate of ^{40}Ar by plate-tectonic activity is written similarly as

$$M_{Ar-p} = \int_0^\tau F_{Ar-p}\, dt$$
$$F_{Ar-p}(t) = f\, C_{Ar}(t)k_p \approx C_{Ar}(t)q_p(t)\xi_p^{-1} \qquad (7)$$

where the subscript p denotes plate tectonics and τ represents the duration of plate tectonics. In estimating supply rate of magma at Martian ridges, we assume three cases for the volume rate k_p (Tajika and Sasaki, in preparation).

In the first case, we use the Earth's production rate $q = 2\times10^{10}$ m^3/yr for Mars and the melt fraction $\xi = 0.2$ of oceanic crust (assuming a higher mantle temperature of the early Mars), together with a surface area correction:

Case 1 $$k_p = 2.0\times10^{10}\,[1/3(R_M/R_E)^2](0.2)^{-1}\,[\text{m}^3/\text{yr}] \qquad (8)$$

where a ratio of planetary radii of Mars and the Earth, R_M/R_E, is 0.53, and the area of plate tectonics is restricted to the northern lowlands (~1/3 of the surface) of Mars [10].

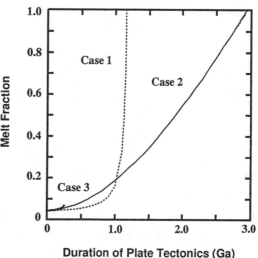

Fig. 3   Summary of the average melt fraction of Martian volcanism as a function of the duration of plate tectonics for Cases 1, 2, and 3. It is noted that in Case 3 the solution is absent over some critical duration of about 0.3Ga.

In the second case, we assume that the Earth's production rate applies to Mars but use the same melt fraction ξ as that of hot spot type activity. Then ξ is a variable parameter also for the plate-tectonic supply.

Case 2 $k_p = 2.0 \times 10^{10} [1/3(R_M/R_E)^2]\xi_h^{-1} \, [\text{m}^3/\text{yr}]$ (9)

In Cases 1 and 2, the melt fraction ξ is an independent variable. But the application of the terrestrial production rate is too rough an assumption. Next we use a model which combines geological parameters [10] and detailed model of the melt generation process beneath oceanic ridges [26]. We use an average plate velocity $v_p = 8$ cm/yr and the length of oceanic ridges of 8000 km [10]. Using the interior solidus curve calculated for Martian low gravity, we can estimate the generated melt thickness d as well as the melt fraction ξ (= the generated melt thickness / the melt generation depth) from the assigned mantle potential temperature T_p, which is the temperature of mantle material expanded adiabatically to the surface pressure. The calculation is based on the model proposed by McKenzie and Bickle [26] (Tajika and Sasaki, in preparation). We have

Case 3 $k_p = 0.08 \times 8 \times 10^6 \, d(T_p) \, \xi(T_p)^{-1} \, [\text{m}^3/\text{yr}]$ (10)

where d and ξ are function of mantle potential temperature T_p. In the above, the mantle potential temperature T_p is an independent variable and $\xi(T_p)$ should also affect the hot spot type eruption rate k_h. For all numerical calculations, time step is taken to be 1.0 Ma.

Our results are shown in Figs. 2 and 3. Since the total degassed ^{40}Ar amount (3.6×10^{14} kg) is fixed, variations of ξ or T_p will change contributions of hot spot and plate tectonics to ^{40}Ar degassing and change the duration of plate tectonics (Figs. 2). The corresponding melt fraction ξ is shown in Fig. 3. In Case 1 (Fig. 2(a)), the hot spot production rate k_h decreases with increasing ξ while the plate production rate k_p is constant. The duration of plate tectonics τ is as long as 1Ga

Table 3   Critical (maximum) plate tectonics duration τ_{cr} and corresponding melting degree ξ_{cr} in Case 3

		τ_{cr}	ξ_{cr}
1	Standard Mars	270Ma	0.086
2	Earth's solidus	440Ma	0.085
3	$[^{40}\text{Ar}] = 2 \times$present $[^{40}\text{Ar}]$	560Ma	0.044
4	$[^{40}\text{Ar}] = 0.5 \times$present $[^{40}\text{Ar}]$	125Ma	0.168
5	$v_p = 0.8$ cm/yr	880Ma	0.084

for $\xi_h \sim 0.2$ (see Fig.3). In Case 2 (Fig. 2(b)), both k_h and k_p decrease with increasing ξ. The duration of plate tectonics would be much longer but plausible ξ_h (≤ 0.2) should keep τ not longer than 1Ga (see Fig.3). In Case 3 (Fig. 2(c)), k_h should decrease with increasing T_p through ξ but k_p should increase with increasing T_p through enhancing melt thickness. Increasing k_p results in a relatively short plate tectonics duration τ (< 0.3Ga) and a critical value for ξ ($\xi_{cr} \sim 0.07$) (see Fig.3). As seen in Fig. 3, no solution exists above the critical value ($\xi > \xi_{cr}$ and $\tau > \tau_{cr}$). Assuming that the start of plate-tectonic activity is at $t = 0$ ($C_{Ar}(0) = 0$), we obtain the longest duration of Martian plate tectonics. If we assume a later time for the beginning of plate tectonics ($t > 0$), then plate activity would outgas more ^{40}Ar and its duration would be shorter.

We obtain plate tectonics duration τ for different constraining parameters in Case 3. The results are shown in Table 3 where the critical duration τ_{cr} is tabulated. Under the terrestrial solidus curve, a smaller volatile accumulation volume results in a smaller degassing rate; we have a longer duration. We have examined this case for different values of the total abundance of atmospheric ^{40}Ar. Jakosky et al. [27] propose that pick-up-ion sputtering by solar winds and photochemical escape should decrease atmospheric species. Using isotopic fractionation of ^{38}Ar/^{36}Ar, about half of ^{40}Ar has been lost from the Martian atmosphere (Jakosky, personal communication). On the other hand, if the initial degassing event was intense ($\alpha \gg 0.1$) and prolonged, early degassed ^{40}Ar would have occupied several tens of percent of the total atmospheric ^{40}Ar, which would decrease the constraining amount of later degassed ^{40}Ar. Then we simulated the cases when the total ^{40}Ar is twice and half as large as the present amount (see Table 3). The former prolongs the plate tectonics duration τ (~ 0.6Ga) and the latter shortens τ (~ 0.1Ga). Finally, we make sure that a slower assumed plate velocity would prolong the duration τ greatly, suppressing the ^{40}Ar degassing rate.

VENUSIAN ^{40}Ar AND VOLCANIC ACTIVITY

It is more difficult to estimate the erupted magmatic volume of Venus because global resurfacing has eliminated any previous record of volcanic activity. Tajika and Matsui [28] assumed that Venusian tectonic history can be divided into two stages: Stage I when plate tectonics prevails and Stage II when volcanic activity is limited to the hot spot type. During Stage I ($0 < t < \tau$) with plate tectonics, we apply the terrestrial magma production rate. According to the study using ^{40}Ar degassing model coupled to the thermal history of the Earth, the seafloor spreading rate averaged over 4.6Ga is constrained to be close to the present value [8], although this result might not guarantee the application of the terrestrial production rate for Venus. During State II ($\tau < t < 4.6 \times 10^9$yr), we use the ^{40}Ar degassing rate $F_{Ar}(t) = C_{Ar}(t)k_h(t)$ (Eq.(6)). Temporal variations of k_h and ξ are estimated from the thermal history calculation; ξ is variable here for the model of Venus although it was constant for the model of Mars. Because the total outgassed ^{40}Ar amount is fixed and equal to the present atmospheric amount, the duration of plate-tectonic activity constrains the average k_h and then the magma eruption rate q_h.

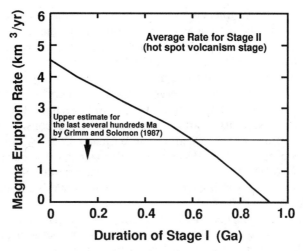

Fig. 4.   The average magma eruption rate of Venus estimated from the [40]Ar degassing model as a function of the duration of Stage I (the presumed plate tectonics stage on Venus). The solid curve represents the magma eruption rate averaged over the Stage II (the stage after plate tectonics ceased). The horizontal line represents the upper bound of melt production activity estimated from crater statistics on Venus by Grimm and Solomon [29].

Our result is shown in Fig. 4. The horizontal axis shows the duration of plate tectonics and the vertical axis corresponds to the average magma generation rate over Stage II. The duration of ancient plate tectonics τ was less than 1Ga for Venus if the ridge spreading rate was the same as that of the Earth over the same time period. If the duration of Stage I, τ, exceeds 1.0Ga, almost all of the [40]Ar degassed into the atmosphere during that stage, suggesting that volcanic activity after t = τ should be nearly "zero" (note that a longer duration is obtained if we assume the lower magma production rate for Stage I). Longer plate-tectonic activity is incompatible with a young surface age of Venus, which suggests recent global volcanic activity. If plate tectonics did not operate on Venus, the magma generation rate averaged over Stage II is estimated to be about 5 km³/yr. Longer plate-tectonic activity, required for degassing the observed amount of [40]Ar, results in an average magma generation rate smaller than 5 km³/yr. Grimm and Solomon [29] considered the Venusian younger surface age inferred from crater statistics to be due to volcanic resurfacing and estimated the recent melt generation rate on Venus to be smaller than 2 km³/yr. If this limit is applicable to the whole of Stage II, some duration of plate-tectonic activity (Stage I) associated with efficient [40]Ar degassing is necessary.

MARTIAN AND VENUSIAN [4]He

The mass spectrometer on board Viking lander did not detect [4]He in the Martian atmosphere [30]. Recently, the EUVE (Extreme Ultraviolet Explorer) satellite observed He airglow (58.4nm resonance scattering) of Mars and detected

43 ± 10R (1R $= 10^6$ photons/cm^2 s) [31]. This corresponds to a ^{4}He abundance of 1.1 ± 0.4ppm, so that the present Martian atmosphere contains 2.8×10^{10} kg of ^{4}He.

In contrast to the heavier ^{129}Xe and ^{40}Ar isotopes, atmospheric ^{4}He, which is more easily lost to space, does not reflect long-term degassing history. In the Earth, ^{4}He escapes along with magnetic field lines from polar regions as "polar wind" [32] and ^{4}He outflow was observed by polar orbiters [33]. The life time of ^{4}He in the atmosphere is estimated to be 10^6yr [34]. Mid oceanic ridges supply about 20-30% of ^{4}He whereas a large part of atmospheric ^{4}He is considered to come from the continental crust where U and Th are enriched [34-36]. Although mid oceanic ridge volcanism can supply ^{4}He continuously over the timescale of 10^6yr, some mechanism such as magmatic transport or water circulation is necessary to maintain continuous degassing from the continental crust because of the slow ^{4}He diffusion rate [34].

Since a dipole magnetic field is absent on Mars, ^{4}He cannot be lost by polar winds. Because of low gravity, however, electron impact ionization followed by solar wind pick-up can cause the escape of the atmospheric ^{4}He on a timescale of 5×10^4yr [31]. Then the detection of ^{4}He implies that the current degassing rate is continuous in this timescale. Since at present Mars does not have an apparent continuous degassing source such as mid oceanic ridges, the existence of Martian ^{4}He requires some other mechanism of supply. We estimated degassing from impact cratering. Using a crater density model and assuming that ^{4}He escapes from volumes ten times as large as that of craters, we have a degassing rate of 10-30 kg/yr. Then the ^{4}He supply timescale is $(1-3) \times 10^9$yr, which is much larger than the estimated lifetime of 5×10^4yr. We speculate that ongoing underground magmatism or hydrothermal circulation is necessary to account for the ^{4}He abundance.

We need more accurate measurements of Martian He. In the EUVE observations, the size of Mars was less than one pixel and there was a high noise level. Future fine mass spectrometric detection by a lander or UV observation by an orbiter will confirm the existence of Martian ^{4}He, which should constrain the current degassing from the interior.

In Venus, the atmospheric ^{4}He abundance is much higher than the Earth: we have $R_{Venus}(^4$He$)/R_{Earth}(^4$He$) = 200$. Because this ratio is even larger than that of Ne or Ar, a large part of atmospheric ^{4}He is not primary and degassed later from the interior. Because of the absence of a magnetic field, the ^{4}He escape rate from the current Venus atmosphere should be much slower than that of the Earth; Prather and McElroy [37] estimated the escape rate and obtained 6×10^8yr for the life time of ^{4}He in the Venusian atmosphere. The lifetime can be also estimated from the degassing rate. If the degassing rate of ^{4}He is compatible with that of ^{40}Ar and if ^{4}He inventory is in steady state in the Earth's and Venusian atmospheres, the life time of atmospheric ^{4}He in Venus t_{He-V} can be approximately estimated from the terrestrial life time t_{He-E} which is better constrained: taking $t_{He-E} \sim 10^6$yr, we have $t_{He-V} \sim [R_{Venus}(^4He)/R_{Venus}(^{40}Ar)] \times [R_{Earth}(^{40}Ar)/R_{Earth}(^4He)] \times t_{He-E} \sim [200 \,/\, 0.29] \times 10^6 \sim 7 \times 10^8$yr. Since this value, which is compatible with the previous estimate, is shorter than 4.6×10^9yr, the present atmospheric ^{4}He abundance in Venus can be explained by a radiogenic degassed component.

ACKNOWLEDGMENTS

We thank Y. Sano and K. Zahnle for discussions. Thoughtful review by D. Porcelli is greatly acknowledged. This work is partly supported by Grant-in-Aid of Ministry of Education (No. 05231103 and 05302026).

REFERENCES

1. H. E. Suess, J. Geol. 57, 600 (1949).
2. H. Brown, In *The Atmospheres of the Earth and Planets*. (ed. Kuiper G. P.) (University of Chicago Press, Chicago, 1952), pp.258.
3. M. A. Lange and T. J. Ahrens, Icarus 51, 96 (1982).
4. W. Rubey, Bull. Geol. Soc. Am. 62, 1111 (1951).
5. T. Staudacher and C. J. Allègre, Earth Planet. Sci. Lett. 60, 389 (1982).
6. Y. Hamano and M. Ozima, In *Terrestrial Rare Gases* (ed. E. C. Alexander Jr. and M. Ozima), (Japan Scientific Society Press, Tokyo, 1978), pp.155.
7. S. Sasaki, In *Noble Gas Geochemistry and Cosmochemistry* (ed. J. Matsuda), (Terra Sci. Publ., Tokyo, 1994), pp.55.
8. E. Tajika and T. Matsui, Geophys. Res. Lett. 20, 851 (1993).
9. R. Greeley, Science 236, 1653 (1987).
10. N. H. Sleep, J. Geophys. Res. 99, 5639 (1994).
11. R. O. Pepin, Icarus 92, 2 (1991).
12. R. O. Pepin and D. Phinney, (famous preprint) (1978).
13. R. O. Pepin, Icarus 111, 289 (1994).
14. K. J. Zahnle, J. Geophys. Res. 98, 10899 (1993).
15. G. Dreibus and H. Wänke, Icarus 71, 225 (1987).
16. D. L. Turcotte and G. Schubert, Icarus 74, 36 (1988).
17. T. A. Scambos and B. M. Jakosky, J. Geophys. Res. 95, 14779 (1990).
18. A. G. W. Cameron, Icarus 56, 195 (1983).
19. J. C. G. Walker, Icarus 68, 87 (1986).
20. H. J. Melosh and A. M. Vickery, Nature 338, 487 (1989).
21. D. S. Musselwhite, M. J. Drake, and T. D. Swindle, Nature 352, 697 (1991).
22. R. Greeley and B. D. Schneid, Science 254, 996 (1991).
23. G. Neukum and D. U. Wise, Science 194, 1381 (1976).
24. W. K. Hartmann et al., In *Basaltic Volcanism on the Terrestrial Planets*. (Pergamon, 1981), pp.1049.
25. K. L. Tanaka, J. Geophys. Res. Suppl. 91, E139 (1986).
26. D. McKenzie and M. J. Bickle, J. Petrol. 29, 713 (1988).
27. B. M. Jakosky et al., Icarus 111, 271 (1994).
28. E. Tajika and T. Matsui, In *Proc. 24th ISAS Lunar Planet. Symp.*, 211 (1991).
29. R. E. Grimm and S. C. Solomon, Geophys. Res. Lett. 14, 538 (1987).
30. A. O. Nier and M. B. McElroy, J. Geophys. Res. 82, 4341 (1977).
31. V. A. Krasnoplosky et al., Icarus 109, 337 (1994).
32. W. I. Axford, J. Geophys. Res. 73, 6855 (1968).
33. T. Abe et al., J. Geophys. Res. 98, 11191 (1993).
34. T. Torgersen, Chemical Geology 79, 1 (1989).
35. R. K. O'Nion and E. R. Oxburgh, Nature 306, 429 (1983).
36. L. H. Kellogg and G. J. Wasserburg, Earth Planet. Sci. Lett. 99, 276 (1990).
37. M. J. Prather and M. B. McElroy, Science 220, 410 (1983).

MERCURY'S ATMOSPHERE AND ITS RELATION TO THE SURFACE

Ann L. Sprague and Donald M. Hunten

Lunar and Planetary Laboratory, The University of Arizona,

Tucson, Arizona 85721

ABSTRACT

Mercury and the Moon have long been considered to be similar objects, but this view was based on limited information at visible wavelengths. It is now known that **real differences exist in the atmospheres and in the rock types** as deduced from visible, near-infrared, mid-infrared, and microwave observations.   At Mercury, Mariner 10 measured H, He, and perhaps O and a number of upper limits for other gases. Less than a decade ago, emissions of Na and K were observed from the ground, At the Moon Ar and perhaps Ne were observed from Apollo landed experiments in spite of a large local background.   An ultraviolet spectrometer (UVS) in orbit placed upper limits on a number of gases including H and O. Na and K were discovered from the ground to have abundances 2 orders of magnitude smaller than on Mercury. A variety of origins is likely for the different components. Impact of meteoroids could supply water, Na, and K, and could also vaporize surface material; Na, K and Ar could be degassed from the crust. H is probably from the solar wind, and He could reasonably come from either degassing or the solar wind.   A substantial loss process for all components is photoionization (or dissociation); ions on one side are swept away into the solar wind, and on the other are swept back to the surface. This process is complicated at Mercury by its permanent magnetic field, but there is little doubt that both loss and recycling still occur. Finally, the light atoms H and He have high enough velocities to escape directly, an effect that may be enhanced by the suprathermal velocity distributions that are likely to be present.   Here we discuss these phenomena and relate Mercury's atmosphere to its surface, especially surface composition.   We then provide a "wish list" of diagnostic observations like a search for atomic S and the OH radical whose presence would help understand the formation history of the planet.

INTRODUCTION

Mercury's atmosphere is not widely studied but it is of special interest because it is potentially diagnostic of its surface composition, the composition of interplanetary dust in the inner solar system, and the morphology of its magnetic field.   An indirect but potentially valuable diagnostic tool for determining surface composition is the identification and measurement of relative abundances of gases sputtered, vaporized, or outgassed from surface materials.   The actual gaseous inventory is probably a combination of surface and meteoritic material. The true surface component must be deduced by applying what is known about the interplanetary dust population, the energetics of impacts into planetary regoliths, and the results of surface remote sensing. Mercury's atmosphere is tenuous; the total mass of all known constituents is approximately 15 orders of magnitude less than Earth's. Ground-based searches for CO_2 resulted only in upper limits. Detection of the atmosphere with measurements of abundances awaited the results of two atmospheric experiments aboard the Mariner 10 spacecraft, which made three encounters with the planet during 1974 and 1975.   The two instruments designed for atmospheric measurements were an occultation experiment (OE), which measured four passbands in the UV, and a ten-passband UV airglow spectrometer (UVS)[1,2]. H, He, and possibly O were detected and upper limits were placed on Ar, Ne, Xe, N_2, H_2O, CO_2 and H_2. Spatial differences are difficult to determine from the ground because of poor spatial resolution, and the coverage by Mariner 10 [1,3] was spotty.   In 1985 and 1986 Na and K were discovered in Mercury's atmosphere [4,5]. Ca, Li, and Fe have been sought [6] at Mercury and Ca, Li, and Ti [7] at the Moon. The abundance of Na appears to be changeable, sometimes appearing in symmetric north-south spots at high latitudes [8] and K shows longitudinal differences [9,10]

Useful comparisons can be made to the Moon, and lunar atmospheric values and upper limits are shown along with those at Mercury in Table 1.   In addition solar

Table 1. ATMOSPHERIC DENSITIES FOR MERCURY, MOON, and EARTH, COMPARED WITH SOLAR VALUES

Substance	Solar $Si = 10^6$	Earth (90 km) cm^{-3}	Mercury cm^{-3}	Moon cm^{-3}	Lifetime (1 AU) 10^4 sec	Wavelength nm
H	-	1.0×10^9	200	<17	2000	122
He	-	-	6000	2000 - 40,000	1400	58
O	2.15×10^7	n.a.	<40,000	<500	250	130
Na	6.0×10^4	1000	20,000	70	5.4	589,590
K	4.2×10^3	67	500	16	3.7	766,770
Li	57	1	< 2	< 0.01	0.37	671
Ar	-	-	$< 3 \times 10^7$	4×10^4	200	105,107
Ca	-	85	< 247	< 6	1.8-2.9	423
Ti	-	-	-	< 2	-	504
S	5×10^7	-	-	-	-	182
OH	-	n.a.	-	-	5.6	306
H_2O	-	n.a.	-	-	88	122

abundances are given for comparison and the abundances of the same elements in the Earth's upper atmosphere. Lifetimes at Mercury are shorter than those at 1 AU by a factor of 4.6 (perihelion) and 10.5 (aphelion). Sources for all values can be found in Ref. 6 with the exception of upper limits for Li, Ca, and Ti at the Moon which can be found in Ref. 7 and Li at Mercury which is a new result of observations we are now preparing for publication. Where comparisons are not applicable, the symbol n.a. is used.

SOURCES, SINKS, TRANSPORT AND RECYCLING

Sources for different species are likely to vary and modeling has shown that several are plausible. At both Mercury and the Moon, H, some He and probably H_2 (which is undetected) come from the **solar wind**. Some He, Ar, and probably Na and K are **degassed**. The case for Na and K is controversial, very dependent on the diffusion coefficient and composition [9,11]. Diffusion coefficients are usually exponentially dependent on temperature and a small difference in temperatures expected at Mercury can make a large difference in flux of Na and K from surface materials. Diffusion fluxes become significant at about 400-500 K. Thus the depth at which diffusion occurs depends critically upon the penetration of the diurnal temperature oscillation or "thermal wave". One of two longitudes on Mercury is beneath the sun every perihelion. One of these passes through Caloris basin; the other is antipodal. These longitudes have been called the "hot longitudes" or, less descriptively, "hot poles". Here the thermal wave penetrates to a few meters depth and a steep gradient conducive to diffusion may be maintained. H_2O could also be degassing from Mercury. **Impact vaporization** of both impactor and surface materials probably supplies Na and K to the atmospheres at both Mercury [12] and the Moon [13]. Likely other constituents yet to be identified include Fe, Ca, O (a marginal detection was made for O by Mariner 10 at Mercury), and of particular interest, H_2O, OH, and S. **Sputtering** and **photo-desorption** are also viable source mechanisms but depletion of the surface layers must be considered [13].

Loss by photoionization and adsorption are certain at both Mercury and the Moon, but the degree of recycling is unknown in each case. Some of the recycling probably occurs by way of the magnetotail by auroral precipitation; the atoms would then diffuse back into the atmosphere shortly after dawn, an effect that seems to be observed for K [9]. Electric fields act on ions created by photionization of neutrals. Photoionization lifetimes at the Moon are on the order of order 10 hours for K and 15 hours for Na; at Mercury they are a factor of 6 shorter (Table 1). At the Moon, ions may be swept away or re-implanted by solar-wind electric field while at Mercury, with its substantial magnetic field, ions may go into the magnetotail either to escape or be redirected back into the midnight sector and re-implanted Some of the fresh ions are picked up by electric fields in the magnetosphere, and a fraction are expected to be accelerated to high energies and to impact the surface [8,9,14]. Impact zones are most likely the auroral zones and regions in the midnight sector where the magnetosheath intersects the surface [14]. The subsequent release of neutrals is highly dependent upon surface temperature; all of the known release processes (evaporation, desorption) have an Arrhenius-type exponential dependence on temperature [9,11,13]. Photodissociation is important for H_2O if it is present at Mercury. Table 1 includes the dissociation time for H_2O, about a day at 1 AU, 2×10^4 sec at Mercury. There could be some recycling of OH back to H_2O. Gravitational escape is important for lighter elements even at Mercury but the Jeans equation cannot apply because there are many suprathermal atoms and the high-velocity tail of the thermal distribution must be truncated. Chemical recombination of O atoms, H atoms, OH radicals, and other constituents is also a likely loss mechanism but only the vaguest of models can be put forward until many more data are available.

Two other phenomena of interest are **transport** and **recycling**. The identified atomic species interact with the surface of Mercury at extremely low pressures (a few femto-bar), providing an opportunity to study gas-surface interactions in a space-based, ultra-high vacuum environment. Atoms move around the surface in a series of ballistic hops – a kind of 2-dimensional diffusion. Na, K, and Ar can condense or adsorb on the night side as might other species yet to be identified. These atoms also evaporate at dawn, with some time delay owing to shadowing or differing adsorption strengths [9,13]. At the Moon at least two velocity components have been seen in the Na and K atmosphere relatively close to the surface [15,16,17]. One (with a 400 K scale height or < 100 km) can be fitted by a temperature close to the surface temperature and a second is at a much higher temperature. A model of competing release mechanisms of thermal and photo-desorption can reproduce observed inventories in these two components at the Moon [16]. Sputtering and photodesorption may have a role. Imaging of the entire Na atmosphere [18] shows the effects of radiation acceleration. An extended corona is seen on the anti-sunward side while the sunside Na atmosphere remains gravitationally bound and displays a relationship with solar zenith angle. The small size of Mercury in the sky and lack of adequate spatial resolution make it impossible to determine if the same multi-component atmosphere exists for Na and K but it was observed for H by Mariner 10 [19]. A major difference is that for H at Mercury, the cold component is at nightside temperatures and a hotter component is at the local subsolar temperature, quite unlike the distributions found [15,16,17] for the Moon.

ATMOSPHERE--SURFACE RELATIONSHIPS

The atmosphere may be diagnostic of surface composition. If Mercury formed in a highly reducing environment, its bulk composition may be similar to that of the enstatite achondrites, rich in Mg-bearing, high-temperature condensates with small but significant residual amounts of Na, K, and Cr sulfides. An anomalously high S abundance in Mercury's atmosphere would thus be an important indicator of the formation history of Mercury. A theoretical calculation of the index of refraction of Mercury's average surface material based on whole disk phase function gives a result higher than that of the Moon [20]. A surface rich in sulfides could explain the derived index of refraction. Alternatively, a S abundance below nominal solar abundance would be a strong indication that Mercury formed in a oxidizing environment similar to that of the Earth and Moon. In this case differentiation following accretion would have partitioned Fe, S, and some O into the core. The remaining, relatively light, silicate crust could be largely composed of feldspars, minerals known to be the low-temperature condensates formed from the residual melt following fractional crystallization and solidification of a magma ocean. Such a process has been shown to have occurred on the Moon[21].

By measuring and interpreting the abundance and distribution of K, Na, and Ca in the context of expected sputtering and volatilization models and our spectrographic measurements, it may be possible to determine the composition of residual feldspar units and differentiate among petrologic types [*e.g.* orthoclase ((Na,K)Si$_3$O$_8$), albite (NaAlSi$_3$O$_8$), labradorite (plagioclase: (Ab$_{50}$-An$_{50}$-Ab$_{30}$An$_{70}$), and anorthite (CaAl$_2$Si$_2$O$_8$)]. A ground-based upper limit for Ca[6] shows that either Ca is released less effectively than Na and K from the regolith or its abundance is below chondritic at Mercury. A null detection of Ca would be an important indicator of the absence of anorthite (Ca-rich plagioclase) on Mercury's surface, and would point toward feldspathoids (Al-rich alkali silicates). It is desirable to detect other atmospheric constituents to better understand sources and surface composition. Particular interest is attached to Li, Ca, Fe, and O for outgassing, diffusion, and oxidation bonding processes.

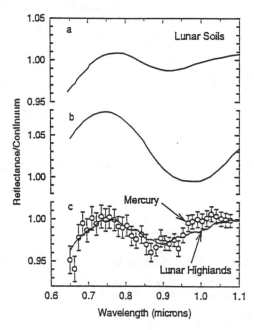

Fig. 1. Laboratory spectra of two lunar soils [22] (a) 62231, noritic, low-Ca orthopyroxene; (b) 12070, basaltic, high-Ca pyroxene; compared to (c) a telescopic spectrum from Mercury [23,24] and lunar highlands soil [23]. The figure shows the shallow FeO absorption in Mercury's spectrum, if it is there at all. Most lunar highlands locations are depleted in FeO relative to maria that typically have absorptions about as deep as those shown in (a) or, not as often, (b).

Besides the indirect measurements of Mercury's surface, **some direct measurements of Mercury's surface materials and properties have been made** despite the observing difficulties. Regolith polarization differences from one surface region to another have been observed [25]. Regolith microwave opacity at Mercury is lower than that on the moon [26], near-infrared measurements point to the apparent absence of FeO in the regolith [23,24,27] and mid-infrared spectroscopic observations show feldspathic rock types at two equatorial locations [28,29]. In contrast, the lunar regolith has up to 8% FeO in the highlands and even higher in the late stage basaltic flows that form some maria [20]. One major spectral difference can be seen in Figure 1. It shows the spectral region in the visible and near-IR spectrograph with the diagnostic olivine and pyroxene features of two lunar soils along with a Mercury spectrum plotted over that of the lunar highlands. Such observations of Mercury can only be made with great difficulty from ground-based observatories. These observations all point to a surface unlike that of the Moon.

FUTURE STUDIES

Several atmospheric constituents have emission lines in the UV, and either cannot be observed or are difficult to observe from the ground (S, 181.6 nm; Fe, 372 and 386 nm; Ca, 422.7 nm). The presence or absence of these gases is really a test of our understanding of the formation of tenuous atmospheres and a tool to determine true source mechanisms from a variety that have been postulated. Their relationship to surface composition has been detailed in the section above, particularly for S, Na, K, and Ca. A particular interest is attached to S and OH. The sulfur resonance triplet at 180.7, 182, and 182.6 nm can be observed only from space, and the 306 nm band of OH can only be observed from the ground with great difficulty. Two considerations are important for making an emission detectable: it must be **bright enough**, and it must have **adequate contrast**: it must stand out above the continuous solar spectrum reflected from the surface. The contrast improves with better spectral resolution, which spreads the continuum out while putting all the flux from a narrow line into a single resolution element. The search for faint atmospheric UV emissions benefits from the darkness of Mercury's surface.

With good signal-to-noise ratio, it is possible to detect emission from a column of 3 $\times$ 10^{11} cm^{-2} OH near 306 nm. This is about twice the amount expected from a continuing supply of H_2O from impacting chondritic material. A search for this band, in the hope of finding evidence for the principal photolysis product (OH) from the icy polar regions on Mercury [30,31,32], is of primary scientific importance. From the brightness of this feature the abundance and degree of exposure of water ice sequestered at Mercury's poles can be inferred. The growth or shrinkage of the polar caps does not depend on the loss rate by photodissociation, but only on the steady-state number density of water vapor and the temperature of the deposit. Meteoritic source estimates have been made: 2.4×10^{24} sec^{-1} [31] and 2×10^{24} sec^{-1} [30]. The number density should then be 2×10^4 cm^{-3} for a global-mean lifetime of 4×10^4 sec. This would condense at any temperature lower than about 115 K and the ice layer would grow at 160 cm per billion years. Degassing could provide a larger source by a factor of 1000 or greater [30]. Using a lifetime for OH [33] of 5.6×10^4 sec and a cometary source rate [34] of 2×10^{-16} g cm^{-2} s^{-1}, we calculate the expected OH abundance at the Moon as 10^9 to 4×10^{11} cm^{-2}.

An ultraviolet spectrograph and telescope in orbit about the Earth, Sun or Mercury could help us understand the formation and condition of Mercury. In particular, it should be equipped with adequate spatial resolution and sensitivity for both the atmospheric measurements described above and for the surface measurements described below. Direct measurement of surface material abundances of magnesium in pyroxene, albite in plagioclase, and ferrous iron in Mercury's surface materials would greatly enhance our understanding of the relationship of Mercury's surface to its atmosphere. Laboratory and petrologic studies of terrestrial rock and meteorite samples in the ultraviolet [35] have shown that reflectance peaks can differentiate between the petrologic type and thus give composition and formation history. Using this spectral region for surface compositional studies is desirable because the near-infrared region so successfully used in mapping the Moon is severely hampered by Mercury's thermal flux which equals the reflected component at wavelengths longer than about 1.7 μm. It is highly desirable to map the extent and chemical composition of feldspar units by studying the spectral region from 135 to 320 nm.   Such spectra can be compared to those from terrestrial analogs measured in the laboratory. These diagnostic features are illustrated in Fig. 2.

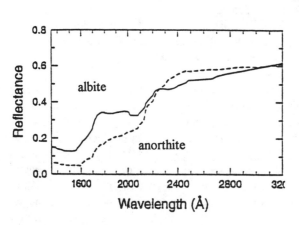

Fig. 2. UV reflectance of two different types of plagioclase feldspar. Albite is the Na-rich end-member and might contribute to the observed Na atmosphere at Mercury. Anorthite is the Ca-rich endmember.   The lunar highlands are anorthositic, yet the upper limit for Ca in the atmosphere is low [7] (see Table 1). At Mercury, the Ca atmospheric upper limit is also low [6] but the type of feldspars present on the surface are unknown.

DISCUSSION

Our growing knowledge of the atmospheres and surfaces strongly suggests that there are substantial differences between the Moon and Mercury. We can study the spatial distribution of a few lunar constituents from the Earth, but the resolution at Mercury is painfully limited. Many emissions of great interest can only be observed from space, which also offers the potential of much better spatial resolution.   Such measurements of Mercury's atmosphere with good time coverage could study magnetospheric effects on atmospheric constituents.   It might be possible to determine the footprint locations of the auroral zones and the magnetosheath on the midnight sector using Na as a tracer and mapping the "burn-off" of neutrals following energetic ion implantation.

Acknowledgments: The manuscript was improved by helpful suggestions from David Mitchell and Faith Vilas.

REFERENCES

1. A.L. Broadfoot, S. Kumar, M.J.S. Belton, and M.B. McElroy, Science **185**, 166 (1974).

2. S. Kumar, Icarus **28**, 579 (1976).

3. A.L. Broadfoot, D. E. Shemansky, and S. Kumar, Geophys.Res.Lett. **3**, 577 (1976).

4. A.E. Potter and T. H. Morgan, Science **229**, 651 (1985).

5. A.E. Potter and T. H. Morgan, Icarus **67**, 336 (1986).

6. A.L. Sprague, R.W.H. Kozlowski, D.M. Hunten, and F.A. Grosse, Icarus **104**, 33 (1993).

7. B.C. Flynn and S.A. Stern, Geophys. Res. Lett., in press (1995).

8. A.E. Potter and T. H. Morgan, Science **248**, 835 (1990).

9. A.L. Sprague, J. Geophys. Res. **97**, 18,257 (1992); correction, J. Geophys. Res. **98**, 1231 (1993).

10. A.L. Sprague, R.W. H. Kozlowski, and D. M. Hunten, Science **249**, 1140 (1990).

11. R. M. Killen and T. H. Morgan, Icarus **101**, 293 (1993).

12. T.H. Morgan, H.A. Zook, and A.E. Potter, Icarus **74**, 156 (1988).

13. D.M. Hunten, T.H. Morgan, and D. Shemansky, in *Mercury*, ed. F.Vilas, C.R. Chapman, and M.S. Matthews (U. of Arizona Press, Tucson, 1988), p. 562.

14. W.H. Ip, Astrophys. J. **418**, 451 (1993).

15. R.W.H. Kozlowski, A.L. Sprague and D.M. Hunten, GRL, **17**, 2253-2256, (1990).

16. A.L. Sprague, R.W.H. Kozlowski, D.M. Hunten, W.K. Wells and F.A. Grosse, Icarus **96**, 2 (1992).

17. S.A. Stern and B.C. Flynn, Astron. J. **109**, 835 (1995).

18. B. Flynn and M. Mendillo, Science **261**, 184 (1993).

19. D.E. Shemansky and A.L. Broadfoot, Rev. Geophys. Space Phys. **15**, 491 (1977).

20. B. Hapke, LPSC XXV Abstracts (1994).

21. S.R.Taylor,   Planetary Science: A Lunar Perspective, (Lunar and Planetary Institute, Houston, 1982).

22. C. Pieters, Lunar Reflectance Spectroscopy (Cambridge University Press, 1993).

23. F. Vilas, M.A. Leake and W. W. Mendell, Icarus **59**, 60 (1984).

24. T.B. McCord and R. N. Clark, J. Geophys. Res. **84**, 7664 (1979).

25. T. Gehrels, R. Landau, and G. V. Coyne, Icarus **71**, 386 (1987).

26. D. Mitchell, Ph.D. thesis (Univ. of Calif., Berkeley, 1993).

27. F. Vilas, Icarus **64**, 133 (1985).

28. A.L. Sprague, R. W. H. Kozlowski, F. C. Witteborn, D. P. Cruikshank and D. Wooden, Icarus **109**, 156 (1994).

29. A.L. Tyler (Sprague), R.W.H. Kozlowski and L.A. Lebofsky, Geophys. Res.Lett. **15**, 808 (1988).

30. B. Butler, D. Muhleman and M. Slade, J. Geophys. Res **98**, 15,003 (1993).

31. J.K. Harmon, M. A. Slade, R.A. Velez, A. Crespo, M. J. Dryer, and J. M. Johnson, Nature **369**, 213 (1994).

32. M. Slade, B. Butler and D. Muhleman, Science **258**, 635 (1992).

33. T.I. Gombosi, A.F. Nagy and T.E. Cravens, Rev. Geophys. **24**, 667 (1986).

34. W.H. Ip and J.A. Fernandez, Icarus **74**, 47 (1988).

35. J.K. Wagner, B. W. Hapke, and E. N. Wells, Icarus **69**, 14 (1987).

3. EXPERIMENTAL STUDIES

STRUCTURE AND BONDING IN HYDROUS MINERALS AT HIGH PRESSURE: RAMAN SPECTROSCOPY OF ALKALINE EARTH HYDROXIDES

T. S. Duffy, R. J. Hemley, and H. K. Mao

Geophysical Laboratory and Center for High-Pressure Research, Carnegie Institution of Washington, 5251 Broad Branch Rd, NW, Washington, DC 20015

ABSTRACT

The high-pressure behavior of the hydrous minerals brucite and portlandite was examined by Raman spectroscopy in a diamond anvil cell. Both the hydrogenated and deuterated forms of brucite [$Mg(OH)_2$, $Mg(OD)_2$] show evidence for structural changes that are first detectable at 4-5 GPa, and these materials remain crystalline to pressures upwards of 30 GPa. Portlandite [$Ca(OH)_2$] becomes amorphous when compressed above 12 GPa. It is proposed that both the structural changes in brucite and amorphization of portlandite are driven by a destabilization of the H substructure as the materials are compressed. In brucite, the hydrogen and deuterium atoms adopt new positions which stabilize the structure to very high pressure. In portlandite, disordering of the hydrogens triggers destabilization of the oxygen substructure, leading to pressure-induced amorphization.

INTRODUCTION

The existence and distribution of volatile species in terrestrial planetary interiors may strongly influence some of the most important physical properties of these bodies, including their viscosity, elasticity, and mineralogy. Hydrogen is particularly significant in this regard, and the possible existence of deep hydrogen reservoirs in the Earth has generated considerable recent interest[1-3]. Little is known, however, about the crystal chemistry and stability of hydrogen-bearing minerals at high pressures and temperatures. In this study, we examine the effects of changes in pressure and chemistry on the behavior of hydrous minerals.

The alkaline earth hydroxides are among the simplest hydrogen-bearing materials, and therefore are useful as prototypes for understanding the role of hydrogen in crystal structures at high pressure. The isostructural hydroxides, brucite, $Mg(OH)_2$, and portlandite, $Ca(OH)_2$, have been the subject of extensive experimental investigation[4-14]. The structural and chemical simplicity of these materials makes them far more amenable to study than the more complex hydrous silicates that have been identified thus far[3]. At the same time, they posses features in common with the more complex phases, including similarities in topology and bonding environments. The crystal structure of brucite and portlandite is shown in Fig. 1.

X-ray diffraction studies[4-7] of these materials have revealed considerable differences in their high-pressure behavior: brucite remains crystalline to 78 GPa at 600 K[4], while portlandite becomes x-ray amorphous when compressed above 12 GPa at room temperature[6] and transforms to a new crystalline structure at pressures as low as 5.6 GPa and 773 K[7]. There are also differences in the infrared spectra of the two materials at high pressure and ambient temperature that are consistent with

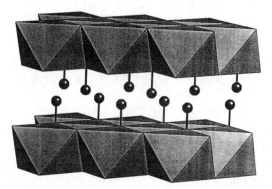

Fig. 1. Crystal structure of brucite and portlandite. Mg/Ca atoms lie at centers of octahedra. Spheres represent hydrogen/deuterium atoms. The hydroxyl bonds are along the c axis.

the x-ray results[8]. Neutron diffraction studies of normal and deuterated brucite to 10 GPa provide evidence for H and D disorder at high pressure[9-10] . We have recently reported measurements of the Raman spectra of brucite to 37 GPa, which document a previously undetected phase transition[11]. In this study, we combine this data with new measurements on portlandite and deuterated brucite to probe how chemical changes affect high-pressure behavior within this system.

EXPERIMENTAL TECHNIQUE

Deuterated brucite was supplied by K. Leinenweber from the same material used in their high-pressure neutron diffraction study[9]. Portlandite samples (>95% purity) were purchased commercially (Aldrich Chemical Co.). The latter were contaminated with a small amount of $Ca(CO)_3$ due to reaction with CO_2 in the atmosphere. Fine-grained samples were loaded into a Mao-Bell diamond anvil cell having 600-μm culets. Raman spectra were excited using an Ar^+ laser operated at 488.0 or 514.5 nm and were recorded using a Dilor XY spectrometer equipped with a CCD detector. Experiments on $Mg(OD)_2$ and $Ca(OH)_2$ were conducted under non-hydrostatic conditions (no pressure medium). Spectra of $Mg(OH)_2$ were recorded both non-hydrostatically and under quasi-hydrostatic conditions using neon as a pressure transmitting medium. The Raman spectra are not strongly affected by degree of hydrostaticity[11]. Pressures were determined from the shift of the fluorescence wavelength of small ruby crystals distributed through the sample volume. Further experimental details can be found in Ref. 11.

RESULTS

Factor group analysis indicates there are four Raman-active modes for brucite and portlandite. Three of these are low-frequency lattice vibrations (external modes) and one is a symmetric OH-stretch vibration (internal mode). Ambient-pressure Raman spectra for $Mg(OH)_2$, $Mg(OD)_2$, and $Ca(OH)_2$ in the low-frequency region are shown in Figure 2a. The ambient-pressure OH-stretch vibrations of $Mg(OH)_2$ and $Ca(OH)_2$ are shown in Figure 2b. For each material, all the expected modes are observed, and agreement with previous data[12] is excellent.

For the translational modes, $E_g(T)$ and $A_{1g}(T)$, there are only small shifts in frequency between the protonated and deuterated forms of brucite. The $E_g(T)$ mode of portlandite is of very low intensity relative to brucite. The librational

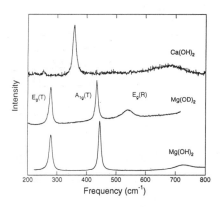
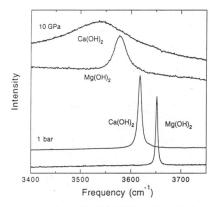

Fig. 2. (a) Ambient-pressure external Raman modes of Ca(OH)$_2$, Mg(OD)$_2$, and Mg(OH)$_2$. E_g (T) and A_{1g} (T) are translational vibrations of the OH or OD groups perpendicular and parallel to the c axis, respectively. The broad E_g (R) mode is a librational motion of the OH or OD unit. (b) OH-stretch modes of brucite and portlandite at ambient pressure and at 10 GPa under non-hydrostatic compression.

modes, E_g(R), are broad and weak for all three materials. In general, the linewidths of all vibrational modes of Ca(OH)$_2$ are about a factor of two larger than those in either form of brucite. This may be consistent with evidence from neutron diffraction for dynamic H disorder in Ca(OH)$_2$ at ambient pressure[13] or it may reflect differences in grain size.

Figure 3 shows the behavior of the external and internal vibrational modes of deuterated brucite under non-hydrostatic compression to pressures of 30 GPa and 24 GPa, respectively. In general, the Mg(OD)$_2$ results are very similar to earlier Mg(OH)$_2$ data[11]. At 5 GPa, a new line is first detectable between the E_g and A_{1g} modes (Fig. 3a). This line subsequently grows at the expense of the E_g mode over a broad pressure interval, with the intensity ratio of the peaks varying approximately exponentially with pressure. The two peaks approach each other and reach a minimum in frequency separation at about 24 GPa (Fig. 4a). These features are characteristic of a Fermi resonance in which two modes of the same symmetry are coupled by an interaction term in the Hamiltonian. The development of the resonance in Mg(OD)$_2$ is similar to the resonance previously observed in Mg(OH)$_2$[11]. The lack of a significant isotope effect is not surprising as the translational modes are not strongly affected by substitution of D for H. The coupling coefficient, estimated by one-half the minimum frequency separation, is ~ 11 cm^{-1} for Mg(OD)$_2$, compared to ~ 9 cm^{-1} for Mg(OH)$_2$.

The high-pressure behavior of the internal mode in deuterated brucite is also similar to Mg(OH)$_2$ (Fig. 3b). In contrast to the external modes, the OH-stretch and OD-stretch vibrations decrease in frequency with increasing pressure. Such behavior is normally associated with hydrogen bonding[8]. A high-frequency sideband on the primary OD stretch is first detected at 5 GPa, consistent with previous detection of a sideband on the OH mode of Mg(OH)$_2$[11] at 4 GPa. Between 14

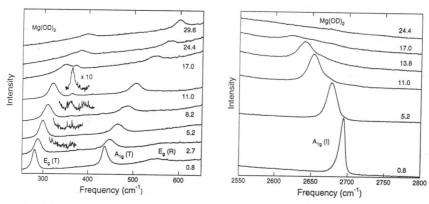

Fig. 3. (a) External modes of Mg(OD)$_2$ under non-hydrostatic compression The pressure (in GPa) is shown next to each trace. An expanded view of the high-frequency side of the E_g (T) mode is shown between 2.7 and 11.0 GPa. (b) OD-stretch vibration of deuterated brucite under non-hydrostatic compression to 24 GPa. A high-frequency sideband is visible at pressures of 5.2 GPa and above.

and 17 GPa, the OD stretch vibration weakens and broadens considerably, and the peak cannot be detected above this pressure. The weakening and disappearance of the symmetric stretch is also observed in Mg(OH)$_2$, but at higher pressures (20-28 GPa).

In general, the substitution of D for H has only minor effects on the high-pressure behavior of the vibrational modes of brucite. New Raman lines are detected in deuterated brucite around 5 GPa, near the pressure (4 GPa) at which similar features were detected in Mg(OH)$_2$. Both materials exhibit a Fermi resonance over a comparable pressure range. The pressure dependence of the vibrational frequencies of Mg(OH)$_2$ and Mg(OD)$_2$ are shown in Figure 4 and Table I.

The vibration modes of Ca(OH)$_2$ were measured to 14 GPa under non-hydrostatic conditions (Fig. 4). No evidence for a crystalline phase transition was found over this range. The E_g(T) mode is difficult to detect in this material, but the E_g(R) mode can be observed at elevated pressures. As with brucite, the OH-vibration decreases in frequency with pressure, but this decrease becomes non-linear above about 5 GPa. The width of the OH-stretching vibration begins to increase strongly above this pressure as well (Fig. 5). At 10.4 GPa, the linewidth of the OH-stretch vibration in portlandite is almost an order of magnitude larger than in brucite (Fig. 2b). Between 10.4 and 12.5 GPa, the amplitude of the OH-stretch vibration decreases dramatically (Fig. 6). Over this same range, the only observable lattice vibration (A_{1g}) similarly becomes weak and broad. This is the same pressure range over which portlandite becomes x-ray amorphous[6], and comparable peak broadening is observed for the infrared-active OH-stretch vibration[8] at these pressures. Upon decompression from 14 GPa, the Raman peaks remain very weak and broad until near 2 GPa, at which point they abruptly recover.

DISCUSSION

The new Raman data for brucite and portlandite complement other recent data[4-14] for these materials and allows for a better understanding of their structural response to compression. In the case of brucite, substitution of D for H produces only relatively minor changes in ambient-pressure structural parameters (Table II). The Raman spectra of $Mg(OH)_2$ and $Mg(OD)_2$ are also similar at high pressures (Figs. 3 and 4), with new peaks appearing in the spectrum of both materials at 4-5 GPa. Due to their weak initial intensity, determination of the pressure at which these peaks first appear is difficult and limited by spectral quality. At ambient pressure, there is no evidence for additional peaks in high-quality spectra for both $Mg(OH)_2$ and $Mg(OD)_2$ (Fig. 2). Therefore, we believe that observed features are due to a phase transition at high pressure rather than due to intensity enhancement of peaks present at ambient pressure.

Table I. Pressure dependence of Raman modes.

Mode	Obs. P range (GPa)	$\nu = A + BP + CP^2$			γ_{i0}[a]
		A (cm^{-1})	B (cm^{-1}/GPa)	C (cm^{-1}/GPa^2)	
		$Mg(OH)_2$			
$E_g(T)$	0-37	280.0	5.40	-0.08	0.81
	4-37	359.6	0.60	0.02	
$A_{1g}(T)$	0-37	444.7	6.93	-0.08	0.65
$E_g(R)$	0	727.5			
$A_{1g}(I)$	0-28	3652.0	-7.68		-0.09
	4-20	3661.3	-5.34		
		$Mg(OD)_2$			
$E_g(T)$	0-30	278.5	4.26	-0.04	0.64
	5-30	346.2	1.30	0.01	
$A_{1g}(T)$	0-30	434.8	6.48	-0.03	0.63
$E_g(R)$	0-3	537.5	11.69		0.91
$A_{1g}(I)$	0-17	2694.4	-4.01		-0.06
	5-17	2709.9	-3.25		
		$Ca(OH)_2$			
$E_g(T)$	0-9	253.5	5.68	0.07	0.85
$A_{1g}(T)$	0-12	356.8	8.78	-0.25	0.94
$E_g(R)$	0-12	671.5	21.34	-1.15	1.21
$A_{1g}(I)$	0-12	3617.5	-3.68	-0.41	-0.04

[a] Ambient pressure mode Grüneisen parameters computed using a bulk modulus of 42 GPa for brucite and 38 GPa for portlandite.

Neutron diffraction data[9] for $Mg(OD)_2$ are consistent with a structural change at 5.4 GPa. Improved fits to the neutron data are obtained at this pressure by allowing the D atoms to be split over three off-axis sites. Such a model is consistent with the Raman spectra for $Mg(OD)_2$, which were obtained on the same

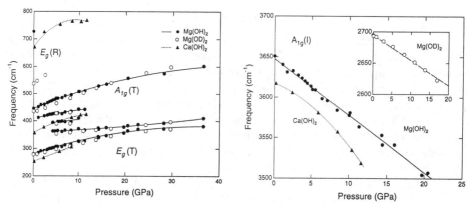

Fig. 4. (a) Pressure-dependence of Raman frequencies for Ca(OH)$_2$, Mg(OH)$_2$, and Mg(OD)$_2$. Lines show fits to data for Ca(OH)$_2$ and Mg(OH)$_2$, but fits are omitted for Mg(OD)$_2$ for clarity. (b) Pressure-dependence of OH/OD-stretch vibrational frequencies.

Fig. 5. Band width of the OH-stretch vibrations under non-hydrostatic compression. FWHM is full width at half maximum.

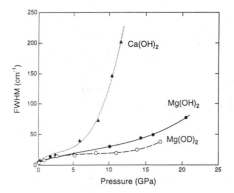

sample material as used in the neutron study. The neutron data[10] for Mg(OH)$_2$ is also consistent with H disordering, but only at a higher pressure of 10.9 GPa. Neutron studies of the two materials also yield different results for the pressure dependence of the O-H and O-D bond lengths. For Mg(OD)$_2$, a slight lengthening of the O-D distance with pressure is reported, whereas for Mg(OH)$_2$ the O-H bond length remains nearly constant or contracts slightly with pressure. One possible interpretation of the neutron results is that disordering occurs more easily and interlayer bonding becomes stronger with pressure in the deuterated form of brucite. However, our Raman measurements indicate that the two materials behave similarly, and that structural changes occur at nearly the same pressure in both forms of brucite. Hence, it is probable that the apparent differences between the two neutron studies arise from differences in experimental technique or data analysis, rather than from intrinsic differences in the high-pressure response of Mg(OH)$_2$ and Mg(OD)$_2$.

All three materials show strong negative frequency shifts (-4 to -8 cm^{-1}/GPa)

Fig. 6. OH-stretch vibration in portlandite between 10.4 and 13.9 GPa. The strong reduction in peak intensity over a narrow pressure interval is evident. Pressures are listed in GPa next to each trace.

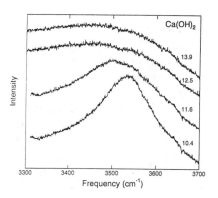

of the Raman-active OH-stretch with pressure (Fig. 4b). The underlying cause of this red-shift of the internal mode frequencies with pressure is unclear from currently available data. It is generally assumed that such behavior arises from decreasing force constants and concomitant lengthening of the O-H/O-D bond as a consequence of increased hydrogen bonding[8]. However, it is important to note that a decreasing force constant does not require bond expansion[15] (i.e., the length could remain unchanged or could even compress).

The ratio of the Raman-active stretch frequencies in $Mg(OH)_2$ and $Mg(OD)_2$ is 1.355 at ambient pressure, and decreases weakly with pressure to 1.341 at 18 GPa. The expected ratio (for a purely harmonic interaction) given by the ratio of reduced masses is 1.376. This indicates there is some anharmonicity which increases slightly with pressure. The lower than expected frequency ratio and its decrease with pressure has been observed in some molecular crystals[16].

A comparison of the behavior of brucite and portlandite at high pressure may yield insights into the process of pressure-induced amorphization. There are a number of differences between our results for $Ca(OH)_2$ and earlier high-pressure Raman data for this material[14], which were obtained with a less sensitive detector (diode array). We have detected all the Raman active modes of portlandite at ambient pressure (Table I) and our results are in good agreement with earlier work[12] that assigned modes on the basis of polarized spectra and deuterated samples. We therefore assign the strong mode at 356.8 cm^{-1} to be the A_{1g} (T) mode rather than the E_g (T) mode as was done previously[14]. Ref. 14 also reports the appearance of new modes near 3650 cm^{-1} at 11 GPa, and near 750 cm^{-1} at 13 GPa. We find no evidence for a new mode near 3650 cm^{-1} up to 14 GPa (Fig. 6). Furthermore, the mode near 750 cm^{-1} may be the E_g (R) mode that we detect near this frequency up to 11.6 GPa (Fig. 4a). Thus, in contrast to Ref. 14, we find no evidence for the appearance of new Raman lines in this material at high pressure. Above 11.6 GPa, we find evidence for only very broad and weak Raman peaks in portlandite, consistent with infrared data[8]. The pressure dependences of the A_{1g} internal and external mode frequencies determined in this study are the same as those found previously[14].

The high-pressure response of $Ca(OH)_2$ can be divided into three regions on the basis of Raman measurements. At low pressures (to 5 GPa) portlandite behaves in a manner analogous to $Mg(OH)_2$ over than same pressure range. From 5-12 GPa, qualitative changes in the Raman spectra are observed. Specifically, the OH-stretch frequency in $Ca(OH)_2$ begins to decrease non-linearly with pressure (Fig. 4b) and its band width increases significantly relative to the band width of the brucite peak (Fig. 5). At about 12 GPa, all portlandite peaks abruptly weaken and merge with the background (Fig. 6). Both IR and x-ray diffraction peaks disappear at a similar pressure, implying that disordering occurs at both spectroscopic and x-ray length scales. It has been suggested that pressure-induced amorphization in portlandite is due to a kinetically inhibited phase transition. A phase transition in $Ca(OH)_2$ has been observed at 5.6 GPa and 773 K[7]. While pressure-induced amorphization occurs at much higher pressure than this, the Raman data provide evidence that structural changes begin to occur near this pressure.

Table II. Ambient-pressure structural parameters.

	$Mg(OH)_2$[a]	$Mg(OD)_2$[b]	$Ca(OH)_2$[c]
a (Å)	3.14979(4)	3.1455(1)	3.5918(3)
c (Å)	4.7702(1)	4.7646(3)	4.9063(7)
c/a	1.51445(4)	1.5147(1)	1.3659(2)
V (Å^3)	40.986(1)	40.831(4)	54.816(12)
ρ (g/cm^3)	2.362	2.453	2.244
z_0	0.2203(3)	0.2218(3)	0.2341(3)
z_H/z_D	0.4130(6)	0.4183(2)	0.4248(6)
Mg-O/Ca-O (Å)	2.1003(6)	2.1012(6)	2.371(1)
O-H/O-D (Å)	0.958(3)	0.956	0.984(4)
H...H'/D...D' (Å)	1.999(2)	1.960	2.202(2)
O...O' (Å)	3.229(2)	3.2133(1)	3.333(2)

[a]Ref. 10, [b]Ref. 17, [c]Ref. 18 (see also Ref. 13).

The three hydroxide compositions examined here provide evidence for a range of high-pressure behavior even within this simple structure. Recently, pressure-induced amorphization has been observed at 11.2 GPa by IR spectroscopy in the isomorphous transition metal hydroxide, $Co(OH)_2$, with evidence for structural changes continuing in the amorphous state to 36 GPa[19]. In brucite, modifications to the H substructure appear to be sufficient to allow the oxygen substructure to remain intact to very high pressures. In portlandite, on the other hand, complete disordering of both the O and H substructure occur at similar pressures. This collapse is preceded by a pressure range over which there is evidence for precursory disordering within the H substructure alone. This disordering begins near the pressure where the H substructure of brucite rearranges itself and where a crystalline phase transition has been observed in portlandite at elevated temperature.

In summary, Raman spectroscopic data for portlandite, and both hydrogenated and deuterated forms of brucite can be interpreted with a simple model that describes the differences in their high-pressure behavior. At low pressures, the H atoms begin to be displaced from their equilibrium positions in these materials due to increased interlayer forces as the structure is compressed. In the case of brucite, the H atoms adopt new positions, possibly in a three-site disordered arrangement, which stabilizes the structure to high pressures. In portlandite, broadening of the Raman peaks between 5 and 12 GPa indicates that the H atoms become increasingly disordered over this range, leading to a range of bond distances, orientations, and strengths. The larger intralayer O-O distances in $Ca(OH)_2$ may prevent the H atoms from achieving stable positions. The disorder in the H substructure triggers disordering of the O sublattice at about 12 GPa. This leads to destabilization of the structure as a whole and relatively extensive amorphization over a narrow pressure interval in this material.

Despite being among the simplest hydrogen-bearing minerals, the alkaline earth hydroxides exhibit complex and varied responses to the application of pressure. Characterization of this behavior requires the use of a variety of in situ experimental probes (spectroscopy, neutron and x-ray diffraction). Such detailed analyses are necessary to achieve an understanding of the crystal chemistry of hydrogen-bearing minerals under Earth mantle conditions.

ACKNOWLEDGEMENTS

K. Leinenweber provided samples of $Mg(OD)_2$. We benefited from discussions and correspondence with Q. Williams, J. Parise, K. Leinenweber, M. Catti, and C. Meade. R. Downs is thanked for technical assistance. This work was supported by the NSF.

REFERENCES

1. A. B. Thompson, Nature **358**, 295 (1992).
2. M. Kanzaki, Phys. Earth Planet. Int. **66**, 307 (1991).
3. L. W. Finger and C. T. Prewitt, Geophys. Res. Lett. **16**, 1395 (1989).
4. Y. Fei and H. K. Mao, J. Geophys. Res. **98**, 11875 (1993).
5. T. S. Duffy, J. Shu, H. K. Mao, and R. J. Hemley, Phys. Chem. Minerals, in press.
6. C. Meade and R. Jeanloz, Geophys. Res. Lett. **17**, 1157 (1990).
7. K. Leinenweber, *Natl. Synchrotron Light Source Activity Rep.* (Brookhaven National Laboratory, Upton, NY, 1993) p. 128; M. Kunz, D. J. Weidner, J. B. Parise, M. Vaughan, and Y. Wang, Eos Trans. AGU, Fall Meeting Suppl., 661 (1994).
8. M. B. Kruger, Q. Williams, and R. Jeanloz, J. Chem. Phys. **91**, 5910 (1989).
9. J. B. Parise, K. Leinenweber, D. J. Weidner, K. Tan, and R. B. Von Dreele, Am. Mineral. **79**, 193 (1994).
10. M. Catti, G. Ferraris, S. Hull, and A. Pavese, Phys. Chem. Minerals, in press.
11. T. S. Duffy, C. Meade, Y. Fei, R. J. Hemley, and H. K. Mao, Am. Mineral. **80**, 222 (1995).
12. P. Dawson, C. D. Hadfield, and G. R. Wilkinson, J. Phys. Chem. Solids **34**, 1217 (1973).
13. L. Desgranges, D. Grebille, G. Calvarin, G. Chevrier, N. Floquet, and J. C. Niepce, Acta Cryst. **B49**, 812 (1993).

14. C. Meade, R. Jeanloz, and R. J. Hemley, in *High-Pressure Research: Applications to Earth and Planetary Science* (Terra Scientific, Tokyo, 1992), p. 485.
15. R. J. Nelmes et al., Phys. Rev. Lett. **71**, 1192 (1993).
16. H. Shimizu, Phys. Rev. **B32**, 4120 (1985).
17. D. E. Partin, M. O'Keefe, and R. B. Von Dreele, J. Appl. Cryst. **27**, 581 (1994).
18. W. R. Busing and H. A. Levy, J. Chem. Phys. **26**, 563 (1957).
19. J. H. Nguyen, M. B. Kruger, and R. Jeanloz, Phys. Rev. B **49**, 3734 (1994).

THERMODYNAMIC STABILITY OF HYDROUS SILICATES: SOME OBSERVATIONS AND IMPLICATIONS FOR WATER IN THE EARTH, VENUS AND MARS

Alexandra Navrotsky and Kunal Bose
Center for High Pressure Research and
Department of Geological and Geophysical Sciences,
Princeton University
Princeton, NJ 08544

INTRODUCTION

Recent high pressure, high temperature experiments in the mantle analog system $MgO-SiO_2-H_2O$ have identified a suite of dense hydrous magnesium silicates (DHMS; Figure 1) that could be conduits to transport water to at least the 660 km discontinuity in the Earth via mature, relatively cold, subducting slabs [1, 2, 3]. In addition, nominally anhydrous silicates, especially wadsleyite, β-Mg_2SiO_4, appear to retain significant amounts of water under mantle conditions [4, 5, 6]. Water released from sequential dehydration of these phases, or melting and/or phase transitions among them, could be responsible for generation of deep focus earthquakes, mantle metasomatism, and other physico-chemical processes in the deep interior of the Earth and other terrestrial planets. It is also plausible that the released water initiates a cycle of hydration - dehydration events in the mantle wedge proximal to a subducting slab; such reactions could affect liquidus temperature, viscosity, and diffusivity of the mantle peridotite.

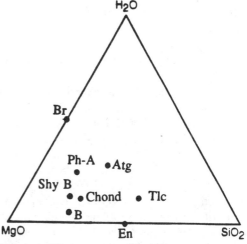

Figure 1: Ternary plot showing location of relevant DHMS phases in the $MgO-SiO_2-H_2O$ system. The locations of brucite, talc and antigorite are included for reference. Abbreviations: Tlc=talc; Atg=antigorite; Br=brucite; Ph-A=phase A; En=enstatite; B=phase B; Shy B=superhydrous B; Chond=chondrodite.

The extent to which such transport of volatiles to, or even beyond, the transition zone actually does occur depends on three factors:

1. The thermodynamic properties of the DHMS and other water-containing phases stable above approximately 70 kbar. What are the details of the P-T stability fields for these phases, both in simple model systems and in multicomponent mantle compositions?

2. The stability fields of "lower" pressure, "lower" temperature hydrous phases such as talc and antigorite. Can they bring water to the depths within the stability fields of the DHMS phases (> ~ 70 kbar), or are they completely dehydrated early in subduction and the source of water all but eliminated at shallow depths?

3. The temperature profiles of subducting slabs. This is receiving considerable attention from geophysicists and geodynamicists at present [7, 8] and is critical in determining the depth to which the hydrous phases can transport water to the mantle. As mineral physicists, we wish to address the first two points and summarize recent work on phase stabilities and thermodynamic data for both low pressure and high pressure phases in the *model* system $MgO-SiO_2-H_2O$, applicable to the hydrated portion of the ultramafic oceanic lithosphere. We also wish to speculate how the very closely balanced energetics of these hydrous phases might provide a mechanism by which rather small differences in thermal gradients (at different times or on different planets) could result in large differences in volatile distribution which might in turn affect the entire style of subduction, convection and evolution of the terrestrial planets.

RECENT THERMODYNAMIC STUDIES

Phase equilibrium experiments on the stability of talc to 50 kbar [9, 10] have established that the equilibrium dehydration boundary (talc = enstatite + quartz/coesite + H_2O) lies ~ 150 °C higher than that calculated from existing data bases [11, 12], the source of discrepancy arising from incorrect assumptions in the data bases about the compressibility of talc. Further, the temperature maximum of the talc dehydration reaction is at ~ 28 kbar, 830 °C instead of at 10 kbar, 800 °C. The experimental data on talc have been combined with thermodynamic analyses of reactions involving talc and antigorite in the $MgO-SiO_2-H_2O$ system to construct a phase diagram (Figure 2) that allows evaluation of the possibility that volatile transport by hydrated phases occurs in subduction zones below the depth (~ 125 km) of generation of arc magmas, to ~ 70 kbar, 700 °C, which marks the lower limit of stability of phase A.

The recent work of Luth [13] on the reaction: phase A + enstatite = forsterite + H_2O places additional topological constraints on the phase diagram. Since the thermodynamic properties of enstatite, forsterite and vapor are reasonably well constrained [14, 15] at these conditions, we have evaluated the free energy of phase A between 70 and 95 kbar from the reversal brackets and used this to calculate the positions of the other four univariant reactions emanating from the quartz, talc absent invariant point. There is uncertainty in exact slopes and/or locations of computed antigorite-bearing equilibria, arising from lack of experimental data (phase equilibria or thermo-elastic properties) of antigorite at these conditions. Additional components, notably substitution of Fe for Mg and dilution of H_2O by other fluids, mainly CO_2, may alter the exact location of the equilibria in P-T space. Since there are no experimental data on partitioning behavior of Fe between the hydrous phases considered, (Phase A, talc, antigorite) and the anhydrous phases participating in the reactions, it is not possible to make any predictions on the extent of the displacement

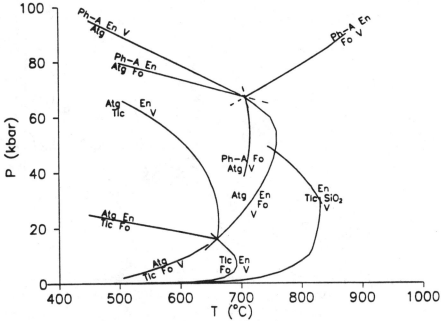

Figure 2: Phase equilibria in the system MgO-SiO$_2$-H$_2$O. The reactions Talc = enstatite + SiO$_2$ + H$_2$0 and phase A + enstatite = forsterite + H$_2$0 have been determined experimentally [9, 13]. The other equilibria shown are calculated. Abbreviations as in *Figure 1*.

of the equilibria due to iron or other minor components. The effect of diluting the vapor phase by 30% CO$_2$ depresses the talc dehydration equilibrium by ~ 60 °C. In spite of these limitations, an interesting feature that arises from a Schreinemakers' analysis [16] in this system is that phase A should be stable at P < 70 kbar and there should be an additional reaction that forms phase A. The high pressure analogue of talc, namely the 10 Å phase, could be a possible candidate, but there is controversy as to its stability [17, 18]. Despite these uncertainties, the phase diagram is instructive in identifying a sequence of key reactions that map out a possible pathway for transporting water to the mantle from near-surface conditions to the onset of the stability of DHMS phases. This pathway in a subducting slab relies on thermodynamically stable phases; thus it does not require assumptions about kinetics of dehydration reactions.

Figure 3 shows the key equilibria in this system along with computed leading-edge temperature profiles of old (50 Ma) and young (5 Ma) subducting oceanic plates[8] descending at velocities ranging from 10 to 3 cm/year with an average subduction angle of 27°. Since there is excess forsterite over talc in the ultramafic section of the oceanic lithosphere, talc that forms at the surface will not be carried beyond ~ 25 kbar pressure unless talc is in the sedimentary pile riding atop the subducted slab and is carried to greater depths as suggested by the ^{10}Be/^{9}Be ratio of mineral separates of arc magmas [19, 20]. In this scenario, talc will be stable to ~ 50 kbar, till the reaction talc = enstatite + coesite + H$_2$O is encountered. In the slab, antigorite

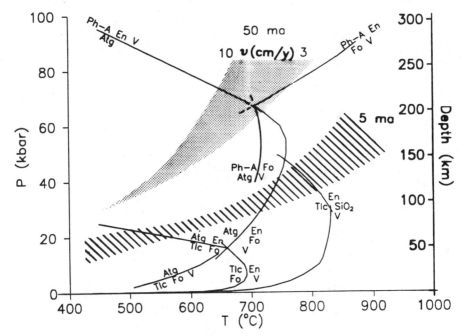

Figure 3: Calculated leading edge temperatures of subducting slab and stability ranges of key equilibria in the system MgO-SiO$_2$-H$_2$O. The shaded band represents the computed temperature ranges[8] for a mature (~ 50 ma old), "cold" slab, plate velocity (υ) of 10 (left extreme) and 3 (right extreme) cm/yr. The diagonally hatched region is for a young, (~ 5 ma old), "hot" slab, with same ranges of velocities. The subduction angle is 27° for both scenarios. Note the major dehydration events corresponding to ~ 70-80 kbar (≈220-240 km depth) for a cold subducting slab. Note also that a number of P-T paths within the "cold" slab maintain hydrous phases continuously from the surface into the stability fields of DHMS.

would form through the vapor conserved reaction talc + forsterite = antigorite + enstatite and would become the dominant hydrous phase. Accordingly, the reaction antigorite + talc = enstatite + H$_2$O is unlikely to take place, and has been left out of Figure 3. The reaction phase phase A + enstatite = antigorite + forsterite, a vapor conserved reaction, is of no consequence for the present purposes.

It is clear that for a "hot" slab, the major dehydration events take place between 20 and 45 kbar corresponding to 60 to 125 km depth. This probably represents the major source of volatile release necessary for island arc magmatism and no hydrous phases are likely to persist beyond this depth in such a thermal regime. In a colder slab however, antigorite can be stable ~ 70 kbar, reacting to form hydrous phase A accompanied by H$_2$O release. This reaction releases some, but not all the H$_2$O present in the antigorite. Provided temperatures remain cold enough, Phase A could at greater depth form the hydrous phase E through the reaction phase A + enstatite + H$_2$O = phase E. Phase E is considered stable to at least 150 kbar, 1400 °C[3] and could potentially carry water to the 440 km discontinuity before it gives rise to the phase superhydrous B which could be stable to considerably greater depths and higher temperatures. It is noteworthy that the antigorite dehydration event corresponds to a depth of ~ 220-250 km, coinciding with the 220 km or Lehmann discontinuity [21]. Could this mark the base of a seismic low velocity zone? Moreover, the antigorite breakdown reaction is characterized by a negative dP/dT slope -- this volume decrease may serve as a crack-healing or anti-crack mechanism. In contrast, the dehydration of phase A at similar conditions is marked by a positive volume change and could serve as a potential deep focus earthquake trigger mechanism. The closely balanced energetics of these two equilibria illustrate how small changes in P-T conditions can give rise to drastically different mineralogical and geophysical scenarios.

The equilibria involving DHMS phases are difficult to determine with satisfactory reversals. In addition, pressure and temperature uncertainties in multi-anvil apparatus at 70-150 kbar, 700-1400 °C are substantial. To better constrain the equilibria shown semi-quantitatively in Figures 2 and 3, we have embarked on a program of measuring the enthalpies of formation of DHMS phases by high temperature oxide melt drop-solution calorimetry. Preliminary data are shown in Table 1 (all data reported in units of kJ/mol).

TABLE I

Enthalpy (kJ/mol)

	$\Delta H_{ds}{}^a$	$\Delta H^f_{elements}$	ΔH^f_{oxides}	$\Delta H_{rxn}{}^{25(b)}$
Phase A[1]	655.6	-6935	-49.9	71.2
B[2]	792	-11,170	-23.9	15.2
Shy-B[3]	718	-9328	-10.9	67.2
Chond[4]	471	-5206	-93.3	-53.2
Talc[5]	554	-5897	-164.7	-125.6

(a): drop solution (b): formation at 25 °C, 1 bar from MgO, SiO_2, and $Mg(OH)_2$.
[1]Phase A: $Mg_7Si_2O_8(OH)_6$; [2]Phase B: $Mg_{12}Si_4O_{19}(OH)_2$; [3]Superhydrous B: $Mg_{10}Si_3O_{14}(OH)_4$; [4]Chondrodite: $Mg_5Si_2O_8(OH)_2$; [5]Talc: $Mg_3Si_4O_{10}(OH)_2$

The second column is the enthalpy of drop solution [22, 23] in molten $PbO.B_2O_3$ at 771°C. Columns 3 and 4 are the data recast in the more conventional, standard form of enthalpies of formation from the oxides and elements at room temperature respectively, obtained by applying the appropriate thermodynamic cycles. The last column is the heat of formation of the phases at ambient conditions from silica, periclase and brucite, indicating that the phases A, B, and superhydrous B are thermodynamically unstable (positive $\Delta H_{rxn}{}^{1, 25}$) at 1 bar, 25°C and confirming that the DHMS are indeed thermodynamically stable in the P-T range in which they are synthesized, thus putting to rest any nagging concern that they may form upon temperature "quench". The excellent agreement of the enthalpy of formation of talc from the elements as obtained in our laboratory with literature value[12] of -5897.4 kJ/mol illustrate the reliability of our calorimetric procedure involving hydrous magnesium silicates. A full thermodynamic analysis is in progress; it is already clear that thermophysical parameters, such as heat capacity, thermal expansion, and compressibility, play a major role in affecting these energetically very closely balanced phase equilibria.

IMPLICATIONS FOR PLANETARY EVOLUTION

The experimental data on stability of hydrous magnesium silicates, thermodynamic calculations, and modelling of thermal structures of subducting slabs presented above indicate that, given favorable temperature distribution with depth, hydrous phases could be stable to the lower mantle in subducting slabs. Recent tomographic images [24, 25, 26] display significant low velocity anomalies almost always proximal to a subducting slab at ~ 220-250 km, 400 km and 650 km depths. Although anisotropy, temperature perturbations, and variations in mineral chemistry may be invoked to account for the negative anomalies, the contribution of volatiles to anelasticity, and partial melting of the hydrated mantle are likely to be the primary causes for this signature. It is tempting to speculate that the "double humped" nature of the low velocity zone at 200-250 km, where two regions of low velocity are separated by a faster region, may represent the two reactions mentioned above, namely antigorite = phase A + enstatite + H_2O and phase A + enstatite = forsterite + H_2O. Furthermore, if a portion of the water does indeed survive to the 670 km discontinuity, then the subduction of a particularly large, wet, cold slab may bring substantial H_2O to the region of slab "pileup" atop the 670 km discontinuity. The effect of H_2O on viscosity or its chemical effects on phase relations could trigger the suggested turnover or avalanche that may periodically mix upper and lower mantle [27, 28]. If the colder parts of the transition zone are wet, overstepping phase transitions appears less likely. Thus water release may trigger the olivine-spinel transition, and may bear an indirect as well as direct relation to deep focus earthquakes. However, to understand whether water plays a pervasive, global role in mantle chemistry and tectonics, the amount of water that could be subducted and subsequently released, and more critically, the lateral extent to which it migrates (and on what time scales), once liberated from the slab, need to be better understood.

Geoid modeling and topographic considerations [29] indicate that neither Venus nor Mars, have Earth-like low viscosity zones in their upper mantle. In addition to being influenced by the heat budget, convection is dependent on viscosity variation with depth which is in turn linked to the volatile inventory. The high surface temperature on present day Venus (~450 °C) essentially precludes common hydrous minerals at the low surface pressure (< 100 bar). The higher surface temperature imposes a hotter "geotherm", and even a "cool" subducting slab would be hotter on Venus than on Earth. These two features make it unlikely that any water can be subducted under present conditions. We speculate that a buildup in surface temperature, occurring on a time scale dictated by the accumulation of the current dense CO_2 atmosphere, perhaps over millions of years, could have resulted in a cessation of subduction of hydrous phases into the Venusian mantle. This would have led to an increase in viscosity, hindering and eventually preventing convective recycling. The changeover from Earth-like to a present day Mars-like tectonic style could have been concomitant with the cessation of the ability of Venus to "fold in" water to its mantle, some 600 my ago [30]. The evidence from SNC meteorites implying a drier, later stage Martian mantle [31] may indicate that with continued outgassing and less effective exchange between surficial and interior water, the plate tectonic cycle could have ceased on Mars approximately 1.3 Gyr ago (if it ever existed).

We thus conclude that it is worth considering whether the three different evolutionary styles of Venus, Earth and Mars, in addition to being influenced by that seminal event, the Moon forming impact on Earth, may derive their current differences from different patterns of water circulation into their mantles. Such significant differences can be triggered by rather small changes in pressure and temperature conditions because of the large number of closely balanced equilibria of hydrous phases. We further conclude that for the present Earth, a significant amount of water can be recycled at least to the 670 km discontinuity, though we are as yet unable to quantify exactly what we mean by the term "significant".

ACKNOWLEDGEMENT

This work was supported by the Center for High Pressure Research, an NSF Science and Technology Center. We thank Jibamitra Ganguly, Guust Nolet and W. Jason Morgan for valuable discussions.

REFERENCES

1. L. Liu, Phys. Earth Planet. Inter., 49, 142, (1987).
2. A. B. Thompson, Nature, 358, 295, (1992).
3. T. Gasparik, J. Geophys. Res., 98, 4287, (1993).
4. T. Inoue, Phys. Earth Planet. Int., 85, 237, (1994)
5. T. E. Young, H.W. Green II, A. M. Hofmeister and D. Walker, Phys. Chem. Min., 19, 409, (1993).
6. J. R. Smyth, Am. Min., 79, 1021, (1994).
7. J. H. Davies and D.J. Stevenson, J. Geophys. Res., 97, 2037, (1992).
8. S. M. Peacock, T. Rushmer and A. B. Thompson, Earth Planet. Sc. Lett., 121, 227, (1994).
9. K. Bose, Ph.D. Dissert., Univ. of Arizona, (1993).
10. K. Bose and J. Ganguly, Geol. Soc. Amer. Abst. w. Prog. 25, 213, (1993).
11. R. G. Berman, Jour. Petrol., 29, 445, (1988).
12. T. J. B. Holland and R. Powell, Jour. Metamor. Geol., 8, 89, (1990).
13. R. W. Luth, accepted for publ. in Geochim. Cosmochim. Acta, (1995).
14. S. K. Saxena, N. Chatterjee, Y. Fei and G. Shen, Thermodynamic Data on Oxides and Silicates, (Springer-Verlag, Heidelberg, 1993) 426 pp..
15. A. Belonoshko, P. F. Shi and S. K. Saxena, Computers and Geosciences, 18, 1267, (1992).
16. E-An Zen, Geol. Soc. Amer. Bull. 1225, 56 pp. (1966).
17. B. Wunder and W. Schreyer, Jour. Petrol., 33, 877, (1992).
18. K. Yamamoto and S. Akimoto, Am. J. Sci., 277, 288, (1977).
19. F. Tera, Geochim. Cosmochim. Acta, 50, 1535, (1986).
20. J. Morris and F. Tera, Geochim. Cosmochim. Acta, 53, 3197, (1989).

21.   I. Lehmann, Geophys. J. Roy. Astron. Soc., 4, 124, (1961).
22.   A. Navrotsky, Phys. Chem. Min., 2, 89, (1977).
23.   S. Circone and A. Navrotsky, Am. Min., 77, 1191, (1992).
24.   R. Van der Hilst, R. Engdahl, W. Spakman and G. Nolet, Nature, 353, 37, (1991).
25.   G. Nolet and A. Zielhuis, J. Geophys. Res., 99, 15813, (1994).
26.   G. Nolet and R. Van der Hilst, subm. for publ. to Science, (1994).
27.   P. J. Tackley, D. J. Stevenson, G. A. Glatzmaier and G. Schubert, Nature, 361, 699, (1993).
28.   P. Machetel and P. Weber, Nature, 350, 55, (1991).
29.   W. S. Kiefer in: Conference on Deep Earth and Planetary Volatiles, LPI Contr. 845, 25, (1994).
30.   W. M. Kaula, Science, 247, 1191, (1990).
31.   G. Dreibus and H. Wänke, Icarus, 71, 225, (1987).

STABILITY OF HYDROUS MINERALS
IN H$_2$O-SATURATED KLB-1 PERIDOTITE UP TO 15 GPA

Tatsuhiko Kawamoto
Department of Geology, Arizona State University, Tempe, AZ 85287

Kurt Leinenweber
Department of Chemistry, Arizona State University, Tempe, AZ 85287

Richard L. Hervig
Center for Solid State Science, Arizona State University, Tempe, AZ 85287

John R. Holloway
Departments of Geology and Chemistry, Arizona State University, Tempe, AZ 85287

ABSTRACT

Stability fields of Ti-chondrodite, Ti-clinohumite, phase A, and Al$_2$O$_3$-bearing phase E were determined in H$_2$O-saturated KLB-1 peridotite at 6 to 15 GPa. Phase E contains 1.5 - 9 wt. % Al$_2$O$_3$ and was found from 800°C, 9 GPa to 1400 °C, 15 GPa. When a downdragged hydrous peridotite follows a hotter P-T path than one passing through 800 °C at 4 GPa, serpentine breaks down, followed by talc, chlorite and pargasite. Beyond the pargasite-out reaction the downdragged peridotite will be almost free of H$_2$O bound in crystals except for a small amount in phlogopite and then K-richterite. This means that the hydrous peridotite should encounter "a choke point" restricting the passage of H$_2$O to greater depth. In subducting basalt, lawsonite is the phase most resistant to pressure induced dehydration. Therefore, dense hydrous magnesium silicates in the downdragged flow of the base of the mantle wedge could absorb H$_2$O from dehydrating lawsonite and become H$_2$O carriers in a very cold subduction zone ($\leq$ 800°C, 9 GPa). However, the scarcity of magmatism above subducting slabs deeper than 250 km suggests that downgoing peridotite and basalt dehydrate at 3-7 GPa to form volcanic arcs.

A calibration line for the secondary ion mass spectrometer determination of H$^+$/^{30}Si$^+$ ion ratios in phase A, chondrodite, and clinohumite has been determined. Its coefficient of H$^+$/^{30}Si$^+$ ion ratios against wt. % H$_2$O concentration is 0.12, which is lower than the 0.21 found for hydrous basaltic glasses and the 0.28 for hydrous rhyolitic glasses.

INTRODUCTION

H$_2$O plays an important role in melting temperatures, melting relations, bulk modulus, and rheological features of terrestrial rocks. Silicate melts can dissolve more than 10 wt. % H$_2$O at pressure higher than 1 GPa [1,2]. The H$_2$O content of a rock depends on the modal abundance of hydrous minerals. Since the pioneering work by Ringwood and Major [3] and Sclar et al.[4], several hydrous phases have been produced in high pressure and temperature experiments and proposed as H$_2$O reservoirs in the mantle: the alphabetical phases, clinohumite, and chondrodite (see Liu [5] and Thompson [6] for comprehensive reviews). These dense hydrous magnesium silicates do not need trace elements like K$_2$O to be stable. Because alphabetical phases were only experimentally produced in the MgO-SiO$_2$- H$_2$O system except for diamond-anvil pressure cell experiments on serpentine [7], it remains to address the stability fields of these hydrous phases in a peridotitic system. Among the dense hydrous magnesium silicates (Figure 1), phase A and phase E seem more likely to exist in the peridotitic upper mantle than the others, because (1) phase A has been shown to coexist with enstatite in the MgO-SiO$_2$- H$_2$O system [8], and (2) phase E can

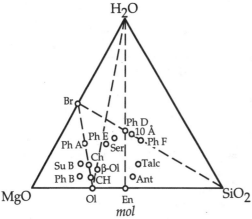

Figure 1, Dense hydrous magnesium silicates in the MgO-SiO$_2$- H$_2$O system [6, 9, 26, 28]. Ol, olivine; β-Ol, β-olivine; En, enstatite; Ph E, phase E; Ph A, phase A; CH, clinohumite; Ch, chondrodite; Br, brucite; Su B, superhydrous phase B; Ph B, phase B; Sep, serpentine; Ph D, phase D; 10 Å, 10 Å phase; Ph F, phase F; Ant, antigorite.

be slightly less ferromagnesian than olivine[9,10], so it should be stable in a peridotitic system. Since most of the other dense hydrous magnesium silicates are more ferromagnesian than olivine, Ringwood [11, 12] suggested that chondrodite, and clinohumite could be stabilized in subducting lithosphere if an SiO$_2$ component was leached by a fluid phase generated through the dehydration of some hydrous mineral [13]. Although it has been shown that clinohumite and chondrodite do not coexist with enstatite in the MgO-SiO$_2$-H$_2$O system [8], recently Kanzaki reported that phase E decomposed into clinohumite, enstatite, and H$_2$O at higher temperature in the Mg$_2$SiO$_4$-H$_2$O system[9]. Therefore, it is important to determine whether these hydrous phases can coexist with pyroxene or not in the multi-component peridotite system.

We report here the results of a series of high pressure and temperature experiments in an H$_2$O-saturated peridotite system and also report a calibration line for the secondary ion mass spectrometer determination of H$^+$/^{30}Si$^+$ ion ratios in phase A, chondrodite, and clinohumite. Finally we will discuss plausible pressure-temperature conditions for hydrous peridotite during subduction of the lithosphere.

EXPERIMENTAL PROCEDURES

As a starting material, we used a gel of KLB-1 peridotite composition [14] in which magnesium was added as brucite (KLB-1-14, 14 wt. % H$_2$O, Table 1). Takahashi [14] suggested that the KLB-1 composition is similar to Ringwood's [11] Pyrolite III (Table 1). We prepared an Mg-absent gel following a method similar to Hamilton and Henderson [15] and heated it using a gas-mixing furnace at 800 °C and 1 atmospheric pressure under the Ni/NiO oxygen buffer for 24 hours. The chemical composition of the gel was determined by Philips PW-1480 XRF analysis following Goto and Tatsumi [16] at the Geological Institute, University of Tokyo. Although the abundance of carbon of the gel was supposed to be negligible, trace amounts of (Mg, Fe)CO$_3$ were found in a few low temperature experiments.

For additional experiments, we used two kinds of mixtures of reagent grade brucite, iron wustite, metallic iron, and SiO$_2$ for the H$_2$O-saturated MgO-FeO-SiO$_2$ system (Mix 1 and Mix 3, Table 1). We found our synthesis results to be consistent with the phase relations of experiments in the FeO-free system [8]. Also we used a mixture of reagent grade brucite, iron wustite, metallic iron, Al$_2$O$_3$ and SiO$_2$ for a 4.4 wt. % Al$_2$O$_3$-bearing phase E composition (Phase E, Table 1).

We used platinum, silver, and gold sample containers. For experiments using platinum capsules in KLB-1 and phase E systems, we added 75 wt. % relative of the initial iron to KLB-1-14 gel and phase E composition as a mixture of iron wustite and metallic iron in order to compensate the iron loss to platinum capsules, to which 40 wt. % relative of the

Table 1 Chemical compositions of Pyrolite III, KLB-1 peridotite, and starting materials

	Pyrolite III	KLB1	KLB-1-14	Mix1	Mix3	Phase E
SiO_2	46.1	44.8	44.7	37.2	58.0	39.7
TiO_2	0.2	0.2	0.2	0.0	0.0	0.0
Al_2O_3	4.3	3.6	3.4	0.0	0.0	4.9
FeO*	8.2	8.2	8.6	10.4	6.9	9.5
MgO	37.6	39.5	39.0	52.4	35.0	45.9
CaO	3.1	3.5	4.0	0.0	0.0	0.0
Na_2O	0.4	0.3	0.3	0.0	0.0	0.0
K_2O	0.0	0.0	0.0	0.0	0.0	0.0
Total	100.0	100.0	100.0	100.0	100.0	100.0
H_2O			13.7	11.5	8.0	12.8

initial FeO was lost in the H_2O-saturated $FeO-MgO-SiO_2$ system after 24 hours at 875-1100 °C and 9.3 GPa.

All experiments were performed in a Walker type multi-anvil apparatus [17] at Arizona State University. The truncation of anvils was 8 mm for 6.3 - 10.3 GPa, and 4 mm for 14.5 - 15.5 GPa. The octahedral pressure medium is cast MgO ceramic (Ceramacast 584). We used two types of cell assembly design, which were described by (1) Pawley [18] and (2) Leinenweber and Parise [19]. The uncertainties for pressures using the 8 mm and 4 mm octahedra are estimated at ± 0.5 and 1 GPa, respectively. We usually put two capsules in a single run using 8 mm assemblies, and temperature was measured with $Pt/Pt_{90}Rh_{10}$ thermocouple at one of the lowest temperature points. For cell assembly (1), a single wrap of Inconel 600 metal was used as a furnace at temperatures $\leq 1000°C$ while Re with $LaCrO_3$ outer sleeve was used at higher temperatures. There is a 150 °C thermal gradient along the samples [18]. In this configuration, a difference of temperatures between the ends of those two capsules was measured using two thermocouples and demonstrated to be less than 50 °C by Kenneth Domanik of Arizona State University (personal communication, 1994). For cell assembly (2), box-shaped graphite was used for the furnace, and the thermal gradient along sample is believed to be smaller. In experiments ≤ 700 °C, the samples were preheated at 1000 °C at the desired pressure for 2 hours prior to lowering to the desired temperature.

After the runs using an 8 mm assembly, H_2O saturation was checked following the method of Luth[20]. Chemical compositions of phases in polished samples were determined with JEOL JXA 8600 electron microprobe using a focused beam. Determination of H concentration was carried out with a secondary ion mass spectrometer (Cameca IMS 3f) at the Center of Solid State Science, Arizona State University, following the analytical procedures of Hervig and Williams [21] and Ihinger et al. [22]. The ion beam diameter was ~20 μm. β-olivine was identified using Raman spectra taken by Andrzej Grzechnik and Paul McMillan at the Department of Chemistry following the similar method described by McMillan and Akaogi [23].

RESULTS

6.3 GPa. No hydrous minerals were found at 550 and 700 °C at 6.3 GPa in the H_2O-saturated peridotite system, although there was phase A and chondrodite in the $MgO-FeO-SiO_2-H_2O$ system (Mix 3 composition) at identical conditions. A trace amount of (Mg, Fe)CO_3 was found at both experimental conditions. Magnesite was identified based on electron microprobe analysis including O but C.

(a)

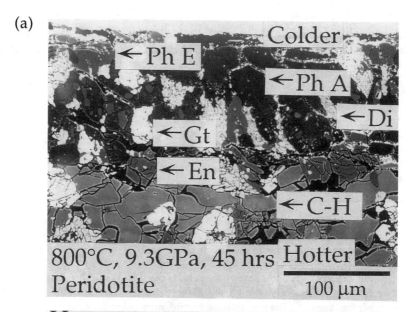

(b)

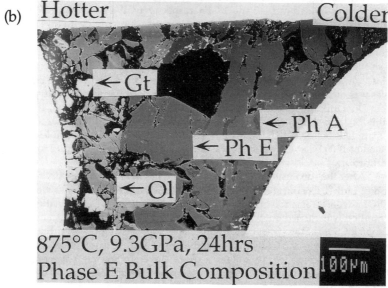

Figure 2, (a) Back scattered electron image of run # 14 at 800 °C and 9.3 GPa. The 800 °C is measured at the cold end, and 950 °C at the hot end. In the colder part of the capsule gray-colored phase E is in contact with dark colored phase A. They are in contact with bright garnet and bright diopside. Phase A is in contact with enstatite and clinohumite in the hotter part (Abbreviations same as in Figure 1).   (b) Back scattered electron image of run # 16 at 875 °C and 9.3 GPa, in a system of H_2O-saturated phase E bulk composition. Phase E is found with phase A in the colder part, and phase E contacts olivine and, olivine contacts garnet in the hotter part.

Table 2 Experimental conditions and mineral assemblages

Run#	Starting material	Furnace	P (GPa)	T (°C)	t (hour)	Capsule	Mineral phases
1	KLB-1-14-FeO	Inc	6.3	1000/550a	2/31	Pt	Ol, En, Di, Gt, Mgs
2	Ph E' FeO	Inc	6.3	1000/550b	2/31	Pt	CH, Gt, Ol
3	KLB-1-14-FeO	Inc	6.3	1000/700a	2/41	Pt	Ol, En, Di, Gt, Mgs
4	Mix 3	Inc	6.3	700b	41	Pt	Ph A, Ch
5	KLB-1-14	Gr	7.7	1000/550	2/37	Ag	CH, Ch, Ol, En, Di, Gt, Mgs
6	KLB-1-14	Gr	7.7	1000/700	2/24	Ag	Ol, En, Di, Gt, Mgs
7	Ph E' FeO	Inc	9.3	750	50	Pt	Ph A, Gt
8	Mix 1	Inc	9.3	800	22	Pt	Ph A, En
9	KLB-1-14	Gr	9.3	800	62	Ag	Ph E, CH, Ol, Gt, En, Di
10	KLB-1-14-FeO	Inc	9.3	875	25	Pt	Ol, En, Di, Gt
11	Mix 1	Inc	9.3	950	3	Pt	Ol, CH
12	Mix 1	Inc	9.3	1100	3	Pt	Ol
13	KLB-1-14	Gr	9.3	1000/800	2/49.5	Ag	Ph E, CH, Gt, En, Di
14	KLB-1-14-FeO	Inc	9.3	800a	45	Pt	Ph E, Ph A, CH, Gt, En, Di
15	KLB-1-14-FeO	Inc	9.3	875a	24	Pt	Ol, En, Di, Gt
16	Ph E' FeO	Inc	9.3	875b	24	Pt	Ph E, Ph A, Gt, Ol
17	KLB-1-14	Inc	9.3	950a	22	Pt	Ol, En, Di, Gt
18	Mix 3	Inc	9.3	950b	22	Pt	Ch, CH
19	KLB-1-14	Gr	10.3	800	42	Ag	Ph A, Ph E, CH, Gt, En, Di, Mgs
20	KLB-1-14	Gr	10.3	900	42	Ag	Ph E, CH, Ol, Gt, En, Di
21	KLB-1-14	Gr	10.3	1000	42	Ag	Ol, En, Di, Gt
22	KLB-1-14-FeO	Inc	14.5	1000	0.2	Ag	Ph E, Gt, En, Di
23	KLB-1-14	Re	14.5	1200	5	Ag	Ph E, Gt, En, Di
24	KLB-1-14	Re	15.5	1400	1	Au	Ph E, β-Ol, Gt, En
25	KLB-1-14	Re	15.5	1500	0.33	Au	β-Ol, Gt, En

FeO, 75 wt. % FeO added. Inc, Inconel 600; Gr, graphite; Re, rhenium; a, above capsule; b, bottom capsule. Ol, olivine; β-Ol, β-olivine; En, enstatite; Di, diopside; Gt, garnet; Ph A, phase A; Ph E, phase E; CH, clinohumite; Ch, chondrodite; Mgs, magnesite.

Table 3 Representative chemical compositions of minerals

Run #		SiO$_2$	TiO$_2$	Al$_2$O$_2$	MgO	FeO*	CaO	Na$_2$O	total	Mg value	M/Si
4	Ch	33.60	0.00	0.00	53.79	6.54	0.00	0.00	93.96	0.94	2.55
	Ph A	25.01	0.00	0.00	57.92	3.21	0.00	0.00	86.16	0.97	3.56
5	En	57.28	0.02	0.26	34.43	5.95	0.12	0.00	98.06	0.91	0.98
	Gt	45.55	0.46	15.64	19.43	8.27	8.58	0.05	97.98	0.81	0.79
	Di	55.25	0.04	0.57	18.55	3.28	20.15	0.21	98.05	0.91	0.94
	Ol	40.42	0.06	0.01	49.30	9.59	0.04	0.00	99.41	0.90	2.02
	Ch	33.30	6.78	0.06	45.50	9.87	0.03	0.00	95.54	0.89	2.44
	CH	36.67	4.41	0.05	47.51	9.70	0.04	0.01	98.38	0.90	2.24
7	Ph A	24.39	0.00	0.16	50.32	13.07	0.00	0.00	87.94	0.87	3.52
	Gt	42.66	0.00	21.99	20.54	15.52	0.10	0.00	100.81	0.70	1.02
11	CH	37.42	0.00	0.00	52.19	7.34	0.00	0.00	96.95	0.93	2.24
	Ol	41.53	0.00	0.00	51.73	6.99	0.00	0.00	100.24	0.93	2.00
13	Ph E	37.61	0.02	4.43	40.31	8.25	0.80	0.04	91.46	0.90	1.92
	CH	37.27	3.08	0.02	46.17	11.30	0.02	0.00	97.86	0.88	2.16
	Ol	41.12	0.06	0.00	49.45	9.24	0.03	0.00	99.90	0.91	1.98
14	Ph A	24.85	0.14	0.41	53.98	6.92	0.52	0.00	86.82	0.93	3.47
	CH	37.94	0.54	0.04	52.42	7.56	0.01	0.00	98.53	0.93	2.24
	Gt	42.09	0.09	17.82	15.91	12.07	10.13	0.00	98.11	0.70	0.80
	En	59.94	0.00	0.11	37.92	3.39	0.04	0.00	101.41	0.95	0.99
	Di	56.23	0.00	0.31	18.58	2.01	24.00	0.22	101.35	0.94	0.97
	Ph E	35.51	0.06	4.41	41.14	8.51	0.05	0.00	89.68	0.90	2.07
16	Ph E	35.74	0.00	5.22	39.88	10.10	0.00	0.00	90.93	0.88	2.07
	Ph A	25.39	0.00	0.15	53.93	9.70	0.00	0.00	89.16	0.91	3.49
	Ol	40.96	0.00	0.00	51.02	7.70	0.00	0.00	99.69	0.92	2.01
	Gt	42.58	0.00	18.63	21.42	14.72	0.00	0.00	97.35	0.72	1.04
18	CH	38.14	0.00	0.00	54.07	6.54	0.00	0.00	98.74	0.94	2.26
	Ch	33.95	0.00	0.00	53.30	6.61	0.00	0.00	93.85	0.93	2.50
19	Ph A	25.22	0.03	0.08	52.80	9.57	0.00	0.00	87.70	0.91	3.44
	Ph E	37.53	0.04	9.17	35.68	8.83	0.04	0.00	91.29	0.88	1.90
	CH	35.96	2.86	0.04	47.36	9.05	0.66	0.00	95.93	0.90	2.23
	Mgs	0.00	0.00	0.00	44.70	4.73	0.13	0.00	49.56	0.94	
20	Ph E	37.97	0.02	7.14	39.62	7.86	0.48	0.01	93.10	0.90	1.95
	CH	37.43	2.09	0.02	47.95	9.12	0.00	0.00	96.60	0.90	2.15
	Ol	41.25	0.04	0.00	49.96	8.76	0.00	0.02	100.02	0.91	1.98
	En	59.40	0.00	0.18	35.92	5.13	0.10	0.00	100.73	0.93	0.97
	Di	57.87	0.02	1.35	34.64	5.35	0.50	0.00	99.74	0.92	0.97
	Gt	42.23	0.26	21.52	17.35	11.31	6.11	0.00	98.78	0.73	0.84
22	Ph E	39.26	0.10	1.98	41.64	8.31	0.03	0.00	91.33	0.90	1.82
24	β-Ol	40.47	0.05	0.30	46.06	8.39	0.03	0.00	95.31	0.91	1.87
	Ph E	37.07	0.41	1.51	40.62	10.04	0.07	0.00	89.72	0.88	1.91
25	β-Ol	41.51	0.03	0.36	46.74	8.57	0.01	0.00	97.21	0.91	1.85
	Gt	46.66	0.13	14.53	24.33	8.31	5.47	0.02	99.43	0.84	0.93

M = Mg+Fe for Ol, β-Ol, En, Ph A, Gt; Mg+Fe+Al for Ph E; Mg+Fe+Ti for Ch, CH; Mg+Fe+Ca for Di

7.7 GPa. At 550 °C, TiO_2-bearing clinohumite and Ti-chondrodite are found with olivine, enstatite, diopside, garnet, and $(Mg, Fe)CO_3$. The TiO_2 contents are 4.3 - 4.5 wt. % in clinohumite and 6.5 - 7.0 wt. % in chondrodite, respectively (#5, Tables 2 and 3). At 700 °C, there are no hydroyus phases, but a trace amount of $(Mg, Fe)CO_3$ is seen.

9.3 GPa. No hydrous phases were found at 875 and 950 °C and 9.3 GPa in the H_2O-saturated peridotite system. An assemblage of phase E, clinohumite, enstatite, garnet, diopside, with phase A or olivine was obtained at 800 °C and 9.3 GPa (# 9, 14, Table 2, Figure 2a). This phase E has 4.4 wt. % Al_2O_3, and has a stoichiometry similar to that described by Kanzaki [9] which is less ferromagnesian than olivine. The clinohumite has 0.4 - 3 wt. % TiO_2 (#14, Table 3). Although there are no hydrous phases in the H_2O-saturated peridotite system at 875 °C and 9.3 GPa, phase E, phase A, olivine, and garnet were found at the identical condition in a system of H_2O-saturated phase E bulk composition (#16, Table 2, Figure 2b). This phase E contains 5 wt. % Al_2O_3 (#16, Table 3).

10.3 GPa. At 800 °C , phase A, phase E, and clinohumite coexist with olivine, enstatite, diopside, garnet and a trace amount of $(Mg, Fe)CO_3$. Phase E possesses about 9 wt. % Al_2O_3 and clinohumite has 3 wt. % TiO_2 at these conditions. At 900 °C, hydrous phases are phase E (7 wt. % Al_2O_3) and clinohumite (2 wt. % TiO_2). No $(Mg, Fe)CO_3$ was found. At 1000 °C, all crystalline phases were anhydrous.

14.5 and 15.5 GPa. A 2 wt. % Al_2O_3 bearing phase E is obtained at 1000 and 1200 °C and 14.5 GPa with garnet, enstatite, and diopside in an H_2O-saturated peridotite system (#22, Table 3). At 1400 °C, 15.5 GPa (#24, Tables 2, 3), phase E coexists with β-olivine, enstatite, and garnet. The β-olivine has a trace amount of Al_2O_3 (0.3 wt. %) and is characterized by a 1.87 $(Mg+Fe^*)/Si$ atomic ratio and low totals (95 wt. %). At 1500 °C, no phase E was found (#25, Tables 2, 3). The crystalline phases are β-olivine, garnet and enstatite. The β-olivine has a trace amount of Al_2O_3 (0.36 wt. %) and is characterized by a 1.85 $(Mg+Fe^*)/Si$ atomic ratio and low totals (97 wt. %). Whether these conditions were above solidus or not remains to be uncertain.

HYDROUS MINERALS IN AN H_2O-SATURATED PERIDOTITE AT 8 - 15 GPA

At 550 °C and 7.7 GPa, chondrodite coexists with clinohumite. Both of them contain significant amounts of TiO_2. Ti-bearing those minerals have been locally found in peridotite nodules [24]. Based on experimental results in the $MgO-SiO_2-H_2O$ system and petrography of the nodules, Aoki et al. suggested that P-T conditions of clinohumite and chondrodite bearing peridotite was 3.3 GPa and 1000 °C. However, our results indicate higher pressures and/or lower temperatures (Figure 4). In simpler systems, Ti-clinohumite can be stable in shallower and hotter regions (e.g., 2 GPa, 700 °C [25]) than in a peridotite system. Also Al_2O_3-bearing phase E is stable in a hotter range in its own bulk composition (#15, Table 2) than in a peridotite system.

At 9.3- 10.3 GPa, clinohumite is the most resistant to temperature induced dehydration, phase E is less thermally stable, and phase A is stable only in lower temperature. At 14.5 - 15.5 GPa, only phase E was found as a hydrous phase but β-olivine is potentially hydrous [26, 27, 28].

Phase E and clinohumite were found with garnet, enstatite, diopside, and phase A or olivine at 800 °C and 9.3 GPa in a peridotite system (#9, 14, Table 2). We suggest that the addition of Al_2O_3 and TiO_2 stabilizes phase E and clinohumite, respectively, in a hydrous peridotite system. We believe this because (1) phase A has less than 0.4 wt. % Al_2O_3 and 0.2 wt. % TiO_2 in both the H_2O-saturated peridotite system and the system of H_2O-saturated phase E bulk composition (Table 3), and (2) phase A coexists with only enstatite in the H_2O-saturated $MgO-FeO-SiO_2$ system at 800 °C and 9.3 GPa (#8, Table 2). In an H_2O-saturated Mg_2SiO_4 system, Kanzaki [9] found phase E at 1000 °C and pressures greater

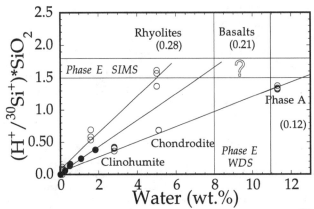

Figure 3, Calibration line for hydrogen intensity normalized to $^{30}Si^+$ plotted against stoicheometric water contents. The slope of the new calibration line for these hydrous minerals is 0.12 which is lower than 0.21 for hydrous basaltic glasses and 0.28 for hydrous rhyolitic glasses.

than 13 GPa. Luth [20] found phase E at 1200 °C and 10 GPa in the same simple system. We found the Al_2O_3 bearing phase E at lower temperature and lower pressure in an H_2O-saturated peridotite system. This demonstrates the value of investigating the stability of hydrous phases in a complex peridotitic system.

Clinohumite, chondrodite, and phase A have 2.8, 5.1, and 11.3 wt. % H_2O, respectively, based on the stoichiometry assuming a 0.9 Mg/(Mg+Fe) atomic ratio and no TiO_2. Phase E is a non-stoichiometric mineral with 1.77 - 2.15 Mg/Si atomic ratios [9,10]. There have been no direct analyses of the H concentration of phase E. In order to obtain precise H abundance in phase E, we measured the $H^+/^{30}Si^+$ ion ratios of olivine, clinohumite, chondrodite, and phase A with a secondary ion mass spectrometer (SIMS). Except for San Carlos olivine, these minerals were obtained through the present experiments in the H_2O-MgO-FeO-SiO_2 system. Their Mg/(Mg+Fe) atomic ratios are in a range between 0.87-0.94. Figure 3 shows H_2O-contents and hydrogen intensity normalized using $^{30}Si^+$ ion after subtraction of background. The slope of the calibration line for these hydrous minerals is 0.12 which is lower than the 0.21 determined for hydrous basaltic glasses and the 0.28 for hydrous rhyolitic glasses. Wavelength dispersive electron microprobe analyses showed that phase E has 89-91 wt. % total oxides, which is higher than the 86-88 wt. % of phase A (Table 3). However, the new SIMS calibration line suggests 12-14 wt. % H_2O, which is wetter than phase A. Using the basaltic glasses calibration line we find 7-9 wt. % H_2O. The short-fall in the WDS analysis suggests values in between. Although we do not have even a qualitative method to assess the matrix effect on $H^+/^{30}Si^+$ intensity ratio in silicates, the chemical composition of this Al_2O_3 bearing phase E is roughly between minerals on the olivine - brucite join and basaltic glasses. Unfortunately, the present result shows that we need another calibration line for phase E in order to determine the matrix effect.

PERSPECTIVE FOR H_2O RECYCLING IN SUBDUCTION ZONES

The most plausible environment for dense hydrous magnesium silicates (DHMS) to exist in the Earth's mantle is in subduction zones, which are characterized by colder and wetter conditions than other regions. Nicholls and Ringwood [29] suggested that subducting basalt will be almost dry beneath the fore-arc region. Therefore, Sakuyama and Nesbitt [30] suggested that induced downgoing mantle will be hydrated through H_2O released by dehydration of the hydrous minerals in the basaltic layer and carry H_2O beneath the volcanic arc. In order to clarify the geological significance of DHMS, we need to address the dehydration processes of downdragged peridotite from shallow to deep level.

Figure 4 shows the stability fields of hydrous minerals in relevant systems for the upper mantle, with schematic P-T paths of the downdragged hydrous mantle [8, 31]. P-T paths (1) and (2) in Figure 4 have been suggested based on a petrological model [32]. These values are also roughly consistent with the model calculations for the surface of subducting slab; ~800 °C, 3 GPa (Furukawa [33]) and ~700 °C, 3 GPa (Peacock [34]). In these models, downdragged flow lines are concentrated at the base of the mantle wedge, and there is a steep thermal gradient. Therefore, the temperature of the downdragged mantle can be higher than the calculated values for the surface of the subducting slab.

When the downdragged hydrous peridotite follows on P-T paths such as (1), (2) or (3) in Figure 4, serpentine breaks down, followed by talc, chlorite and finally pargasite. Beyond the pargasite-out reaction the subducting lithosphere will be almost free of H_2O bound in crystals except for a small amount in phlogopite at 3.5-4.5 GPa. This means that when downdragged hydrous peridotite goes on paths like these, the hydrous minerals should encounter "a choke point" like that shown in Figure 4. The downdragged hydrous

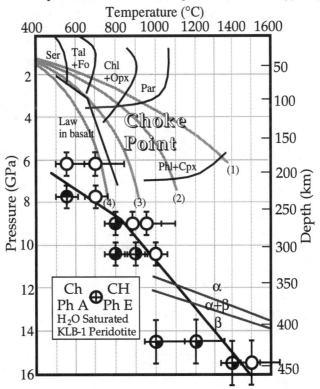

Figure 4, A P-T projection of the stability of dense hydrous magnesium silicates in H_2O-saturated KLB-1 peridotite (this study), lawsonite in a basaltic system [37], and hydrous minerals in relevant systems for peridotite [31] with schematic P-T paths of the downdragged hydrous mantle. Phase boundaries among α, $\alpha + \beta$, and β $(Mg_{0.9}, Fe_{0.1})_2SiO_4$ are after Katsura and Ito [38]. See text.

peridotite will carry a small amount of H in phlogopite into the deeper mantle. At 6-11 GPa, the phlogopite breaks down into K-richterite, which has an equal H/K atomic ratio and is stable at least up to 13 GPa [35]. In subducting basaltic crust, lawsonite is one of the most resistant hydrous minerals [36, 18, 37]. Lawsonite in a subducting basaltic crust contains ~11 wt. % H_2O and can cover some region of the choke point (Figure 4). Therefore, if the lower pressure stability of DHMS can overlap with that of lawsonite, DHMS in the downdragged flow of the base of the mantle wedge can take H_2O from decomposing lawsonite in the basaltic layer and become H_2O carriers in a very cold subduction zone like path (4) to the deeper mantle beyond the choke point. However, the scarcity of magmatism above subducting slabs deeper than 250 km[32] indicates that downgoing peridotite and basalt dehydrate at 3-7 GPa (the choke point) to form volcanic arcs.

There is an alternative candidate for an H_2O reservoir in subducting slabs. Nicholls and Ringwood [29] suggested that the serpentinite in the subducting peridotite layer can carry H_2O into the mantle beneath the volcanic arc. Furukawa [33] and Peacock [34] suggested that the temperature of the subducting peridotite layer 10 km below the surface of the slab will be 600-700 °C at 3.5 GPa (like paths (3) and (4) in Figure 4). In order to understand the fate of the hydrous minerals in the serpentinite of the slab, it is required to experimentally investigate the low temperature region. However, if the Clapeyron curves of the dehydration reactions of serpentine, talc + forsterite, chlorite + orthopyroxene, and pargasite are negative there, the subducting serpentinite is unlikely to play an important role, because it cannot receive the H_2O from the decomposing lawsonite of the overlying basaltic layer.

CONCLUSIONS

(1) Stability fields of Ti-chondrodite, Ti-clinohumite, phase A, and phase E were determined in H_2O-saturated KLB-1 peridotite at 550 - 1500 °C, 6 - 15 GPa (Figure 4).

(2) Phase E contains 1.5 - 9 wt. % Al_2O_3 and clinohumite contains 0.5-4 wt. % TiO_2 in a peridotite system (Table 3). We suggest that the addition of Al_2O_3 and TiO_2 stabilizes phase E and clinohumite, respectively, in a hydrous peridotite system.

(3) We suggest that downdragged hydrous peridotite has "a choke point" restricting the passage of H_2O to greater depth beyond 3-7 GPa (Figure 4).

(4) Dense hydrous magnesium silicates in the downdragged flow of the base of the mantle wedge could absorb H_2O from dehydrating lawsonite and become H_2O carriers in a very cold subduction zone (800°C, 9 GPa). However, the scarcity of magmatism above subducting slabs deeper than 250 km indicates that downgoing peridotite and basalt dehydrate at 3-7 GPa to form volcanic arcs.

(5) A calibration line for the secondary ion mass spectrometer determination of $H^+/^{30}Si^+$ ion ratios in phase A, chondrodite, and clinohumite has been determined (Figure 3). Its coefficient of $H^+/^{30}Si^+$ ion ratios against wt. % H_2O concentration is 0.12, which is lower than the 0.21 found for hydrous basaltic glasses and the 0.28 for hydrous rhyolitic glasses.

Acknowledgment We thank Jim Clark for assistance with electron micro probe analysis, and Andrzej Grzechnik and Paul McMillan for taking Raman spectra. Kei Hirose kindly helped TK to cook a gel. Bob Liebermann and Eiji Ito kindly provided us with LaCrO3. Stefano Poli and Max Schmidt permitted us to refer to their unpublished data. Technical assistance by Ken Domanik and Rosemary Gerald Pacalo and comments by Xiaoyuan Li, Simon Peacock, Yoshiyuki Tatsumi, Atsushi Yasuda, and an anonymous reviewer were helpful. This work was sponsored by NSF DMR 9121570 for multi-anvil apparatus, NSF EAR 8408163 for electron microprobe analysis, NSF EAR 9305201 for SIMS, and NSF EAR 9312498 for laboratory support at ASU, and Grant-in Aid for Scientific Research of Japanese Ministry of Education, Science, and Culture # 02402018 for XRF at the Geological Institute, University of Tokyo. Japanese Society for the Promotion of Science postdoctoral fellowship enables TK to stay at ASU.

REFERENCES

1. D.L. Hamilton, C.W. Burnham, and E.F. Osborn, J. Petrol. **5**, 21 (1964).
2. P.F. McMillan and J.R. Holloway, Contrib. Mineral. Petrol. **97**, 320 (1987).
3. A.E. Ringwood and A. Major, Earth Planet Sci. Lett. **2**, 106 (1967).
4. C.B. Sclar, L. C. Carrison, and O. M. Stewart, Trans. Am. Geophys. Union **48**, 226 (1967).
5. L.-G. Liu, Phys. Earth Planet. Interiors **49**, 142 (1987).
6. A.B. Thompson, Nature **358**, 295 (1992).
7. L.-G. Liu, Phys. Earth Planet. Inter. **42**, 255 (1986).
8. K. Yamamoto, and S. Akimoto, Am. J. Sci. **277**, 288 (1977).
9. M. Kanzaki, Phys. Earth Planet. Inter. **66**, 307 (1991).
10. Y. Kudoh, L. W. Finger, R. M. Hazen, C. T. Prewitt, M. Kanzaki, and D. R. Veblen, Phys. Chem. Minerals **19**, 357 (1993).
11. A.E. Ringwood, Composition and petrology of the Earth's mantle (McGraw Hill, New York, 1975), p. 618.
12. A.E. Ringwood, Tectonophysics **32**, 129 (1976).
13. Y. Nakamura, and I. Kushiro, Carnegie Inst. Wash. Yearb. **73**, 255 (1974).
14. E. Takahashi, J. Geophys. Res. **91**, 9367 (1986).
15. D.L. Hamilton and C.M.B. Henderson, Mineral. Mag. **36**, 832 (1968).
16. A. Goto and Y. Tatsumi, Rigaku J. **22**, 28 (in Japanese) (1991).
17. D. Walker, M.A. Carpenter, and C.M. Hitch, Am. Mineral. **75**, 1020 (1990).
18. A. Pawley, Contrib. Mineral. Petrol. **118**, 99 (1994).
19. K. Leinenweber and J. Parise, J. Solid State Chem. **114**, 277 (1995).
20. R.W. Luth, Geophys. Res. Lett. **20**, 233 (1993).
21. R.L. Hervig and P. Williams, in Secondary Ion Mass Spectrometry, SIMS VI (eds. Benninghoven A, H., A. M., Werner, H. W., J. Wiley and Sons, New York, 1988) 961.
22. P.D. Ihinger, R. L. Hervig, and P. F. McMillan, in Volatiles in magmas. M. S. A. Reviews in mineralogy (eds. M. R. Carrol and J. R. Holloway, Mineralogical Society of America, Washington D. C., 1994) 67.
23. P.F. McMillan and M. Akaogi, Am. Mineral. **72**, 361 (1987).
24. K. Aoki, K. Fujino, and M. Akaogi, Contrib. Mineral. Petrol. **56**, 243 (1976).
25. M. Engi and D.H. Lindsley, Contrib. Mineral. Petrol. **72**, 415 (1980).
26. J.R. Smyth, Am. Mineral. **72**, 1051 (1987).
27. D.L. Kohlstedt, H. Keppler, D. C. Rubie, Trans. Am. Geophys. Union, **75**, 652 (1994).
28. T. Inoue, H. Yurimoto, and Y. Kudoh, Geophys. Res. Lett. in press, (1995).
29. I.A. Nicholls, and A. E. Ringwood, J. Geol. **81**, 285 (1973).
30. M. Sakuyama and R. W. Nesbitt, J. Volcanol. Geotherm. Res. **29**, 413 (1986).
31. Y. Tatsumi, J. Geophys. Res. **94**, 4697 (1989).
32. Y. Tatsumi and S. Eggins, Subduction zone magmatism (Blackwell, in press).
33. Y. Furukawa, J. Geophys. Res. **98**, 8309 (1993).
34. S. M. Peacock, Chem. Geol. **108**, 49 (1993).
35. A. Sudo and Y. Tatsumi, Geophys. Res. Lett. **17**, 29 (1990).
36. A.R. Pawley and J.R. Holloway, Science **260**, 664 (1993).
37. S. Poli and M. W. Schmidt, submitted to J. Geophys. Res. (1995).
38. T. Katsura and E. Ito, J. Geophys. Res. **94**, 15,663 (1989).

CO$_2$ AND H$_2$O IN THE DEEP EARTH: AN EXPERIMENTAL STUDY USING THE LASER-HEATED DIAMOND CELL

Xiaoyuan Li, Jeffrey H. Nguyen and Raymond Jeanloz
University of California, Berkeley, CA 94720-4767

ABSTRACT

Experiments with the laser-heated diamond cell show that H$_2$O and CO$_2$ can be stabilized within crystalline mineral structures of the lower-mantle, and hence can be present as relatively non-volatile components of the Earth's deep interior. Free CO$_2$ reacts both with ferromagnesian silicates and with metallic Fe at high pressures and temperatures to form (Fe,Mg)CO$_3$-containing assemblages, the MgCO$_3$-FeCO$_3$ solid solution being stable and coexisting with (Mg,Fe)SiO$_3$ perovskite to at least 30-40 GPa and ~1500-2000 K. Fluid H$_2$O combines with Mg-Fe silicate to form (Mg,Fe)SiH$_2$O$_4$ phase D along with (Mg,Fe)SiO$_3$ perovskite; if enough water is present, phase D can become the predominant phase in the MgSiO$_3$-H$_2$O system at lower-mantle conditions. The results of high-pressure experiments suggest that both H$_2$O and CO$_2$ can be abundant in the Earth's lower mantle, being present in stable hydroxysilicate and carbonate phases.

INTRODUCTION

CO$_2$ and H$_2$O are major volatile constituents of the terrestrial planets. In particular, knowing the concentration and distribution of these components within the entire Earth System is important for understanding the evolution of both the atmosphere and interior of the planet. The distribution of CO$_2$ and H$_2$O early in geological history also influenced the development of life on Earth.

With these motivations, we investigate the chemical reactions between Earth materials (i. e., ferromagnesian silicates and metallic iron) and either free CO$_2$ or free H$_2$O at high pressures and temperatures. Our purpose is to investigate the degree to which these nominally volatile components might be locked up deep inside the planetary interior.

Our experiments are carried out using the laser-heated diamond-anvil cell, in combination with gas loading, *in situ* high-pressure FTIR (Fourier-Transform Infrared) spectroscopy, X-ray diffraction and film digitizing techniques. The ultrahigh pressure–temperature experiments show that ferromagnesian carbonate and hydrous silicate minerals are stable at lower-mantle conditions, thus reinforcing and extending previous results indicating that CO$_2$ and H$_2$O lose their volatile character deep in the Earth. To our knowledge, the present work is the first study of the high pressure-temperature reactions between CO$_2$ and Earth materials, with free CO$_2$ as a starting material.

EXPERIMENTAL METHODS

Natural enstatite, siderite and magnesite samples are used as starting materials in our work. Of these, the enstatite, with a composition of (Mg$_{0.88}$Fe$_{0.12}$)SiO$_3$, is from Bamble, Norway, and has been utilized in previous investigations (e.g., Refs. 1 and 12). The FeCO$_3$ siderite is from Ivigtwt, Greenland, and the MgCO$_3$ magnesite from Snarum, Norway. The identification and purity of these starting materials have been verified by x-ray diffraction as described in the following sections.

The diamond cell used in the present study is of the Mao-Bell type, and is complemented by a continuous Nd:YAG laser-heating system.[2,3] The culets of the diamond anvils are 350 µm in diameter. In order to hold the sample, spring-steel gaskets (from McMaster-Carr Company, Los Angeles, CA) are used in the diamond cell. The spring-steel gasket, containing 0.01-1.5 wt % carbon, is also used as a starting material for studying reactions between CO$_2$ and Fe.

To avoid contamination, no ruby or pressure medium has been included with the sample. Instead, the sample pressures are calibrated using the spring length of the diamond cell.[4] The reliability of this calibration is confirmed for CO$_2$-containing samples by FTIR spectroscopy, via the pressure dependency of the CO$_2$ v_1+v_3 vibrational frequency.[5-7] As shown in Fig. 1, the frequency–pressure relation determined in the present study is in good agreement with the results of previous work. Based on these data, we estimate the uncertainties in sample pressures to be less than 3 GPa.

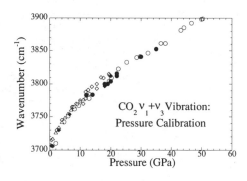

Fig. 1 Pressure dependence of the $\nu_1 + \nu_3$ frequency used for pressure calibration, with ν_1 and ν_3 representing the symmetric and asymmetric stretching modes of CO_2. Solid symbols are from the present study on compression (circles) and decompression (triangles). Open symbols are from refs. 5 (triangles), 6 (diamonds) and 7 (circles), and the kink at ~11 GPa is due to a phase transition from cubic to orthorhombic structures in solid CO_2.[8]

Fluid CO_2 and H_2O are loaded into the sample chamber at temperatures between 77 and 195 K, and sealed at pressures less than ~10 GPa. After loading the sample, the diamond cell is warmed up to room temperature and the sample is compressed gradually to higher pressures. The sample is then laser-heated to high temperatures, and chemical reactions within the sample are observed visually during the heating. Both on compression and on decompression, high-pressure infrared absorption spectra of the samples are measured *in-situ* (at 300 K) with a Bruker IFS 66v Fourier-transform infrared spectrometer equipped with L-N$_2$ cooled MCT and InSb detectors, and operating at a resolution of 4 cm^{-1}. The reaction products are finally unloaded and examined at ambient conditions by means of X-ray diffraction, and the resultant diffraction patterns are analyzed with the film-scanning techniques described by Nguyen and Jeanloz.[9]

RESULTS AND DISCUSSION

Experiments in this study involve reactions between CO_2 and Fe, between CO_2 and $(Mg,Fe)SiO_3$, between H_2O and $(Mg,Fe)SiO_3$, between $FeCO_3$ and $MgCO_3$, and between $FeCO_3$ and $(Mg,Fe)SiO_3$, all at high pressures. The results are summarized in Table 1. A more detailed discussion of these experimental results is given in the following paragraphs.

Table 1. Experimental conditions and results

Starting Materials	P (GPa)	T (K)	Quenched Products
En + CO_2 (saturated)	14	~2500	Ma-Sd + St
Fe + CO_2 (saturated)	22	~2500	Sd + Fe$_x$C
Sd + Ma (1:1 mol ratio)	38	1500-2000	Ma-Sd
En + Sd (20 wt %)	40	1500-2000	Pv + Sd
En + H_2O	40	1800-2000	Pv + D

En = enstatite; Sd = siderite; Ma = magnesite; Ma-Sd = magnesite-siderite solid solution; Pv = perovskite; D = hydrous silicate phase D (compositions of these phases are given in the text).

Reactions between CO_2 and $(Mg,Fe)SiO_3$ silicate

A mixture of free CO_2 and $(Mg,Fe)SiO_3$ enstatite was laser-heated to about 2500 K at 14 GPa. Visually, the enstatite grains are initially transparent in the CO_2 matrix and become opaque after laser-heating, suggesting that a chemical reaction occurs between the silicate and CO_2 at high temperatures.

This reaction is further documented by the appearance of carbonate modes in the mid-infrared absorption spectrum of the sample (Fig. 2a). Before heating, only CO_2 and silicate vibrational modes are present. However, after laser heating, three carbonate modes appear at 759 (±9), 876 (±9) and 1477 (±9) cm^{-1}, and their intensities increase with increased duration of heating (increased degree of reaction), as shown in the figure. By referring to the vibrational spectrum of $CaCO_3$,[10] the three carbonate modes can be assigned to in-plane bending, out-of-plane bending and asymmetric stretching motions, respectively. The spectra thus demonstrate the formation of carbonate in the high-pressure chemical reaction between CO_2 and $(Mg,Fe)SiO_3$.

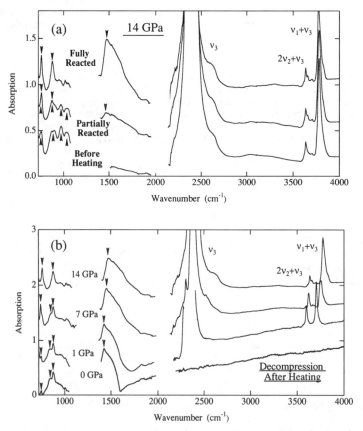

Fig. 2 Infrared absorption spectra of a sample that is initially a mixture of free CO_2 and $(Mg,Fe)SiO_3$ enstatite. (a) Spectra at 14 GPa, before and after heating: the down-pointing arrows indicate carbonate absorption bands, while the up-pointing arrows indicate absorption peaks due to the starting enstatite. (b) IR absorption patterns upon decompression, after heating. The symbols, v_1, v_2, v_3 and their combinations represent vibrational modes of CO_2, with v_2 being the bending mode, and v_1 and v_3 being symmetric and asymmetric stretching modes. No spectrum is shown at ~1000-1400 cm^{-1} and 2000-2200 cm^{-1}, where impurity and 2-phonon absorption by the diamond anvils blocks the signal.

Fig. 2b shows the IR spectrum of the sample upon decompression, after laser-heating at 14 GPa. The carbonate peaks remain present at ambient conditions (frequencies of 744 ±9, 876 ±9 and 1427 ±9 cm^{-1}), by which point the CO_2 has escaped from the sample chamber. In addition, a new carbonate peak appears at pressures ≤ 7 GPa, and is characterized by a frequency of 845 (±12) cm^{-1} at ambient conditions. In comparison with the vibrational frequencies of end-member $MgCO_3$ (748, 856, 886 and 1446 cm^{-1}, respectively, at ambient pressure),[11] the present frequencies are slightly but systematically smaller. This is reasonable because our sample contains iron, and increasing the $FeCO_3$ content reduces the vibrational frequencies of the $MgCO_3$-$FeCO_3$ solid solution.[18]

To further identify the phases of the reaction product, we have collected an X-ray diffraction pattern from the decompressed sample (Fig. 3). More than 16 diffraction lines are resolved in the pattern. Still, there is no indication of any enstatite phase in the X-ray pattern, demonstrating that this starting material has been totally used up in the reaction. Most of the lines are from either SiO_2 stishovite (denoted as St) or $(Mg,Fe)CO_3$ magnesite (denoted as Ma). Only 1 line (indicated by question marks in Fig. 3) cannot be identified, and may be due to contamination.

Based on these experimental results we conclude that at mantle conditions of pressure and temperature, CO_2 and $(Mg,Fe)SiO_3$ react according to:

$$CO_2 + (Mg,Fe)SiO_3 \rightarrow (Mg,Fe)CO_3 + SiO_2. \tag{1}$$

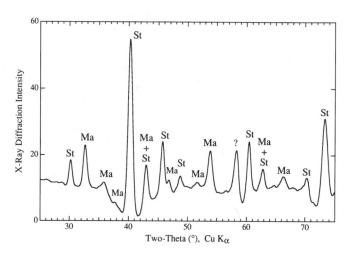

Fig. 3 X-ray diffraction pattern of the reaction products of CO_2 and $(Mg,Fe)SiO_3$ enstatite, after laser heating at 14 GPa and being quenched to ambient conditions. The diffraction peaks of the resultant $(Mg,Fe)CO_3$ magnesite (Ma) and SiO_2 stishovite (St) are shown, and question marks indicate unidentified diffraction peaks.

Since X-ray diffraction has only been carried out on quenched samples, the crystal structure of the carbonate at deep-mantle pressures is not clear from the present work. It can be seen from Fig. 2b that the IR data exhibit some differences in the high- and low-pressure absorption spectra of the carbonate. For example, two peaks appear as out-of-plane bending modes at pressures ≤7 GPa while only one occurs at 14 GPa (876 cm^{-1}), and we cannot tell if this difference corresponds to a phase transition.

Previous experimental results on the high-pressure crystal structure of magnesite have been controversial. The X-ray diffraction results of Fiquet, et al.[19] suggest that a phase transition takes place at 25 GPa, yet no phase transition was observed in the X-ray diffraction studies of Katsura et al.[20] to 55 GPa, in the shock experiment of Kalashnikov et al.[21] to 120 GPa, and in the Raman measurements of Williams et al.[10], Gillet,[22] and Gillet et al.[23] to ≥ 30 GPa. Thus, further work is required to clarify the structure of $(Mg,Fe)CO_3$ at deep mantle conditions.

Reactions between CO$_2$ and Fe

Fig. 4a shows the IR spectrum of CO$_2$ contained within a steel gasket at 22 GPa. The spectrum on the top is obtained after heating the interface of the gasket in contact with the CO$_2$ to about 2500 K. The sample changes from being clear to having a brown color, indicating that the CO$_2$ and steel have reacted. It can also be seen from Fig. 4a that a carbonate peak appears at about 1755 cm^{-1} after heating, demonstrating that Fe-carbonate is a product of the high-pressure chemical reaction.

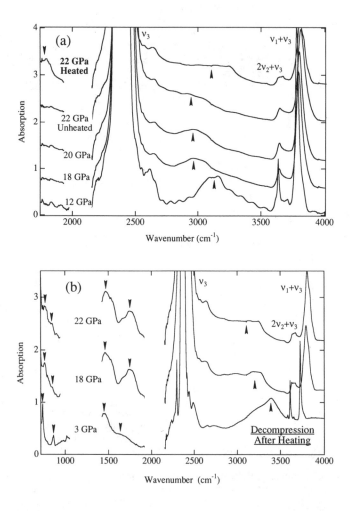

Fig. 4 Infrared absorption spectra of a sample starting with free CO$_2$ contained in a steel gasket. (a) Patterns measured on compression using an InSb detector: the down-pointing arrows indicate the carbonate absorption peak formed after laser heating at 22 GPa, and the up-pointing arrows indicate peaks corresponding to condensed H$_2$O (see text). (b) Patterns measured during decompression using an MCT detector. The labels ν_1, ν_2 and ν_3 have the same meaning as in Figs. 1 and 2. Impurity and 2-phonon absorption by the diamond anvils blocks the signal at ~1000-1400 cm^{-1} and 2000-2200 cm^{-1}.

Fig. 4b shows the IR spectrum of the sample on decompression, measured here using an MCT detector. More carbonate vibrations can be observed with this detector than with the InSb detector used for Fig. 4a. At 3 GPa, the frequencies of the four absorption bands of iron carbonate are 737 ±4 (in-plane bend), 864 ±5 (out-of-plane bend), and 1446 ±27 and 1676 ±52 cm⁻¹ (both are asymmetric stretches). The first three frequencies are in good agreement with the values reported in Ref. 11 (739, 867 and 1422 cm⁻¹). Note that in the IR spectrum of both Refs. 11 and 18, there is a broad band centered at 1422 cm⁻¹ and covering the whole frequency range from 1100 to 1700 cm⁻¹, such that the present observations (including the absorption at 1676 cm⁻¹, which is weak at low pressures) are fully consistent with these published results.   X-ray diffraction of the quenched sample proves that FeCO₃ siderite is the major reaction product (Sd peaks in Fig. 5). Based on the chemical formulae of the known reactants and product phases, one or more of the following chemical reactions can have taken place at elevated pressures and temperatures:

$$(2+x)\ Fe + 3\ CO_2\ \rightarrow\ 2\ FeCO_3 + Fe_xC, \tag{2}$$

$$(4+x)\ Fe + 6\ CO_2\ \rightarrow\ 4\ FeCO_3 + C\ (diamond) + Fe_xC, \tag{3}$$

$$Fe + 2\ CO_2\ \rightarrow\ FeCO_3 + CO, \tag{4}$$

$$2\ Fe + 2\ CO_2\ \rightarrow\ FeCO_3 + C\ (diamond) + FeO. \tag{5}$$

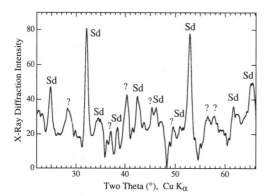

Fig. 5 X-ray diffraction pattern of the reaction product of CO_2 and the spring-steel gasket, heated to 2400 K at 22 GPa and quenched to ambient conditions. Siderite is denoted by Sd, and question marks indicate unidentified diffraction peaks.

We find no evidence in the X-ray diffraction pattern for the presence of either diamond or FeO, so (3) and (5) are not likely to have been significant reactions in our experiments. The ~2100 cm⁻¹ spectral region characteristic of the infrared-active CO stretching vibration[26] is blocked by 2-phonon absorption in the diamond anvils (Fig. 4), and therefore cannot be used to document reaction (4). It is possible that the change in sample color observed after heating is due to (4) because photoreaction of CO at high pressures has been reported to induce similar color changes.[27] Yet because we find no evidence by X-ray diffraction or FTIR spectroscopy of photoreaction products as described by Mills, et al.,[27] we conclude that (4) is improbable at the conditions of our experiments.

Consequently, (2) is considered the most likely chemical reaction to dominate at the high pressures and temperatures of our study. This conclusion is supported by the x-ray diffraction observation of the reaction products. As shown in Fig. 5, there are some unidentified peaks in the x-ray diffraction pattern, demonstrating that a new crystalline phase has been quenched together with siderite. We found that the new diffraction peaks are consistent with those of the high-pressure reaction product of Fe + C.[28] Therefore, these peaks demonstrate that some iron carbide phase is formed in the present experiments.

It should be noted that IR absorption due to H$_2$O is observed in our CO$_2$-Fe sample (Fig. 4). Water ice apparently condensed into the sample chamber as the CO$_2$ was loaded at low temperatures. We do not know what role the H$_2$O plays in the ultrahigh pressure–temperature experiments, except to say that it does not prevent the carbonation reaction from taking place. Thus, further work is required to characterize the observed reactions.

Stability of the Carbonate at Lower Mantle Conditions

To further investigate the stability of carbonate at the conditions of the deep Earth, we compress a 1:1 molar ratio mixture of siderite and magnesite to 38 GPa and heat it to about 1600-2000 K. After heating, the sample turns from a brownish color to black. The X-ray diffraction pattern of the quenched sample, shown in Fig. 6, exhibits no change in crystal structure after treatment at lower-mantle pressures and temperatures, reinforcing the conclusion that carbonate is a stable phase at these conditions. The change in color during heating is interpreted as the siderite combining with the magnesite to form an FeCO$_3$-MgCO$_3$ solid solution, (Mg,Fe)CO$_3$ magnesite.

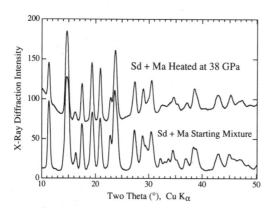

Fig. 6   X-ray diffraction pattern of a 1:1 molar ratio mixture of siderite (Sd) and magnesite (Ma) before the experiment (bottom), and after heating to 1500-2000 K at 38 GPa and quenched back to ambient conditions (top).

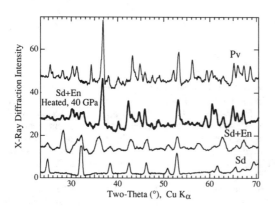

Fig. 7   X-ray diffraction pattern of initial (Mg,Fe)SiO$_3$ enstatite (En) + 20 wt % siderite (Sd) mixture. Heavy line denotes the pattern obtained after heating to 1500-2000 K at 40 GPa, and quenching back to ambient conditions. Two patterns at bottom are from the starting siderite and enstatite + siderite mixture. Top pattern is for (Mg,Fe)SiO$_3$ perovskite converted from En at about 40 GPa and 2000 K.

The stability of the iron carbonate in the presence of ferromagnesian silicate is documented by the following experiment. We compress a mixture of Bamble enstatite with 20 wt % siderite to 40 GPa and heat it to 1500-2000 K. The X-ray diffraction pattern after quenching to ambient conditions shows that the enstatite transforms to the high-pressure (Mg,Fe)SiO$_3$ perovskite phase while the carbonate diffraction pattern remains unchanged (Fig. 7). This demonstrates that $(Fe,Mg)CO_3$ can coexist with $(Mg,Fe)SiO_3$ perovskite at lower-mantle conditions of pressure and temperature.

Previous experimental results of Katsura and Ito[16] have shown that $MgCO_3$ is stable and coexists with $MgSiO_3$ phases to pressures of 8-26 GPa, corresponding to a depth of 140-720 km in the mantle. Biellmann et al.[24] have further shown that $CaMg(CO_3)_2$ can react with ferromagnesian silicates to form stable assemblages of $(Mg,Fe)CO_3$, $CaSiO_3$ and $(Mg,Fe)SiO_3$ perovskites, and $(Mg,Fe)O$ magnesiowüstite at pressures up to 50 GPa and temperatures of 1500-2500 K. Based on all of the experimental results, both present and previous, we conclude that the crystalline $FeCO_3$-$MgCO_3$ system is a stable solid solution under the chemical and thermodynamic (P and T) conditions of the lower mantle.

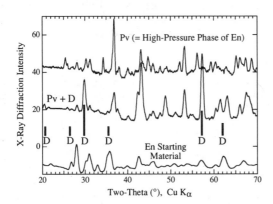

Fig. 8 X-ray diffraction pattern of $(Mg,Fe)SiO_3$ perovskite (Pv) and $(Mg,Fe)SiH_2O_4$ phase D (middle) obtained by heating a mixture of $(Mg,Fe)SiO_3$ enstatite (En) + H_2O starting material (bottom) to 1800-2000 K at 40 GPa, as described in the text. The vertical bars (D) show the diffraction lines of phase D reported by Liu.[14] The enstatite starting material and the perovskite (Pv) sample are the same as in Fig. 7.

Reactions between H_2O and $(Mg,Fe)SiO_3$

In the present work, we have also studied a mixture of H_2O and $(Mg,Fe)SiO_3$. Water ice is condensed from the gas phase into the sample at less than 200 K and, after pressurization to 40 GPa, heated to 1800-2000 K. The sample is then examined by X-ray diffraction after quenching to ambient conditions (Fig. 8). The resultant product is a mixture of $(Mg,Fe)SiO_3$ perovskite (Pv) and hydrous silicate phase D. Also shown in Fig. 8 are the X-ray diffraction patterns of the enstatite starting material, as well as the high-pressure perovskite and D phases. Clearly, D is the primary mineral phase in our quenched sample, as it is this phase which has the strongest diffraction peak (Fig. 8).

The hydrous phase D, with an estimated composition of $(Mg,Fe)SiH_2O_4$, was first found by Liu[13,14] at 22 GPa, and was then shown by Li and Jeanloz[12] to coexist with $(Mg,Fe)SiO_3$ perovskite at lower-mantle conditions. The present results confirm these earlier findings and demonstrate that phase D can become the predominant phase in the $MgSiO_3$-H_2O system at lower-mantle conditions if enough water is present.

Implications for the Earth's Interior

Our experiments show that free CO_2 and H_2O combine with Earth materials at high pressures and temperatures to form assemblages containing carbonate phases or hydrous silicate phases. The resultant crystalline carbonates and hydroxysilicates are stable under the chemical and pressure-temperature conditions of the transition zone and lower mantle, consistent with the experimental results of Newton and Sharp,[15] Katsura and Ito,[16] Kraft et al.,[17] Biellmann et al.,[24] Redfern et al.[25] and Li and Jeanloz.[12]

The present work extends previous high-pressure studies on Fe-bearing compositions, and provides the first information on reactions involving free CO_2 at the pressures of the lower mantle. In particular, whereas experiments with carbonate starting material show no evidence of a separate CO_2 phase being formed at high pressures and temperatures,[24] our work offers a reversal of the reaction (in composition) in showing that any free CO_2 that is present combines with silicate to form carbonate at lower-mantle pressures.

Taken together, the high-pressure experiments demonstrate that it is carbonate rather than carbon dioxide that is stable in the presence of silicates, even at the highest temperatures studied. This result can be understood in terms of the relatively large molar volume of CO_2 at elevated pressures,[8] which stabilizes the carbonate despite counteracting effects of entropy at high temperatures. We believe that the dense hydrous-silicate phases are similarly stabilized at elevated pressures.

In aggregate, the high-pressure experiments suggest that CO_2 and H_2O lose their volatile character at depths below the uppermost mantle. Thus, it is possible that through the formation of carbonate and hydroxysilicate phases at depth, CO_2 and H_2O have been locked inside the planet since early in Earth history. If so, CO_2 and H_2O may be more abundant in the mantle and core than has previously been appreciated, and could significantly participate in geodynamic processes of the planetary interior.

Acknowledgments

We thank A. Kavner for her assistance in the temperature measurement and R. Lu for informing us of his work on CO_2. We also thank T. J. Ahrens and H. K. Mao for their valuable comments on the manuscript. This work was supported by NASA and NSF.

REFERENCES

1.   E. Knittle and R. Jeanloz, Science, 235, 668 (1987).
2.   D. Heinz and R. Jeanloz, High Pressure Research in Mineral Physics (Terra Scientific Publishing, Tokyo, 1987), p. 113.
3.   B. O'Neill and R. Jeanloz, EOS, Transactions of American Geophysical Union. 71, 1611 (1990).
4.   X. Li and R. Jeanloz, Geophys. Res. Lett. 14, 1075 (1987).
5.   R. C. Hanson and L. H. Jones, J. Chem. Phys. 75, 1102 (1981).
6.   K. Aoki, H. Yamawaki and M. Sakashita, Phys. Rev. B48, 9231 (1993).
7.   R. Lu and A. M. Hofmeister, Phys. Rev. B, submitted (1994).
8.   K. Aoki, H. Yamawaki, M. Sakashita, Y. Gotoh and K. Takemura, Science, 263, 356 (1994).
9.   J. H. Nguyen and R. Jeanloz, Rev. Sci. Instrum. 64, 3456 (1993).
10. Q. Williams, B. Collerson and E. Knittle, Amer. Mineral. 77, 1158 (1992).
11. G. C. Jones and B. Jackson, Infrared Transmission Spectra of Carbonate Minerals (Chapman and Hall, New York, 1993).
12. X. Li and R. Jeanloz, Nature 350, 332 (1991).
13. L.-G. Liu, Phys. Earth Planet. Int. 42, 255 (1986).
14. L.-G. Liu, Phys. Earth Planet. Int. 49, 142 (1987).
15. R. C. Newton and W. E. Sharp, Earth Planet. Sci. Lett. 26, 239 (1975).
16. T. Katsura and E. Ito, Earth Planet. Sci. Lett. 99, 110 (1990).
17. S. Kraft, E. Knittle and Q. Williams, J. Geophys. Res. 96, 17997 (1991).
18. J. V. Dubrawski, A-L. Channon and S. St. J. Warne, Amer. Mineral. 74, 187 (1989).
19. G. Fiquet, F. Guyot and J. Itie, Amer. Mineral. 79, 15 (1994).
20. T. Katsura, Y. Tsuchida, E. Ito, T. Yagi, W. Utsumi and S. Akimoto, Proceedings of Japan Academy, 67B, 57 (1991).
21. N. G. Kalashnikov, M. N. Pavlovski, G. V. Simakov and R. F. Trunin, Phys. Solid Earth 2, 23 (1973).
22. P. Gillet, Amer. Mineral. 78, 1328 (1993).
23. P. Gillet, C. Biellmann, B. Reynard and P. F. McMillan, Phys. Chem. Minerals 20, 1 (1993).

24. C. Biellmann, P. Gillet, F. Guyot, J. Peyronneau and B. Reynard, Earth Planet. Sci. Lett., 118, 31 (1993).
25. S. A. T. Redfern, B. J. Wood and C. M. B. Henderson, Geophys. Res. Lett. 20, 2099 (1993).
26. K. Nakamoto, Infrared Spectra of Inorganic and Coordination Compounds, 2nd ed. (Wiley, New York, 1970).
27. R. L. Mills, D. Schiferl, A. I. Katz and B. W. Olinger, J. Phys. (Paris) C8, 187 (1984).
28. X. Li and R. Jeanloz, unpublished data (1995).

DENSE HYDROGEN IN THE OUTER SOLAR SYSTEM: IMPLICATIONS FROM RECENT HIGH-PRESSURE EXPERIMENTS

R. J. Hemley, H. K. Mao, T. S. Duffy, J. H. Eggert, A. F. Goncharov,
M. Hanfland, M. Li, M. Somayazulu, W. L. Vos, and C. S. Zha
Geophysical Laboratory and Center for High-Pressure Research,
Carnegie Institution of Washington, 5251 Broad Branch Rd. N.W.,
Washington, DC 20015

ABSTRACT

Recent high-pressure experiments on hydrogen and related low-Z materials from ~1 GPa to 300 GPa have uncovered a range of phenomena relevant to understanding the interiors of the outer planets and their satellites. The new results in the lower pressure range (<50 GPa) include the observation of new compounds (clathrates and van der Waals compounds), accurate determination of equations of state, and direct measurements of sound velocities. A key result at the highest pressures is evidence for stability of the molecular bond to ~250 GPa (at 77 K). However, the bond begins to weaken well below this pressure, and this weakening could be enhanced with increasing temperature. Experimental data also indicate a significant increase in absorption in the near infrared above 150 GPa and in the visible-near ultraviolet above 250 GPa.

INTRODUCTION

Hydrogen is the most abundant material in the solar system. As a result, the chemistry and physics of this material is crucial for numerous problems in planetary science, particularly for the outer solar system. There has been rapid growth in understanding the behavior of hydrogen and related low-Z materials at low density in the outer planets and their satellites from observational data[1]. However, there has been little direct information on the behavior at high densities, for example in deep interior conditions. For many years, the chemistry of hydrogen-rich bodies of the solar system was the subject of speculation based on extrapolations from low-pressure experiments or theoretical calculations[2].

This has changed as a result of recent breakthroughs in high-pressure experimentation, primarily static compression techniques[3]. At lower pressures (<50 GPa), for example, the equation of state of the solid (cold compression isotherms) has been accurately determined by diffraction techniques. Sound velocities have also been measured over this pressure range. Finally, an entirely new chemistry of hydrogen and other materials has been uncovered. At higher pressures — well into the megabar range (>100 GPa) — spectroscopic studies have been carried out to characterize the state of bonding of hydrogen, identify phase transitions, determine the subsolidus phase diagram, characterize optical properties (visible and infrared) and elucidate electronic properties of the material. Such conditions correspond to densities approaching 1 g/cm^3 (at > 200 GPa). By comparison, the mean density of Jupiter and Saturn are 1.2 and 0.6 g/cm^3, respectively. At the highest pressures (250–300 GPa), such spectroscopic studies have put key bounds on the range of stability of the molecular bond (H$_2$). The astrophysical implications of some of these results have recently been reviewed[4].

EQUATION OF STATE

The low temperature (e.g., room-temperature) isotherm is crucial for planetary modeling because it provides an important reference for higher temperature equation of state determinations, including the density change along planetary adiabats[2]. Early work relied on interpolation between low-pressure static compression data (<2.5 GPa) and quantum statistical (Thomas-Fermi-Dirac) calculations at very high pressures (> 1 TPa). This intermediate pressure-density region includes much of the molecular envelopes of the outer planets. A number of interior models for the giant planets have been constructed[5–7]. Uncertainty in the equation of state of dense molecular hydrogen is one of the principal factors limiting their accuracy. Recently, it has been shown that through diffraction techniques, accurate equation of state data can be obtained for H$_2$ to very high compression at room temperature[8]. Recent x-ray diffraction results are shown in Fig. 1[9].

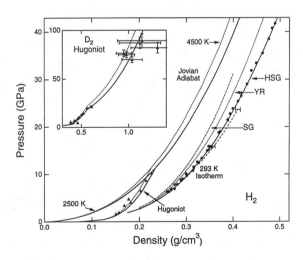

Fig. 1. Equation of state data for H$_2$ and D$_2$. Comparisons with potential models are made for 293 K isotherm data from x-ray diffraction (circles), shock compression data (triangles) for both deuterium (inset) and hydrogen, and a Jovian adiabat[7,17]. The long dashed line shows the 293 K isotherm determined independently from Brillouin scattering data[18].

Due to their high temperatures, the interiors of the giant planets are in a fluid state. The application of shock-wave methods to hydrogen produce very high temperatures along the Hugoniot, as a result of the high compressibility of the material (Fig. 1), and thus probe the equation of state of the fluid[10] (although the temperatures along the Hugoniot do not follow planetary adiabats). Hence both static and dynamic experimental results require extrapolation for determining the pressure-density relations under conditions appropriate to planetary interiors. For simplicity in treatment of thermal properties, these calculations are typically based on effective pair potentials. The results of calculations using a series of different pair potential models are summarized in Fig. 1. The pair potential models are based on fits to lower pressure data. Hence, the measured densities are greater than predicted from the extrapolation of lower pressure data (which is also true for other phenomenological equation-of-state models[3]).

SOUND VELOCITY

Seismic observations of the giant planets offer the promise of radically improving our understanding of the interior structure of these bodies, much as helioseismology has opened a new window into the solar interior. Significant theoretical progress in understanding Jovian oscillations has been reported[11]. Observational programs based on impact sources[12] and ring perturbations[13] have been discussed, and in the case of the former, carried out during the impacts of Shoemaker-Levy 9 into Jupiter in July, 1994. Other ground-based searches for low-degree global oscillations have been conducted[14,15] and the first successful observations reported[15]. The observations are intriguing because the observed fundamental frequency differs strongly from that predicted from current interior models[5-7]. Direct measurement of sound velocities in planetary materials at elevated pressure are needed to better understand the seismic response of Jupiter and other gaseous planets.

During the past several years, we have developed techniques for measuring sound velocity in planetary materials, including single crystals, polycrystals, and fluids, to very high pressures. This represents an extension over our previous methods[16] in the following way: (1) single crystals grown and full anisotropy of the sound velocities determined; (2) technique combined with single-crystal synchrotron x-ray diffraction to determine the orientation of the single crystal; (3) higher sensitivity in the measurement (6-pass tandem Fabry-Perot spectrometer); (4) improved diamond anvil cells to access wider $P - T$ range. In recent work, the new technique has been applied to determine the compressional and shear wave velocities of single-crystal hydrogen to 24 GPa at room temperature[17,18]. From such measurements, the single-crystal elastic moduli were determined[18]. Together with Debye theory, the thermodynamic properties of H_2 — including the Grüneisen constant — were calculated.

Figure 2 shows the hydrodynamic or bulk sound velocities for a polycrystalline aggregate obtained from the measurements by averaging over the elastic constants[17]. We also compare the results of thermodynamic perturbation calculations for the fluid phase using several effective pair potentials. As found from x-ray diffraction work, there is a need for softening the equation of state at higher pressures relative to that predicted from pair potentials fit to lower pressure data. The improved potential (HSG) was used to calculate the fundamental oscillation frequency of Jupiter under a wide range of conditions The frequency of the fundamental mode is given by $v_0 = [\int_0^R dr/v(r)]^{-1}$, where R is the radius of the planet and v(r) is the sound speed. The analysis indicates that the reported frequency is inconsistent with that expected given our current knowledge of the physical properties of the constituents of Jupiter, assuming existing compositional models[17]. Hence, either there is a need for a significant revision of current interior models or the reported observations are incorrect. This is also in agreement with recent analysis based on shock-wave measurements for pure hydrogen[19].

Fig. 2. Sound velocity of hydrogen as a function of pressure. The filled symbols are experimental data[17]. Lines show potential model calculations as in Fig. 1. The vertical line shows the 293 K fluid-solid phase boundary for H_2. The inset shows experimental fluid phase data for hydrogen (solid symbols and full line) and helium (full line). Open symbols are for a equimolar mixture of hydrogen and helium, and the dashed line shows the calculated sound velocities in the mixture[17].

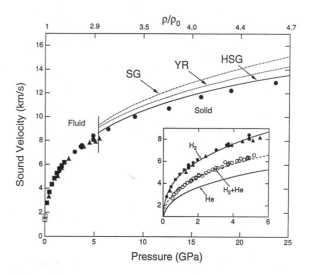

VIBRATIONAL SPECTROSCOPY

A major outstanding problem in planetary science is the pressure of the theoretically predicted molecular dissociation transition in hydrogen. The classic question considered here concerns the transition in the low temperature limit (i.e, to the Wigner-Huntington monatomic state[20]), first predicted to be as low as 25 GPa. Raman, and infrared spectroscopy have provided crucial experimental constraints on the transition. The Raman measurements indicate that the molecular bond is stable to at least ∼250 GPa (at liquid nitrogen temperatures and below)[21,22]. The measurements reveal that the frequency of the intramolecular vibration observed in both Raman scattering and infrared absorption decreases with increasing pressure, indicating that the molecular bond is becoming destabilized with increasing density. The critical pressure at which the frequency decrease is observed differs in Raman and infrared as a result of intramolecular vibrational coupling, which increases dramatically with pressure (Fig. 3)[3,23]. Thus, the results indicate that in addition there is increase in intramolecular interactions such as charge transfer. Measurements conducted to date above 250-300 GPa by Raman scattering have failed to observe the signature of the molecular bond. The loss of the Raman signal suggests dissociation, but this remains speculative because of technical difficulties associated with the Raman measurement.

Moreover, Raman and infrared vibrational spectroscopic measurements reveal a rich phase diagram for the solid at megabar pressures. These measurements involve principally the observation of discontinuities in the vibrational frequency across phase transitions. A generalized phase diagram for deuterium, showing $P - T - v$ along with proposed invariant points p_i in the system, is presented in Fig. 4. Three phases persist to megabar pressures (> 100 GPa)[3]: phase I, the rotationally disordered lower pressure phase with the hcp structure; phase II, the low-temperature intermediate phase, and phase III, which is stable above 150 GPa. The latter is characterized by a large discontinuity in the Raman and IR vibrons[21,24,25], a dramatic increase of the infrared vibron oscillator strength[25,26], and sharp decrease of the vibron discontinuity at a possible invariant point p_2[27,28]. In studies over a narrow temperature range, the I-II phase line has been extended to megabar pressures and found to intersect the phase III boundary at a triple point p_1[3,27-29]. Although these results pertain directly to the solid in a lower temperature domain, the results are important for planetary applications because they show the properties of the material at such pressures (100-300 GPa) differ markedly from those at lower pressures.

Fig. 3. Pressure dependence of the Raman and infrared intramolecular modes (vibrons) in solid hydrogen at room temperature.

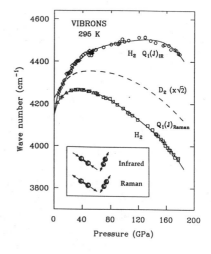

OPTICAL PROPERTIES

Recently it has been proposed that radiative layers may exist near the surfaces of Jupiter, Saturn, and Uranus at temperatures of ~2000 K[30]. If true, the interiors of these bodies would not be fully convective and their interior structure may be quite different from what is generally assumed. For example, interior temperatures would be significantly lower than predicted by completely convective models[31], although the effect is not likely to resolve discrepances in seismic observations[17]. Among the significant uncertainties in determining the opacity of the giant planets is the effect of pressure on the collision-induced absorption in the H_2-He system[30]. Determination of infrared absorption is therefore crucial for models of radiative heat transfer and optical opacity as a function of depth within the planets. In the isolated molecule, the internal stretching mode is dipole-forbidden but becomes weakly allowed by collision-induced processes, an effect which increases with increasing density[32]. Similar induced processes occur in the dense solid, but with the added constraint imposed by crystalline symmetry. Mid- to near-infrared absorption measurements have been conducted as a function of pressure. The first set of measurements to very high pressure was a study to 54 GPa at room temperature using diamond-cell techniques and a conventional Fourier-Transform spectrometer[33]. This study included measurements in the fluid state (i.e., to 5.4 GPa). More recently, the development of high-pressure synchrotron infrared spectroscopy has permitted such measurements to be extended to above 200 GPa[23,25].

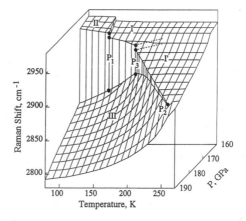

Fig. 4. Generalized phase diagram for deuterium in the megabar pressure range based on Raman scattering measurements.

Measurements have been conducted as a function of temperature between 77 and 300 K over this pressure range. In the fluid and solid phase (phase I) the intensity in the region of the H-H stretching fundamental (vibron) increases with increasing density ρ as ρ^2, or as $1/r^6$ where r is the average intermolecular separation. This is a larger increase than is found in lower pressure calculations of collision-induced absorption, which is based on low density gas-phase data[32]. Moreover, recent experiments reveal a marked increase in absorption in the high-pressure solid phase (phase III), which is stable above 150 GPa[25], as discussed above. The pressure or density dependence of the band gap of hydrogen is the central question associated with the electronic properties of the material in the molecular state[3]. There is now strong experimental evidence for a decrease in the band gap of hydrogen under pressure. Measurements of the dispersion of the refractive index at lower pressures can be used to determine an effective oscillator frequency which tracks the band edge (see, Ref. 3). These measurements (for the solid) are in excellent agreement with shock-wave electrical conductivity measurements as well as theoretical calculations (beyond the local density approximation)[3,34,35]. In addition, there is evidence for direct observation of absorption at visible wavelengths at the highest pressures reached in static experiments on hydrogen ($\sim$300 GPa)[22].

HYDROGEN CHEMISTRY

The gaseous planets are multicomponent systems, and it is generally assumed that these systems can be modeled using only end-member properties. This assumption has not been tested under conditions relevant to the interiors of these bodies, and the properties of mixtures of planetary materials are just beginning to be studied. One of the most exciting results of this field is the discovery of new high-pressure compounds involving hydrogen and other low-Z elements or simple molecular materials, in particular in the C-H-O-N system. Thus, there is a strong indication of compound formation in the cooler regions of the large planets and surfaces of the satellites. Although the deeper interiors of the Jovian planets are thought to consist primarily of fluids due to the high temperatures that prevail, the chemical interactions which drive compound formation should influence phenomena in this higher temperature regime as well (e.g., fluid-fluid separation). The properties of mixtures of He, H_2O, NH_3, CH_4, and H_2 are of critical importance for progress on a variety of issues relating to planetary interiors[36]. In particular, recent studies of Saturn's moon Titan suggest that its interior may contain complex mixtures of methane, ammonia, and water, including a possible methane clathrate hydrate[37].

In the small number of mixtures studied to date, compound formation has been the rule rather than the exception. The behavior was first observed in the N_2-He system, where evidence for $He(N_2)_{11}$ was found[38]. For hydrogen mixtures, one of the most interesting studied has been the H_2-H_2O system, where two new clathrates have been discovered (Fig. 5)[39,40]. The higher pressure one is stable to at least 60 GPa. Studies of the Ar-H_2 system reveal evidence for $Ar(H_2)_2$; spectroscopic studies suggest unusual optical and electronic behavior above 170 GPa[41]. More recent work has revealed evidence for new compounds in the H_2-CH_4[42] and H_2-O_2[43] systems. A particularly important system is H_2-He. No compound formation has yet been documented in this system[44]; moreover, recent work indicates no measurable solubility of He in H_2 above 100 GPa (<300 K)[45]. Compound formation has also been observed in related low-Z systems[46,47].

The reactivity of hydrogen with other materials such as silicates, oxides, and metals at very high pressure is important for constraining the amount of hydrogen in planetary cores and rocky layers. The structure, stability, and physical properties of high-pressure hydrous silicates and oxides, and their relevance to terrestrial planetary interiors, are examined elsewhere in this volume. Here, we note that recent work has documented the formation and stability of metal hydrides to very high pressures. In particular, Fe-H has been shown to be stable to pressures approaching the megabar range (at room temperature)[48]. Moreover, there is strong evidence that hydrogen reacts and forms stable dense hydrides with nominally inert metals such as rhenium at high pressures (e.g., > 10 GPa)[49]. Such work suggests that hydrogen could exist in substantial quantities in planetary cores, and that hydrides could have formed in the early stages of planetary evolution. The presence of hydrogen (as well as other light elements) significantly depresses melting relations (e.g., Ref. 50) and this could have a profound effect on the thermal state of planetary interiors.

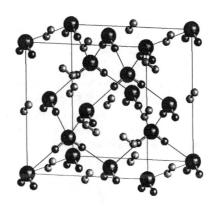

Fig. 5. Crystal structure of the high-pressure H_2-H_2O clathrate (idealized proton-ordered form)[39].

HIGHER $P - T$ BEHAVIOR

Most static compression studies of planetary materials have been performed at lower temperatures ($\leq$ 300 K) than those relevant for planetary interiors. One question therefore is what is the temperature effect on phase transitions in hydrogen? For example, the I-III phase boundary has a positive slope, but has not been explored above room temperature[27]. There is also the question of metallization in the molecular phase associated with band-overlap (in a band picture) or the convergence of a mobility edges (in a disordered system having no well defined bands). This will lead to metallic conductivity in the molecular state in the limit of $T \rightarrow 0$ K. Both of these are in general consistent with the decrease in band gap obtained from recent experiments. Thermal excitation of electrons at temperatures that prevail within the molecular layer of the large planets is expected to give rise to a conducting fluid prior to both formal gap closure and dissociation. The new data on mixtures leads to the possibility that impurity ionization could also play an important role; i.e., pressure- and temperature-induced ionization of soluble components (such as H_2O, NH_3, and O_2) could contribute conducting electrons from impurity bands into the gap prior to gap closure.

The thermal effects on the Wigner-Huntington transition require examination. In particular, recent theoretical work predicts the existence of a plasma phase transition[51]. Hence, experimental checks on this transition, and establishing its possible connection, in both a thermodynamic and structural sense, to the low-temperature Wigner-Huntington transition remain to be explored. Thermally induced dissociation, and concomitant density increase, at high $P - T$ conditions was suggested several years ago[52] on the basis of the decrease in the intramolecular stretching frequency with pressure documented for hydrogen and by analogy to proposed behavior of shocked nitrogen[53]. This would further enhance the electrical conductivity at shallower depths. Interestingly, recent high-temperature quantum simulations predict rapid exchange of atoms between molecules in the fluid state at pressures below those at which the molecules fully dissociate (150 GPa)[54]. Recent measurements and analyses of shock-temperatures have been interpreted in terms of a continuous dissociation transition in the high-density fluid state[19]. This conclusion is based on the low temperatures measured relative to those predicted from effective pair potential calculations. Such a conclusion is complicated, however, by uncertainties associated with fitting pair potentials appropriate to the high-density fluid state; recent work has shown that pair potentials that most accurately describe the x-ray and sound velocity measurements for the solid also predict lower shock temperatures along the Hugoniot (i.e., without invoking dissociation)[17]. Nevertheless, such a continuous increase in conductivity with depth cannot be ruled out, and needs to be examined within current interior models (i.e., influence on magnetic fields).

CONCLUSIONS

The results of recent static high-pressure experiments thus provide a growing experimental database for understanding the composition, chemistry, global processes, and evolution of the outer planets and their satellites. The dominant themes emerging from this work are the new phenomena in hydrogen ranging from lower pressure phenomena from pressures of < 1 GPa to the 300 GPa range. As a result, a great deal of new information has been obtained which is relevant to planetary surfaces and atmospheres, including the observation of new compounds (clathrates and van der Waals compounds), highly accurate determination of equations of state, and direct measurements on sound velocities under pressure. However, a complete microscopic description based on accurate many-body potentials and applicable over a wide $P - T$ range is not yet in hand. The challenge for future work will be the direct determination of physical and chemical properties of hydrogen and related systems at combined high $P - T$ conditions.

This work was supported by the N.S.F. and N.A.S.A.

REFERENCES

[1] B. Fegley and K. Lodders, Icarus **110**, 117 (1994).
[2] V. N. Zharkov, *Interior Structure of the Earth and Planets* (Harwood, New York, 1986).

[3] H. K. Mao and R. J. Hemley, Rev. Mod. Phys. **66**, 671 (1994).

[4] P. Loubeyre, in *Equations of State in Astrophysics,* edited by G. Chabrier and F. Schatzman, in press.

[5] T. V. Gudkova, V. N. Zharkov, and V. V. Leont'ev, Sov. Astron. Lett. **14**, 157 (1988).

[6] W. B. Hubbard and M. S. Marley, Icarus **78**, 102 (1989).

[7] G. Chabrier, D. Saumon, W. B. Hubbard, and J. I. Lunine, Astrophys. J. **391**, 817 (1992).

[8] H. K. Mao, et al., Science **239**, 1141 (1988); V. P. Glazkov, et al., JETP **47**, 661 (1988).

[9] J. Z. Hu, H. K. Mao, J. F. Shu, and R. J. Hemley, in *High-Pressure Science and Technology - 1994,* edited by S. C. Schmidt et al. (AIP, New York, 1994) p. 441.

[10] W. J. Nellis, A. Mitchell, M. Ross, J. Chem. Phys. **79**, 1480 (1983).

[11] D. Bercovici and G. Schubert, Icarus **69**, 557 (1987); S. V. Vorontsov, T. V. Gudkova, V. N. Zharkov, Sov. Astron. Lett. **15**, 278 (1989); U. Lee, Astrophys. J. **405**, 359 (1993); J. Provost, B. Mosser, and G. Berthomieu, Astron. Astrophys. **274**, 595 (1993).

[12] H. Kanamori, Geophys. Res. Lett. **20**, 2921 (1993); M. S. Marley, Astrophys. J. **427**, L63 (1994); P. Lognonne, B. Mosser, and F. A. Dahlen, Icarus **110**, 180 (1994); D. Deming, Geophs. Res. Lett. **21**, 1095 (1994); U. Lee and H. M. Van Horn, Astrophys. J **428**, L41 (1994).

[13] M. S. Marley and C. C. Porco, Icarus **106**, 508 (1993).

[14] D. Deming, M. J. Mumma, F. Espenak, D. E. Jennings, T. Kostiuk, G. Wiedemann, R. Loewenstein, and J. Piscitelli, Astrophys. J. **343**, 456 (1989).

[15] B. Mosser, F. X. Schmider, Ph. Delache, and D. Gautier, Astron. Astrophys. **251**, 356 (1991); B. Mosser, et al., ibid. **267**, 604 (1993).

[16] H. Shimizu, E. M. Brody, H. K. Mao, and P. M. Bell, Phys. Rev. Lett. **26**, 128 (1981).

[17] T. S. Duffy, W. L. Vos, C. S. Zha, R. J. Hemley, and H. K. Mao, Science **263**, 1590 (1994).

[18] C. S. Zha, T. S. Duffy, H. K. Mao, and R. J. Hemley, Phys. Rev. B **48**, 9246 (1993).

[19] W. J. Nellis, M. Ross, and N. C. Holmes, submitted.

[20] E. Wigner and H. B. Huntington, J. Chem. Phys. **3**, 764 (1935).

[21] R. J. Hemley and H. K. Mao, Phys. Rev. Lett. **61**, 857 (1988).

[22] H. K. Mao, and R. J. Hemley, Science **244**, 1462 (1989).

[23] M. Hanfland, R. J. Hemley, H. K. Mao, and G. P. Williams, Phys. Rev. Lett. **69**, 1129 (1992).

[24] H. E. Lorenzana, I. F. Silvera and K. A. Goettel, Phys. Rev. Lett. **63**, 2080 (1989).

[25] M. Hanfland, R. J. Hemley and H. K. Mao, Phys. Rev. Lett. **70**, 3760 (1993).

[26] R. J. Hemley, Z. G. Soos, M. Hanfland and H. K. Mao, Nature **369**, 384 (1994).

[27] A. F. Goncharov, I. I. Mazin, J. H. Eggert, R. J. Hemley, and H. K. Mao, to be published.

[28] R. J. Hemley and H. K. Mao, in *Elementary Processes in Dense Plasmas,* edited

by S. Ichimaru and S. Ogata (Addison-Wesley, New York, 1995) p. 269.

[29] L. Cui, N. H. Chen, S. J. Jeon and I. F. Silvera, Phys. Rev. Lett. **72**, 3048 (1994).

[30] T. Guillot, D. Gautier, G. Chabrier, and B. Mosser, Icarus **112**, 337 (1994).

[31] T. Guillot, G. Chabrier, P. Morel, and D. Gautier, Icarus **112**, 354 (1994).

[32] J. van Kranendonk (editor), *Intermolecular Spectroscopy and Dynamical Properties of Dense Systems* (North-Holland, New York, 1980).

[33] H. K. Mao, J. A. Xu, and P. M. Bell, in *High-Pressure in Science and Technology*, edited by C. Homan, et al. (North-Holland New York, 1984) p. 327.

[34] W. J. Nellis, et al., Phys. Rev. Lett. **68**, 2937 (1992).

[35] H. Chacham and S. G. Louie, Phys. Rev. Lett. **66**, 64 (1991).

[36] D. J. Stevenson, in *Shock Waves in Condensed Matter 1987*, edited by S. C. Schmidt and N. C. Holmes (Elsevier, New York, 1988).

[37] J. I. Lunine, Rev. Geophys. **31**, 133 (1993).

[38] W. L. Vos, et al., Nature **358**, 46 (1992).

[39] W. L. Vos, L. W. Finger, R. J. Hemley, and H. K. Mao, Phys. Rev. Lett. **71**, 3150 (1993).

[40] W. L. Vos, L. W. Finger, R. J. Hemley, H. K. Mao, and H. S. Yoder, in *High-Pressure Science and Technology - 1994*, edited by S. C. Schmidt et al. (AIP, New York, 1994) p. 857.

[41] P. Loubeyre, R. LeToullec, and J. P. Pinceaux, Phys. Rev. Lett. **70**, 2106 (1993).

[42] M. Somayazulu, R. J. Hemley, H. K. Mao, and L. W. Finger, to be published.

[43] P. Loubeyre and R. LeToullec, to be published.

[44] W. L. Vos, R. J. Hemley, and H. K. Mao, to be published.

[45] P. Loubeyre, private communication.

[46] W. L. Vos and J. A. Schouten, Low Temp. Phys. **19**, 338 (1993).

[47] J. A. Schouten, J. Phys.: Condens. Matter **7**, 469 (1995).

[48] J. V. Badding, R. J. Hemley, and H. K. Mao, Science, **253**, 421 (1991).

[49] J. V. Badding, D. C. Nesting, and R. B. Baron, in *High-Pressure Science and Technology - 1994*, edited by S. C. Schmidt et al. (AIP, New York, 1994) p. 1317.

[50] T. Yagi, et al., in *High-Pressure Science and Technology - 1994*, edited by S. C. Schmidt, et al. (AIP, New York, 1994) p. 943.

[51] G. Chabrier and D. Saumon, Phys. Rev. Lett. **62**, 2397 (1989).

[52] R. J. Hemley, Eos Trans. Am. Geophys. Union **69**, 159 (1988).

[53] H. B. Radousky, W. J. Nellis, M. Ross, D. C. Hamilton, and A. C. Mitchell, Phys. Rev. Letts. **57**, 2419 (1988).

[54] D. Hohl, V. Natoli, D. M. Ceperley, and R. M. Martin, Phys. Rev. Lett. **71**, 541 (1993).

RECYCLING OF HYDRATED BASALT OF THE OCEANIC CRUST AND GROWTH OF THE EARLY CONTINENTS

R. P. Rapp

Center for High Pressure Research (CHiPR) and Mineral Physics Institute
Department of Earth and Space Sciences
State University of New York at Stony Brook
Stony Brook, NY 11794

ABSTRACT

Experimentally-determined phase relations for a number of amphibolitized MORB compositions indicate that large volumes of high-Al_2O_3 trondhjemitic-tonalitic liquids (sodic-granitoids) result from melting near and just beyond the high-temperature limb of the amphibole-out phase boundary at 0.8-3.2 GPa, leaving dry and dense eclogitic residues. These liquids are strikingly similar in their major, minor and trace element chemical characteristics to the tonalite-trondhjemite-granodiorite (TTG) suite of rocks that dominate early Archean high-grade gneiss and granite-greenstone terrains. The temperatures required for complete dehydration of metabasalt by partial melting (~1000°C) imply subduction of anomalously hot (young) oceanic lithosphere; as a consequence, "slab melting" is rare in modern island arcs. Colder (older) oceanic lithosphere in "normal" subduction zones dehydrates before intersecting the wet basalt solidus, and water-rich fluids produced in these subsolidus reactions infiltrate the overlying wedge of mantle lithosphere. Partial melting of this metasomatized mantle produces high-Mg andesites and basalt. Intracrustal fractionation of these primary magmas, accompanied by varying extents of assimilation of crustal material, produces granitoids that are geochemically distinct from slab-derived, sodic-granitoids (TTG). In the Archean, shallow subduction and intraoceanic obduction may have led to imbricate thrust stacking of hot, buoyant oceanic lithosphere in compressional zones, partial melting of metabasalt and TTG magmatism. In the post-Archean, deeper subduction of older, colder oceanic lithosphere results in (incomplete?) predominantly subsolidus slab dehydration, and metasomatism and partial melting of the mantle wedge to produce high-Mg andesites and basalts. Andesite-dominated arc magmatism is the final result of protracted fractionation of these magmas. Deep infusion of oceanic lithosphere, and water, into the mantle may be restricted to the subduction of cold slabs, and the post-Archean.

INTRODUCTION

Secondary basaltic crusts are a common feature of the terrestrial planets, yet only on Earth is tertiary or continental crust conspicuous[1]. The most active sites of basaltic volcanism occur along divergent plate boundaries (e.g., mid-ocean ridges), where hot, new oceanic crust is created and hydrothermally altered. The most active sites of felsic magmatism occur along convergent plate margins (e.g., subduction zones), where old, cold basaltic crust dehydrates as it is cycled back into the mantle. As oceanic crust is transported to increasing depth by subduction, hydrous minerals formed in the basaltic oceanic crust at the mid-ocean ridge exceed their pressure-temperature (P-T) stability limits, releasing water-rich fluids which are ultimately responsible for much of arc magmatism, mass transfer between crust and mantle, and the incremental growth of the continents. The most important hydrous phase at moderate to high pressures in metamorphosed basalt is amphibole[2]; however, the conditions under which amphibole dehydration takes place during subduction, and the nature of the decomposition reaction (specifically, whether it occurs at subsolidus conditions or involves partial melting), remain poorly understood. However, it is reasonable to expect that the geochemical characteristics of a fluid formed by subsolidus dehydration of subducted oceanic crust will be distinctly different from a melt formed by amphibole dehydration-melting.

The debate over the ultimate origin of continental crust has been revitalized by the suggestion that certain (albeit rare) island arc magmas represent "slab melts", derived directly by anatexis of subducted oceanic crust[3,4], or, alternatively, result from partial melting of underplated sub-arc basalt, previously emplaced at the crust-mantle interface[5]. Volumetrically, volcanism along island arcs and continental margins is dominated overwhelmingly by andesitic magmas, which are considered to be derived by intracrustal fractional crystallization of primary high-Mg basaltic magmas[6,7] that result from partial melting of a mantle wedge that has been metasomatized by slab-derived fluids. Despite the dominance of andesitic volcanism, the net flux of material from the mantle wedge to the arc crust is nevertheless basaltic[8].

Whether or not slab-derived fluids in subduction zones are water-rich, supercritical fluids resulting from sub-solidus dehydration of the slab, or hydrous melts produced by amphibole dehydration, will depend upon the orientation of the *amphibole-out* phase boundary in metabasalt with respect to the prevailing thermal structure[9]. In addition to their geochemical implications, these considerations will be relevant to the issue of crust-to-mantle recycling of water, since the extent of slab dehydration will limit the amount of water returned to the upper mantle via subduction.

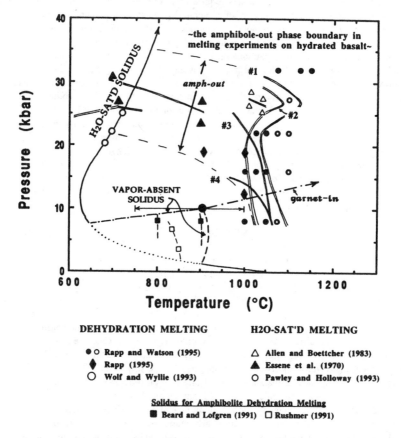

Fig. 1. Position of the amphibole-out phase boundary in melting experiments on hydrated metabasalt, with reference to the water-saturated basalt solidus (ref. 11), and the vapor-absent solidus for dehydration melting (ref. 12). Amphibole-out curves #1-#4 from ref. 17. Solid circles and diamonds denote quartz-normative partial melts.

THE AMPHIBOLE-OUT PHASE BOUNDARY IN MELTED METABASALT

1. Amphibole Stability Limits

A number of experimental studies, summarized in Figure 1 (from ref. 10) and presented with reference to the water-saturated basalt solidus of Lambert and Wyllie[11] and the vapor-absent solidus of Wolf and Wyllie[12], have helped delineate various portions of the *amphibole-out* phase boundary. It is clear that the high-pressure limb of *amphibole-out*, both above and below the water-saturated solidus, is only loosely constrained by the <u>water-saturated</u> studies of Allen and Boettcher[13], Essene et al.[14], and Pawley and Holloway[15]. The emphasis in Fig. 1 is on dehydration melting under vapor-absent conditions (i.e., no free fluid phase), since most interstitial pore fluids in subducted sediments and oceanic crust are probably expelled at shallow depths in subduction zones[16]. A summary of amphibole stability in water-saturated basalts is given in Helz[2]. In general, both the water-saturated and vapor-absent melting studies indicate that there is a broad field over which amphibole coexists with partial melt, but that the exact position of the *amphibole-out* phase boundary is primarily a function of the bulk composition of the basalt (e.g., silica activity[2]), and is affected to a lesser extent by such factors as oxygen fugacity and water activity (*ibid.*). In dehydration melting experiments on four natural, amphibolitized olivine tholeiite basalts, Rapp and Watson[17] focused on the location of *amphibole-out* (curves #1-#4 in Fig. 1) and its effect on the composition of coexisting partial melt, and found that compositional parameters other than silica-activity influence phase relations. Amphibolite dehydration melting experiments at 0.3-0.8 GPa substantiate the composition-dependence of *amphibole-out* below the garnet-in phase boundary[18,19], and the relationship between the backbent segment of the vapor-absent, amphibole dehydration melting solidus and the garnet-in phase boundary has been discussed by Wolf and Wyllie[12] and Wyllie and Wolf[20].

2. Partial Melt Compositions Near *Amphibole-Out*

Electron micro-probe analyses of the quenched glass (i.e., liquid) in experiments below and slightly above the amphibole-out phase boundary, from 0.8 to 3.2 GPa, reveal that 10-40 volume % melting produces quartz-normative, sodium-rich granitic liquids[21,22] (trondhjemite-tonalite in terms of ternary feldspar components Ab-An-Or). Depend-ing upon the exact pressure-temperature conditions, residual crystalline assemblages may consist of amphibolite, garnet-amphibolite, granulite, garnet-granulite, or eclogite (Fig. 1.). The composition of liquids produced by partial melting above the wet basalt solidus is also strongly dependent on the P-T conditions under which melting takes place relative to the *amphibole-out* phase boundary. Partial melts between the wet solidus and *amphibole-out* are peraluminous (i.e., A/CNK, molar [Al2O3/(CaO+Na2O+K2O)] > 1.0) quartz-normative granitoids, but liquids at and just beyond this phase boundary are meta-luminous (i.e., A/CNK < 1.0)[10]. In terms of major element oxide components, the experimental liquids bear close resemblance to the Archean trondhjemite-tonalite-granodiorite suite of grani-toids, which are the dominant felsic igneous rock in high grade gneiss and granite-greenstone terrains.

The trondhjemitic-tonalitic liquids produced in several of the basalt melting experiments of Rapp and Watson[17] were analyzed by ion microprobe for a number of trace elements, including the rare-earth elements (REEs), Sr, Y and Cr. Chondrite-normalized rare-earth patterns for TTG partial melts at 1.2 GPa and 1000°C, 2.2 GPa and 1050°C, and 3.2 GPa and 1100-1150°C, were found to be strongly fractionated and depleted in the heavy rare-earth elements (HREE)[23]. Rare earth element patterns with relatively high La/Yb ratios at low total Yb contents appear to be typical of Archean sodic granitoids (TTG), a feature that distinguishes them from their post-Archean, potassic granite counterparts (Fig. 2, after Martin[24]). The experimental tonalite-trondhjemite liquids also possess high Sr (~400-1200 ppm), and low Cr (5-10 ppm) and Y (2-8 ppm) contents[23].

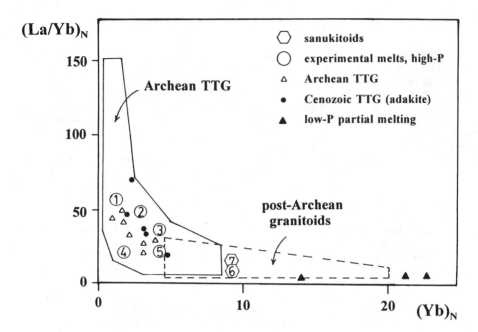

Fig. 2. Chondrite-normalized La/Yb versus Yb for Archean TTG granitoids and post-Archean granitoids, sanukitoids from the Superior province[27] (#7), the Setouchi volcanic belt, SW Japan[28] (#6), and experimental high-pressure partial melts of metabasalt[17,23] at 3.2 GPa, 1000°C (#1), 3.2 GPa, 1100°C (#2), 3.2 GPa, 1150°C (#3), 2.2 GPa, 1050°C (#4), all coexisting with eclogite residue; and at 1.2 GPa, 1000°C, garnet-amphibolite residue. Also shown are average Cenozoic adakite and Archean TTG, and low-pressure partial melts of amphibolite[3].

SLAB-DERIVED VERSUS MANTLE-WEDGE DERIVED GRANITOIDS: ARCHEAN TO MODERN EXAMPLES

Residual (crystalline) assemblages coexisting with the experimental tonalite-trondhjemite liquids are dominated by eclogite minerals (e.g., garnet and clinopyroxene), with minor but significant quantities of amphibole and plagioclase present below 1000°C at ~1-2.5 GPa. The strongly fractionated and HREE-depleted rare-earth patterns of the experimental melts support early interpretations, based on trace element distribution patterns in Archean granitoids from the North Atlantic Craton, southern Africa, and the Canadian Shield, that the TTG suite was derived by partial melting of an amphibole- and/or garnet-bearing mafic crustal source (e.g.,ref. 25). The Superior Province of Canada, like most Archean granite-greenstone terrains, appears to be dominated by sodium-rich trondhjemites and tonalites. Recently, however, more comprehensive and detailed field mapping, and a greater number of geochemical analyses of granitoids, has revealed that another distinct suite of silica-saturated rocks are present in parts of the Superior Province in subordinate but significant proportions (circa 20%) (e.g., refs. 26, 27). This suite of rocks, termed the sanukitoid suite due to their striking resemblance geochemically to high-magnesian andesite volcanics (sanukitoids) in the Setouchi Belt of SW Japan[28], are distinguished by their high Mg#'s (i.e.,molar Mg/(Mg+Fe)), high Ni and Cr contents, and large-ion lithophile element (LILE) enrichment. Sanukitoids are interpreted to be primary magmas generated by partial melting of hydrous, LILE-enriched metasomatized mantle peridotite[26,27]. Considerable controversy revolved around attempts to demonstrate experimentally that primary andesitic magmas could be formed by partial melting of peridotite under hydrous conditions[29-34]; much of the debate centered on the question of the composition of equilibrium liquids[34]. There appeared to be general agreement, however, that the important geochemical characteristics of orogenic andesite were inconsistent with a model in which <u>unaltered</u> peridotite was the source rock for the primary partial melt[35,36,37]. However, liquidus experiments on high-magnesian andesites from the Setouchi volcanic belt, SW Japan (the type-locality for sanukitoids) demonstrated that under hydrous conditions these magmas could have been in equilibrium with either lherzolitic or harzburgitic peridotite assemblages at pressures above 1.0 GPa, depending upon the exact temperature, pressure, and amount of water present[38,39,40]. In any case, sanukitoids seem to require a mantle source metasomatized by a LILE-enriched fluid or melt; the origin of this fluid, and the nature of the enrichment process, remains poorly constrained.

Compared to both experimentally-produced tonalite-trondhjemite liquids and Archean TTG granitoids, sanukitoids in general have lower La/Yb ratios and somewhat higher Yb contents (see Fig. 2). In terms of their Sr/Y versus Y characteristics, natural and experimental TTG granitoids are intermediate to boninites from the Circum-Pacific (Marianas[41], Bonin Isles[41], New Guinea[42]), which possess relatively low Sr/Y ratios and low Y contents, and high-Mg andesites from the American Cordillera[43] which possess equivalent Sr/Y ratios at higher Y contents (Fig. 3a). Sanukitoids appear to be hybrid products with characteristics intermediate to these groups (Archean sanukitoids and hybrid granitoids of the Superior Province[44,45] in Fig. 3a). However, when the Cr/Y versus Y characteristics of these same samples are plotted (Fig. 3b), mantle-derived arc magmas (i.e., high-Mg andesites and boninites) are easily distinguished from TTG magmas derived from melting of a mafic crustal source (i.e., trondhjemite-tonalite suite), and sanukitoids appear to more like the mantle-derived magmas. Note that a tonalite-granodiorite-granite-quartz monzonite (TGGM) suite of metaluminous granitoids from the Superior Province of Canada[44] plot on Figs. 3a and 3b in the vicinity of the experimental TTG liquids that coexist with eclogite residues at 3.2 GPa, as do average Archean TTG[3], Cenozoic-Mesozoic adakite[4], and Cenozoic sodic granites from the Cordillera Blanca[5]. In contrast, a tonalite-trondhjemite-granodiorite (TTG) suite of peraluminous granitoids from the same study in the Superior Province, as well as larger sample suites from the Archean of East Antartica[46] and the Kaapvaal Craton[47,48] to varying extents exhibit trends of significant increases in Y content correlated with more gradual decreases in Sr/Y and Cr/Y ratios Fig. 3a and 3b). These geochemical features (i.e., peraluminosity, higher Y content relative to eclogite-derived trondhjemite-tonalite) can be attributed to derivation from a mafic source dominated by residual amphibole with little or no garnet, whereas a mafic source dominated by residual garnet results in metaluminous TTGs with significantly lower Y contents, due to high partition coefficients for Y in garnet coexisting with TTG liquids. Partial melting in the latter case would be expected to have proceeded beyond the *amphibole-out* phase boundary. The distinctive features of Archean TTG, as exemplified by the data of Fig. 2 and Fig. 3, can be attributed to the presence of a significant proportion of residual garnet in their mafic source. Using the same criteria, modern arc magmas with the same geochemical characteristics can be identified (e.g., adakites[4], and sodic granitoids of the Cordillera Blanca[5]), suggesting their derivation by melting of mafic crust within the garnet stability field. Within a modern plate tectonic framework, therefore, Archean TTGs are slab-derived granitoids with a mafic crustal source, and sanukitoids are mantle wedge-derived granitoids (Table I).

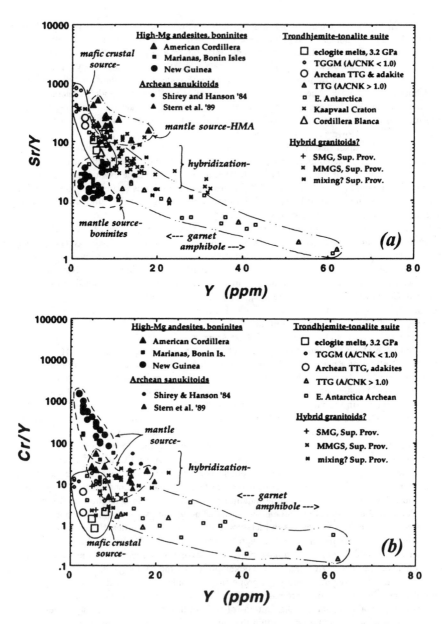

Fig. 3. (a). Relationship between Sr/Y ratio and Y content in high-Mg andesites[42,43] and boninites[41,42] from the Circum-Pacific, Archean sanukitoids[26,27], both experimental[17,23] and Archean[3,46-48] to modern[4,5] TTG, and "hybrid" granitoids from the Superior Province[44,45] with intermediate characteristics. (b). Relationship between Cr/Y ratio and Y content, for same set of samples.

Table I. Source regions for Archean granitoids and modern analogues.

source	amph. and/or garnet-bearing mafic crust	hydrous, LILE-enriched mantle
composition	high-Al trondhjemites and tonalites	sanukitoids (trachyandesite) and monzodiorites
geochemical features	fractionated REEs, strongly HREE-depleted; high LILEs, Mg# < 0.5, low Cr and Ni high Sr/Y, low Cr/Y, low Y	fractionated REEs, moderately HREE-depleted; high LILEs, Mg# > 0.6, high Cr and Ni high Sr/Y and Cr/Y, moderate Y
analagous arc magmas	adakites, Na-granites low-K dacites, tonalites	high-Mg andesites sanukites (boninites?)

DISCUSSION

1. Granitoid Petrogenesis and Slab Geotherms

The question of whether silica-saturated arc magmas are predominantly derived by hydrous melting of the oceanic crust of the subducted slab or by melting of the overlying mantle wedge, metasomatized by a water-rich, slab-derived fluid, ultimately addresses the issue of whether granitoid magmas which form the continental crust originate by melting of mafic crustal rocks (i.e., tertiary crust from secondary crust), or are primary melts of mantle ultramafic rocks. Much effort has been devoted recently to the development of numerical models to describe the thermal regime of subduction zones[49], in order to ascertain whether or not slab geotherms intersect the wet basalt solidus, and under what circumstances. These models suggest that under extreme conditions (e.g., subduction of young, hot oceanic lithosphere; ridge subduction) melting of the subducting slab is possible[49,50], although constraints from such models are liberal.

The products of magmatism along convergent plate margins provide the most direct evidence of melting processes in subduction zones, and therefore arc magmas themselves may be the best indicators of the thermal state of the slab. Slab-derived melts should in principle be distinguish-able from mantle-derived arc magmas on the basis of Cr/Y and Sr/Y ratios, and Cr, Ni and Y contents (Fig. 3a,b). Realistic models would accomodate intermediate products of end-member granitoid-generating processes (e.g., see Fig. 3a,b), by some measure of source heterogeneity[51], assimilation between quartz-normative magmas (TTG) and mantle peridotite[36,37,7], or both.

2. Implications of Slab Melting in the Archean

The overwhelming abundance of TTG granitoids in Archean cratons testifies to the likelihood that younger, hotter, and therefore relatively buoyant, hydrated oceanic crust was transported or displaced to depths of at least 40-100 km, where partial melting to produce large volumes of trondhjemite-tonalite magmas left dry and dense eclogitic residues. Subduction of hot oceanic crust in the early Archean, if it occurred, would be shallow, and intraoceanic obduction might in fact dominate, as suggested for the Kaapvaal Craton by deWit et al.[52]. Imbricate thrust-stacking of oceanic lithosphere could conceivably lead to load-induced subsidence and dehydration melting of oceanic crust. The eclogite residue of melting would sink into the underlying sub-continental mantle "keel" at depths of ~100-300 km, where it would later be sampled by kimberlite magmas[53]. As the Earth cooled, reduced heat flow would result in older and colder oceanic lithosphere, and slab geotherms during subduction would intersect the wet basalt solidus less and less frequently. Subsolidus dehydration of the slab, metasomatism of the mantle wedge, and andesite-dominated arc magmatism would become an important crust-generating process. Oceanic lithosphere would be denser, and deep transport of water via subduction and the penetration of slabs into the lower mantle would only then be possible[54].

Geochemical evidence for an evolution in granitoid petrogenesis during the late Archean (~2.75-2.67 Ga) is found in the Abitibi granite-greenstone terrain of the Superior Province, Canada. Feng and Kerrich[44] identify four distinct granitic magma series, developed systematically in space and time, and marking the transition from pre- and syn-tectonic TTG magmas derived by anatexis of mafic crust, to late- to post-tectonic, mantle-derived sanukitoids.

ACKNOWLEDGEMENTS

This research was supported by NSF grant number EAR-8920239 to the Center for High Pressure Research. The author wishes to thank the Lunar and Planetary Institute and the organizing committee for the Conference on Deep Earth and Planetary Volatiles, especially Tom Ahrens and Ken Farley, for inviting this contribution. A review by Dana Johnston greatly improved the manuscript, and is much appreciated.

REFERENCES

1. S. R. Taylor, Tectonophysics 161, 147-156 (1989).
2. R. T. Helz, In: Veblen, D.R. and Ribbe, P.H. (Eds.) *Amphibole:Petrology and Experimental Phase Relations.* Reviews in Mineralogy Vol. 9B, Mineral. Soc. America, 229-353.
3. M. S. Drummond and M. J. Defant, J. Geophys. Res. 95, 21503-21521 (1990).
4. M. J. Defant and M. J. Drummond, Nature 347, 662-665 (1990).
5. M. P. Atherton and N. Petford, Nature 362, 144-146 (1993).
6. T. L. Grove and R. J. Kinzler, Ann. Rev. Earth Planet. Sci. 14, 417-454 (1986).
7. P. B. Kelemen, Mineral. Mag. 58A, 464-465 (1994).
8. R. M. Ellam and C. J. Hawkesworth, Geology 16, 314-317 (1988).
9. A. B. Thompson, Nature 358, 295-302 (1992).
10. R. P. Rapp, J. Geophys. Res. (in press).
11. I. B. Lambert and P. J. Wyllie, J. Geology 80, 693-708 (1972).
12. M. B. Wolf and P. J. Wyllie, Contrib. Mineral. Petrol. 115, 369-383 (1994).
13. J. C. Allen and A. L. Boettcher, Am. Mineral. 68, 307-314 (1983).
14. E. J. Essene, B. J. Hensen and D. H. Green, Phys. Earth Planet. Int. 3, 378-384 (1971).
15. A. R. Pawley and J. R. Holloway, Science 260, 664-667 (1993).
16. M. G. Langseth and J. C. Moore, J. Geophys. Res. 95, 8737-8741 (1990).
17. R. P. Rapp and E. B. Watson, J. Petrology (in press).
18. J. S. Beard and G. E. Lofgren, J. Petrology 32, 365-401 (1991).
19. T. Rushmer, Contrib. Mineral. Petrol. 107, 41-59 (1991)
20. P. J. Wyllie and M. B. Wolf, In: H. M. Pritchard, T. Alabaster, N. B. W. Harris and C. R. Neary (Eds.) *Magmatic Processes and Plate Tectonics*, Geol. Soc. Spec. Pap. 76 (London) p. 405-416 (1994).
21. R. P. Rapp, E. B. Watson and C. F. Miller, Precambrian Res. 51, 1-25 (1991).
22. K. T. Winther and R. C. Newton, Bull. Geol. Soc. Denmark 39, 213-228 (1991).
23. R. P. Rapp and N. Shimizu, IUGG XXI General Assembly abstracts (in press).
24. H. Martin, Geology 14, 753-756 (1986).
25. F. Barker and J. G. Arth, Geology 4, 596-600 (1976).
26. S. B. Shirey and G. N. Hanson, Nature 310, 222-224 (1984)
27. R. A. Stern, G. N. Hanson and S. B. Shirey, Can. J. Earth Sci. 26, 1688-1712 (1989).
28. Y. Tatsumi and K. Ishizaka, Earth Planet. Sci. Lett. 60, 293-304 (1982).
29. I. Kushiro, J. Petrology 13, 311-334 (1972).

30. I. A. Nicholls and A. E. Ringwood, Earth Planet. Sci. Lett. 17, 243-246 (1972).
31. D. H. Green, Earth Planet. Sci. Lett. 19, 37-53 (1973).
32. I. A. Nicholls, Contrib. Mineral. Petrol. 45, 289-316 (1974).
33. B. O. Mysen and A. L. Boettcher, J. Petrolgy 16, 549-593 (1975).
34. D. H. Green, Can. Mineral. 14, 255-268 (1976).
35. B. O. Mysen, In: R. S. Thorpe (Ed.) *Andesites* (Wiley and Sons, London) 489-522 (1982).
36. A. D. Johnston and P. J. Wyllie, Contrib. Mineral. Petrol. 102, 257-264 (1989).
37. M. R. Carroll and P. J. Wyllie, J. Petrology 30, 1351-1382 (1989).
38. Y. Tatsumi, Earth Planet. Sci. Lett. 54, 357-365 (1981).
39. Y. Tatsumi, Earth Planet. Sci. Lett. 60, 305-317 (1982).
40. T. Kawasaki, K. Okusako and T. Nishiyama, The Island Arc 2, 228-237 (1993).
41. R. L. Hickey and F. A. Frey, Geochim. Cosmochim. Acta 46, 2099-2115 (1982).
42. G. A. Jenner, Chem. Geology 33, 307-332 (1981).
43. G. Rogers and A. D. Saunders, In: A. J. Crawford (Ed.) *Boninites*, 416-443 (Unwin Hyman, London).
44. R. Feng and R. Kerrich, Chem. Geology 98, 23-70 (1992).
45. R. H. Sutcliffe, A. R. Smith, W. Doherty and R. L. Barnett, Contrib. Mineral. Petrol. 105, 255-274 (1989).
46. J. W. Sheraton and L. P. Black, Lithos 16, 273-296 (1983).
47. D. R. Hunter, F. Barker and H.T. Millard, Jr., Precambrian Res. 24, 131-155 (1984).
48. D.R. Hunter, R. G. Smith and D. W. W. Sleigh, J. African Earth Sci. 15, 127-151 (1992).
49. S. M. Peacock, T. Rushmer and A. B. Thompson, Earth Planet. Sci. Lett. 121, 227-244 (1993).
50. S. M. Peacock, Science 248, 329-337 (1990).
51. R. P. Rapp, In: L. D. Ashwal and M. J. deWit (Eds.) *The Tectonic Evolution of Greenstone Belts* (Oxford Univ. Press, London) (in press).
52. M. J. deWit, C. Roering, R. J. Hart, A. Armstrong, C. E. J. de Ronde, R. W. E. Green, M. Tredoux, E. Peberdy, and R. A. Hart, Nature 357, 553-562 (1992).
53. H. Helmstaedt and D. J. Schulze, GSA Spec. Publ. 14, 358-368.
54. M. T. McCulloch, Earth Planet. Sci. Lett. 115, 89-100 (1993).

IMPACT-INDUCED DEGASSING OF NOBLE GASES FROM OLIVINE

Shun-ichi Azuma, Hajime Hiyagon
Department of Earth and Planetary Physics, University of Tokyo
Yayoi 2-11-16, Bunkyo-ku, Tokyo 113, Japan.

Tsuneji Futagami
Cosmic Ray Research Institute, University of Tokyo
Tanashi 3-2-1, Tokyo 188, Japan.

Yasuhiko Syono and Kiyoto Fukuoka
Institute for Materials Research, Tohoku University
Katahira 2-1-1, Aoba-ku, Sendai 980-77, Japan.

ABSTRACT

Shock-recovery experiments were performed on ^{4}He-^{40}Ar-irradiated olivine samples. The estimated peak pressures were ~30, ~40 and ~50 GPa. Noble gases were extracted from the recovered samples (OL-30, -40 and -50) and an unshocked sample (UOL2) by stepwise heating and analyzed with a sector-type mass spectrometer. Three peaks were observed in the ^{40}Ar release pattern for UOL2 at 700-800°C, 1100-1300°C and 1500-1600°C, suggesting three different trapping sites of Ar in olivine. In spite of significant loss of ^{40}Ar from OL-30 and OL-40 (62-86%), they also showed three release peaks, very similar to those for UOL2. If thermally-induced diffusion was responsible for ^{40}Ar loss, preferential loss from the lower temperature peaks must have occurred, which was not the case. Furthermore, degassed fractions of ^{40}Ar and ^{4}He were very similar to each other: 62% and 67% for OL-30, and 86% and 89% for OL-40, respectively. These observations are consistent with the idea of "all or nothing"-type degassing, where some portion was severely destroyed with the shock and released the gas completely while the other portion had little damage and retained most of the gas.

INTRODUCTION

Impact-induced degassing is one of the important candidates for the sudden degassing from the early Earth[1]. Ahrens and his colleagues have reported impact-induced degassing of H_2O and CO_2 from hydrous minerals (e.g. serpentine) or carbonate minerals[2,3,4]. Noble gases have been used as important tracers in considering the Earth's degassing processes[1], but their degassing conditions and degassing mechanisms are not well understood because of relatively small number of data[5,6]. In the present study, we performed shock recovery experiments on olivine samples irradiated with ^{4}He and ^{40}Ar. In the present report, we briefly summarize the results of noble gas analyses and some observations of XRD and TEM analyses. More details of the results including XRD, SEM and TEM observations will be shown elsewhere[7].

EXPERIMENTAL

A thinly sliced (~0.4 mm thick) olivine sample (from San Carlos) was irradiated with ^{4}He and ^{40}Ar ions at Takasaki Radiation Chemistry Research Establishment, Japan

270 © 1995 American Institute of Physics

Atomic Energy Research Institute. The irradiation doses, irradiation energies and the estimated penetration depths of ^{4}He and ^{40}Ar ions were 9.9×10^{14} ions/cm^2 and 4.9×10^{14} ions/cm^2, 40keV and 200keV and 1470Å and 2800Å, respectively. The procedure for ion implantation was similar to that described by Futagami et al.[8] The irradiated sample as well as an unirradiated sample were powdered (grain size: 63-125 μm), mixed together, and used for the starting material (UOL2). The concentrations of ^{4}He and excess ^{40}Ar in UOL2 were 2.12×10^{-5} cm^{3}STP/g and 1.28×10^{-5} cm^{3}STP/g, respectively. Contribution of noble gases originally contained in San Carlos olivine was negligibly small. By using excess ^{40}Ar, we could avoid the problem of atmospheric contamination. The powder was pressed to make pellets (5 mm in diameter x 1 mm in thickness, ~50 mg in weight, and ~20% porosity) and used for targets. The sample was put in a stainless steel (SUS304) container having a ventilating path for the released volatiles. The container was shot with a polycarbonate projectile with an iron flyer accelerated by a single-stage powder gun at Institute for Materials Research (IMR), Tohoku University. Three shock experiments were performed. The estimated peak pressures for the three samples (OL-30, OL-40 and OL-50) were 30.4GPa, 39.6GPa and 48.3GPa, respectively. The peak temperatures, on the other hand, could not be estimated, because temperature increase in the powdered samples having large porosities must be very heterogeneous. The recovered fractions for OL-30 and OL-40 were about 80% and 50%, respectively, while only a small fraction (<10%) was recovered from OL-50 due to severe destruction of the container. The recovered samples composed of relatively large grains (similar to the original grain size) as well as much smaller grains (a few μm or less). Noble gases were extracted by heating the samples stepwise from 100°C to 1800°C or 1860°C with 100°C steps for UOL2, OL-30 and OL-40 and analyzed with a sector-type mass spectrometer. For OL-50, noble gases were extracted with 300°C steps because of the small sample size (~5 mg). The recovered samples were also analyzed with an X-ray diffractometer (XRD), a scanning electron microscope (SEM), and a transmission electron microscope (TEM).

RESULTS AND DISCUSSION

Degassed fractions of ^{4}He and ^{40}Ar are 67% and 62% for OL-30, 89% and 86% for OL-40, and 95% and 79% for OL-50, respectively (see Fig.1). Except for OL-50, the degassed fractions of ^{4}He and ^{40}Ar are very similar to each other within experimental uncertainties (~5%). If thermally-induced diffusion was responsible for the loss of ^{4}He and ^{40}Ar, it would have caused significant fractionation between them due to large difference in their diffusivities, which however was not the case. Therefore, some mechanisms other than diffusion must be responsible for the loss of ^{4}He and ^{40}Ar.

It is curious that loss of ^{40}Ar for OL-50 is smaller than that for OL-40, while loss of ^{4}He for OL-50 is larger than that for OL-40. The cause for this is not clear at present. One possibility is that some part of olivine was melted at ~50GPa and ^{40}Ar once liberated from the crystal structure was trapped in the melt and fixed when the melt cooled to form glass or quench crystals. On the other hand, once liberated ^{4}He might be lost from the sample due to its high diffusivity. Another possibility is that ^{40}Ar once liberated from the sample at the passage of the first shock wave was again shock-implanted into the sample during repeated reflections of the shock wave in the sample, while once liberated ^{4}He was lost from the sample due to its high diffusivity. At present, we can not judge which is really the case. Further study is needed to fully understand this observation.

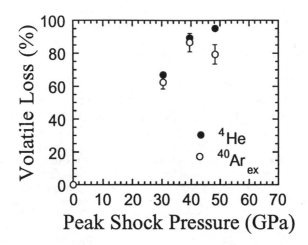

Fig.1 Degassed fractions of ⁴He and excess ⁴⁰Ar.

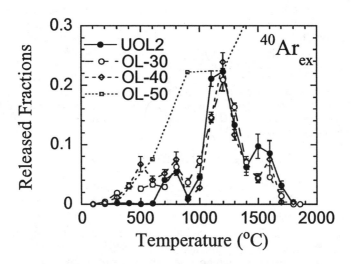

Fig.2 Thermal release patterns of ⁴⁰Ar.

The thermal release patterns of ^{40}Ar from the shocked and unshocked samples are shown in Fig.2. Some important features are observed in the figure. First, for the unshocked sample (UOL2), ^{40}Ar was released at three different temperature ranges, 700-800°C, 1100-1300°C and 1500-1600°C, which is consistent with the previous result[8]. This suggests that there are three different trapping sites of Ar in olivine, though details about these trapping sites are not well known. Second, in spite of significant loss of ^{40}Ar from OL-30 (62%) and from OL-40 (86%), they preserved all the three peaks and the release patterns were quite similar to that for UOL2 (except for minor but notable gas release below 600°C). If thermally-induced diffusion was responsible for ^{40}Ar loss, preferential loss from the lower temperature peaks must have occurred, which was not the case. This suggests that loss of ^{40}Ar from the shocked samples was not caused by simple diffusional processes.

Note that for OL-50 the gas was extracted with 300°C steps instead of 100°C steps and hence a direct comparison cannot be made between the release pattern for OL-50 and those for the other samples. However, ^{40}Ar release at >1200°C for OL-50 is apparently higher than those for the other samples. For OL-50, some process must have operated by which ^{40}Ar was trapped very tightly in the sample, which however is not known from the present result only.

Both of the above observations (shown in Fig.1 and Fig.2) are not consistent with thermally-induced diffusional loss of noble gases. Instead, they seem to support the idea that some portion of olivine was severely destroyed with the shock and released the gas completely while the other portion had little damage and retained all the gas without any loss. We may call this kind of degassing as the "all or nothing"-type degassing.

It should be noted here that noble gases in the present samples existed only in the uppermost 1000-3000Å layer of the original surface, and hence, the observed thermal release patterns of noble gases were almost independent of the grain size of the recovered samples. If the above interpretation (i.e., "all or nothing"-type degassing) is correct, the observed loss of ^{40}Ar (and ^{4}He) from the shocked sample may indicate the fraction of the damaged portion of the (original) surface layer of olivine grains.

Judging from the almost unfractionated degassing of ^{4}He and ^{40}Ar from OL-30 and OL-40 (Fig. 1), it seems clear that "all or nothing"-type degassing might also occur for ^{4}He. However, as shown in Fig.3, the release patterns of ^{4}He were not identical among the shocked and unshocked samples. The peak release temperatures were 500°C for UOL2, 500-600°C for OL-30, 600-700°C for OL-40 and ~900°C for OL-50, that is, the peak release temperature of ^{4}He became higher as the shock pressure increased. (Here, also note that the release pattern for OL-50 cannot be compared directly with the others because of different temperature steps.)

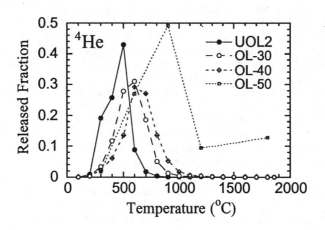

Fig.3 Thermal release patterns of 4He.

One possible explanation for the shift of the peak is diffusion of ^{4}He towards both deeper and shallower portion of the sample surface but some ^{4}He in the shallower portion was lost from the sample possibly due to residual heat. This would effectively result in the shift of the average depth of ^{4}He from the surface and hence result in the shift of the release peak to a higher temperature. If this is the case, some ^{4}He must be lost even from the relatively undamaged portion of the shocked samples, implying diffusional loss must be considered for ^{4}He in addition to the "all or nothing"-type degassing. Diffusional loss of ^{4}He, however, must not be too large judging from the almost unfractionated loss of ^{4}He and ^{40}Ar (Fig.1).

In order to examine whether or not there were any changes in the crystal structures caused by the shock, an X-ray diffraction analysis (XRD) were performed on the recovered samples (OL-30 and OL-40) at the Radio-Isotope Centre, University of Tokyo. The X-ray spectra of the shocked samples were very similar to that of the unshocked sample (UOL2) except for the decrease in the peak intensities. This suggests that olivine was not decomposed into other minerals by the shock and the crystal structure was retained at least in some portions of the shocked samples. The decrease in the peak intensities shows a possibility that some portion of olivine was converted to amorphous structure. It should be noted, however, that an X-ray diffraction analysis gives only qualitative information and we cannot derive such a conclusion from the X-ray analysis only.

In order to further examine whether an amorphous structure was formed in the shocked olivine, we conducted a TEM analysis on OL-30 and OL-40 at Mineralogical Institute, University of Tokyo. It was found that a large portion of olivine grains retained the original structure but that some portion became aggregates of very fine grains or even became almost amorphous judging from the halo image of TEM electron diffraction pattern. The TEM observations suggests that melting probably did not occur for OL-30 and OL-40. Unfortunately it was not possible from the TEM analysis to estimate the fraction of the amorphous structure. For OL-50, on the other hand, we observed by SEM a few-μm-sized olivine spheres just at the boundary of the sample and the container, which apparently suggests some part was melted at the shock.

The X-ray diffraction analysis and TEM and SEM observations show that the impact-induced destruction of the crystal was heterogeneous, that is, some portion was severely destructed to form (almost) amorphous structure but the other portion was not damaged and retained the original structure. These observations are very consistent with the "heterogeneous yielding model" [9,10] and with the idea of the "all or nothing"-type degassing suggested by the noble gas results. Noble gases might be almost completely released from the severely damaged (amorphous) portion but only a negligible release occurred from the (relatively) undamaged portion.

The present study clearly shows that the "all or nothing"-type degassing associated with heterogeneous destruction of the crystal is an important degassing process for noble gases. In this context, it is important to understand the mechanism controlling the size and distribution of the destructed regions, which however remains unanswered. The effect of elemental fractionation between ^{4}He and ^{40}Ar was very minor in the present experimental conditions. However, some effect of diffusion was suggested by the shift in the release temperature of ^{4}He from the shocked samples (Fig.3). Therefore, it is expected that the effect of elemental fractionation will be more important in the case of natural impact phenomena with km-sized bodies, because of longer duration of the peak temperatures (~1 sec) compared with that for the present experiments (~1 μsec).

ACKNOWLEDGMENT

We thank Masana Morioka for help in XRD analysis and Osamu Tachikawa for TEM analysis. We thank Naoji Sugiura and Ichiro Kaneoka for fruitful discussions. We also express thanks to Linda Lowan for valuable comments in the review, which improved the manuscript. This work was performed under the inter-university research program of IMR (Institute of Materials Research), Tohoku University, and financially supported in part by Society for Promotion of Space Science, Japan.

REFERENCES

1. Y. Hamano and M. Ozima, *Adv. Earth Planet. Sci.*, **3**, 155 (1978).
2. M. A. Lange and T. J. Ahrens, *Icarus*, **51**, 96 (1982).
3. T. J. Ahrens, J. D. O'Keefe and M. A. Lange, *Origin and Evolution of Planetary and Satellite Atmospheres* (Univ. Arizona, Tucson), 328 (1989).
4. J. A. Tyburczy, R. V. Krishnamurthy, S. Epstein and T. J. Ahrens, *Earth Planet. Sci. Lett.*, **98**, 245 (1990).
5. C. Gazis and T. J. Ahrens, *Earth Planet. Sci. Lett.*, **104**, 337 (1990).
6. T. J. Ahrens, L. Rowan and W. Yang, *LPSC* **XXIII** (abstract), 3 (1992).
7. S. Azuma et al., (in preparation).
8. T. Futagami, M. Ozima, S. Nagai and Y. Aoki, *Geochim. Cosmochim. Acta*, **57**, 3177 (1993).
9. D.E.Grady, W.J.Murri and P.S.DeCarli, *J. Geophys. Res.* **80**, 4857 (1975).
10. D.E.Grady, *J. Geophys. Res.* **85**, 913 (1980).

SOLUBILITIES OF NITROGEN AND ARGON IN BASALT MELT UNDER OXIDIZING CONDITIONS

Akiko Miyazaki, Hajime Hiyagon and Naoji Sugiura
Department of Earth and Planetary Physics, University of Tokyo, Tokyo 113, Japan.

ABSTRACT

We measured solubilities of nitrogen and argon in basalt melt at 1300°C under highly oxidizing conditions ($P(O_2)$=0.16-0.37 atm) using $^{15}N^{15}N$-labeled air (P_{total}= 1 atm). The obtained solubilities (Henry's constants) are $K(N_2)$= (1.3-2.9)x10^{-9} mol/g/atm and $K(Ar)$= (2.6-4.5)x10^{-9} mol/g/atm, respectively, showing that the solubility of nitrogen is slightly lower than that of Ar under highly oxidizing conditions.

The $^{15}N^{15}N$-labeled gas used in the solubility experiment was not isotopically in equilibrium but highly enriched in $^{15}N^{15}N$, that is, the ratios of $^{14}N^{14}N$: $^{14}N^{15}N$: $^{15}N^{15}N$ in the gas were not 1: 2r: r^2, where r=$^{15}N_{total}/^{14}N_{total}$. The recovered gas from the synthetic basalt glass also showed similar isotopic disequilibrium. This suggests that isotopic exchange among N_2 molecules did not occur during dissolution of the gas in the melt. In other words, nitrogen dissolved in the basalt melt as molecules in the present (highly oxidizing) experimental conditions.

INTRODUCTION

Gas/melt and melt/solid partitioning would be important factors controlling degassing of volatile elements from inside the Earth. However, compared with noble gases, which have been used to construct various Earth degassing models, relatively little data exist about the physico-chemical properties of nitrogen.

Solubilities of noble gases in silicate melts at high temperatures have been studied by many authors[1,2,3,4,5] and it is known that: (1) solubilities are in the order He>Ne>Ar>Kr>Xe, (2) they depend on the composition of the melt, and (3) they show small positive dependence on temperature. As noble gases are chemically inactive, they are thought to dissolve physically in melts.

On the other hand, solubility of nitrogen in silicate melt depends not only on temperature, pressure, and melt composition, but also on oxygen fugacity. Mulfinger[6] measured nitrogen solubility in some glass melts under various conditions and found that nitrogen solubilities under highly reducing conditions are more than four orders of magnitude higher than those under oxidizing conditions. Shilobreeva et al.[7] measured solubilities of nitrogen in basaltic and albitic melts at 1250°C at pressures of 1-3 kbar under rather reducing conditions (fO$_2$=IW or NNO). Their results also show that nitrogen solubility in basaltic melt increases with decreasing oxygen fugacity, though almost no fO$_2$ dependence is observed for albitic melt. Solubility of nitrogen in E chondritic melt under reducing conditions (fO$_2$=CCO) obtained by Fogel[8] is as high as 2 mol% at 1500°C. Such high nitrogen solubilities under reducing conditions suggest that nitrogen chemically dissolves in silicate melts, that is, dissolves as anions or replaces oxygen in silicate networks. However, solubility and dissolution mechanisms of nitrogen under oxidizing conditions are not well understood. Javoy and Pineau[9] reported a relatively low nitrogen solubility in basaltic melt (~3.6x10^{-9} mol/g/atm), but experimental details are not given in the report.

To obtain nitrogen solubility in silicate melts under various oxygen fugacity and to understand its mechanisms, we conducted solubility experiments by using a $^{15}N^{15}N$-labeled gas. By analyzing excess ^{15}N relative to air in the sample and comparing it with that in the gas phase, we obtained the solubility (the Henry's constant) without the problem of contamination with atmospheric nitrogen. It has also been proved that this method has a great advantage in understanding the dissolution mechanism of nitrogen. Since the $^{15}N^{15}N$-labeled nitrogen gas is isotpically in disequilibrium (see text below), dissolution of nitrogen as atoms or anions would result in isotopic exchange among nitrogen molecules and hence result in isotopic equilibration of the gas, while dissolution of nitrogen as molecules would preserve the isotopic disequilibrium.

EXPERIMENTAL

A chunk of basalt, collected at Izu-Oshima, Japan, was crushed and completely melted in a Mo crucible at 1300-1400°C under high vacuum, and gases originally contained in the basalt were degassed. After a few hours heating, the melt was quenched to make glass. Molybdenum adhering to the surface of the glass was carefully removed, and the glass was used as a starting material for the solubility experiment. The chemical composition of the glass is shown in Table 1.

Table 1. Chemical composition (wt %) of the starting material obtained by EPMA analysis.

SiO_2	Al_2O_3	Na_2O	MgO	K_2O	CaO	TiO_2	Fe_2O_3	MnO	total
53.56	14.70	2.43	5.45	0.53	9.33	1.04	11.65	0.00	98.68

The basalt glass was cut into thin plates ~0.5 mm thick. A few pieces (total weight ~600 mg) of the glass plates were placed on a piece of platinum mesh put on the top of a platinum crucible and loaded in a reaction chamber made of quartz glass (~4 cm in diameter x ~35 cm in length).

The $^{15}N^{15}N$ gas was synthesized in another glass chamber of ~70 cm^3 by heating and decomposing 10-20 mg of ^{15}N-labeled ammonium sulfate, $(^{15}NH_4)_2SO_4$ (containing 98 atom % ^{15}N), in an oxygen atmosphere or in air at ~400°C. The SO_2 and H_2O gases produced in this process were condensed and removed by using liquid nitrogen and only the $^{15}N^{15}N$ gas was transported to the reaction chamber. Air was further introduced into the reaction chamber so that the total pressure became to be about 0.3-0.4 x the atmospheric pressure. The lower 10 cm of the reaction chamber, where the samples were placed, was put in an electric furnace and kept at 1300°C for 3 to 6 hours. The total pressure inside the chamber reached about one atmosphere during heating. The samples were melted on the platinum mesh to form a thin melt layer, which was expected to equilibrate with the $^{15}N^{15}N$-labeled gas during heating. Finally the reaction chamber was pulled out from the furnace and samples were quenched to form glass. Experimental conditions and compositions of the gas for the solubility experiments are listed in Table 2.

The basaltic glass samples thus synthesized were loaded in vacuum lines, VL1 or VL2, for extraction of the dissolved gases. The two lines were very similar to each other. The description of VL1 is given by Hashizume and Sugiura[10]. Gases were extracted from the samples by heating stepwise from 600°C to 1200°C in a quartz chamber by using a resistance-heated furnace under ~1 torr (~10^2 Pa) of oxygen atmosphere (stepwise combustion method). Contaminating gases, such as CO_2 or H_2O, were removed by several cold traps, and about half of the purified gas was used

Table 2. Experimental conditions for the solubility experiments.

Run #	Temp. (°C)	Time (hr)	P(O$_2$) (atm)	P(^{28}N$_2$) (atm)	P(^{30}N$_2$) (atm)	P(^{40}Ar) (atm)
SLB-1	1300	6	0.19	0.804	0.0061	0.0096
SLB-2	1300	6	0.37	0.417	0.015	0.0050
SLB-3	1300	3.3	0.16	0.823	0.0072	0.0098

for nitrogen analysis. The rest of the gas was further exposed to a furnace containing titanium foil (heated at ~300°C for VL1, or kept at room temperature for VL2) to remove nitrogen, and used for argon analysis. Procedures for extraction, purification, and analysis of nitrogen with a QMS are presented in detail by Hashizume and Sugiura[10].

In order to extract the gas completely, a laser extraction was further applied to the samples. The samples were loaded in another sample container (connected to VL1), which had a glass window (~4 cm in diameter) on the top of it, and the samples were shot through the window with a Nd-Yag laser operated at a continuous mode. It was difficult to estimate the heating temperatures during laser extraction, but the observation with an optical pyrometer gave approximate temperatures of 1100-1800°C depending on the applied power.

The extracted nitrogen and argon were analyzed with quadrupole-type mass spectrometers (QMS1 for VL1, or QMS2 for VL2). A typical tungsten filament was used in QMS1, while an yttria (Y$_2$O$_3$)-coated iridium filament was used in QMS2. The latter filament has a lower work function for emitting electrons and hence can be operated at lower temperatures. This suppressed substantially the decomposition rate (or isotopic exchange rate) of nitrogen molecules during the analysis with the QMS. This point will be further discussed later.

In each nitrogen measurement, masses (m/e) 28, 29, 30 and 26 were analyzed repeatedly for 20 cycles. Mass 26 was used for monitoring and correcting for hydrocarbons on the mass range of 28-30. Contribution of CO was not corrected for the present results because of difficulties in its estimation. However, judging from the peak intensity of mass 31 (^{13}C^{18}O + hydrocarbon), which was measured at the beginning of the analysis for SLB-2-2 and SLB-3, contribution of CO might not be serious for most of the sample runs (except for the 600°C fraction of SLB-3).

Decrease of nitrogen peaks during the analysis was observed both for QMS1 (~4.6 %/min) and QMS2 (~2 %/min). In the case of QMS2 (using an yttria-coated iridium filament), changes of the peak intensities were slow and the effect of isotopic exchange was small. We obtained the gas composition by extrapolating the peak intensities of masses 28, 29 and 30 to 'time zero', when the gas was admitted to the QMS. In the case of QMS1 (using a tungsten filament), however, a rapid isotopic exchange was observed for sample gases having high excess ^{15}N in mass 30. The ^{30}N$_2$ peak decreased and the ^{29}N$_2$ peak increased rapidly during the analysis, and the ratios of ^{28}N$_2$: ^{29}N$_2$: ^{30}N$_2$ approached the equilibrium ratios by the end of the analysis (in ~6.5 minutes). Because of such rapid changes in the mass composition of nitrogen molecules, it was difficult to extrapolate the peak intensities to 'time zero', so we calculated the gas composition by adopting the observed composition (^{28}N$_2$: ^{29}N$_2$: ^{30}N$_2$) for the first cycle and correcting them for the 'decay' of the peak from t=0 using the decay rate for normal nitrogen (~4.6 %/min). For this reason, the nitrogen data (^{28}N$_2$, ^{29}N$_2$ and ^{30}N$_2$) for SLB-3 listed in Table 3 do not represent the compositions at t=0. Note, however, that this treatment would not change the results for the calculation of total excess [15]N.

RESULTS AND DISCUSSION

Before calculating solubilities of nitrogen and argon from the present results, we must examine whether or not equilibrium was attained between the gas and the basaltic melt under the present experimental conditions. The diffusion coefficient of Ar in tholeiitic basalt melt at 1350°C is reported to be ~6x10[-6] cm^2/s [4]. The diffusion coefficient of nitrogen is not yet known for basalt melt, but for synthetic glass melts (soda-lime-silica glass and alkali-barium-silica glass) at ~1300°C, diffusivities were reported to be (1.8-4.0) x10[-6] cm^2/s [11]. Hence, we assume that diffusion coefficients for Ar and nitrogen are on the order of 10[-6] cm^2/s. If this is the case, equilibrium may safely be attained between the gas and the melt within 3-6 hours, because the expected diffusion length (a few mm) is much larger than the thickness of the melt (~0.5 mm).

The analytical data are listed in Table 3. Temperature fractions from 600°C to 1200°C indicated as C are the results for the typical stepwise combustion and the others (1100-1800°C fractions indicated as L) are those for laser extraction. Two duplicate analyses were conducted for the sample SLB-2. For SLB-2-2, an accidental electricity shut down happened after the first 1000°C fraction, and we restarted the analysis from another 600°C fraction.

The data of 600°C and 800°C fractions were omitted from the present calculation because they are mostly atmospheric contamination. In fact, negligible excess [15]N (and argon) was detected for these fractions except for SLB-3. High nitrogen release was observed at 600°C (and to some extent at 800°C) fraction(s) of SLB-3, but it was most likely due to contamination of organic nitrogen, judging from very low release of argon in these fractions. Before the analysis of nitrogen for the 600°C fraction of SLB-3, we pumped out by mistake the vacuum line before closing a valve of a cold trap which contained the sample gas. Although the cold trap was kept at liquid nitrogen temperature, the sample gas might be partly lost from the system, which might have caused mass fractionation in preference to the heavier molecules. Apparently high release of excess ^{15}N calculated for this fraction is, therefore, at least partly due to this mass fractionation effect. Contribution of CO might also be present in this fraction judging from the relatively high intensity of mass 31 ([13]C[18]O + hydrocarbon). However, data for these fractions (600-800°C fractions of SLB-3) were not used in the solubility calculation, anyway.

From the observed concentrations of masses 28, 29 and 30 (corresponding to [14]N[14]N, [14]N[15]N and [15]N[15]N), we calculated the amount of excess ^{15}N relative to air. Excess ^{15}N existing as [14]N[15]N and as [15]N[15]N were calculated as follows:

$$\text{Excess } ^{15}\text{N in mass 29} = \left\{ \left(\frac{^{29}\text{N}_2}{^{28}\text{N}_2} \right)_{obs} - \left(\frac{^{29}\text{N}_2}{^{28}\text{N}_2} \right)_{air} \right\} \times \left[^{28}\text{N}_2 \right]_{obs} \quad \cdots\cdots\cdots\cdots (1)$$

$$\text{Excess } ^{15}\text{N in mass 30} = \left\{ \left(\frac{^{30}\text{N}_2}{^{28}\text{N}_2} \right)_{obs} - \left(\frac{^{30}\text{N}_2}{^{28}\text{N}_2} \right)_{air} \right\} \times \left[^{28}\text{N}_2 \right]_{obs} \times 2, \quad \cdots\cdots (2)$$

where 'obs' represents the observed ratios or concentration, and 'air' represents the atmospheric values: $(^{29}\text{N}_2/^{28}\text{N}_2)_{air} = 7.34 \times 10^{-3}$ and $(^{30}\text{N}_2/^{28}\text{N}_2)_{air} = 1.35 \times 10^{-5}$.

Comparing the total excess ^{15}N thus calculated and the partial pressure of excess ^{15}N^{15}N (excess ^{30}N$_2$) in the gas phase in the reaction chamber, we may calculate the solubility (Henry's constant) of nitrogen in basalt melt as

$$K \text{ (mol/g/atm)} = \frac{\text{Excess } ^{15}\text{N in basalt melt} \times (1/2) \text{ (mol/g)}}{\text{Partial pressure of excess } ^{15}\text{N}^{15}\text{N (atm)}} . \quad \cdots\cdots\cdots\cdots(3)$$

It should be noted that this treatment is valid only for the case in which nitrogen dissolves in the melt as molecules (Case (4), below).

$$N_2(\text{gas}) \leftrightarrow N_2(\text{melt}) \quad \cdots\cdots\cdots\cdots\cdots\cdots\cdots\cdots\cdots\cdots\cdots\cdots\cdots\cdots(4)$$

$$N_2(\text{gas}) \leftrightarrow N(\text{melt}) + N(\text{melt}) \quad \cdots\cdots\cdots\cdots\cdots\cdots\cdots\cdots\cdots\cdots(5)$$

A simple consideration on thermodynamic equilibrium shows that, in the case of (5), where nitrogen dissolves in the melt as atoms or anions, Henry's Law would no longer be valid and concentration of nitrogen would be proportional to $\sqrt{P(N_2)}$ instead of $P(N_2)$. In the present study, it has been proved that nitrogen dissolved in the basalt melt as molecules (see below). Hence, we may safely assume the Henry's constants calculated by Eq.(3) are representative of the solubility of molecular nitrogen in basalt melt under oxidizing conditions.

The solubility of nitrogen (as N$_2$) thus obtained is $(1.30-2.86) \times 10^{-9}$ mol/g/atm or $(2.91-6.41) \times 10^{-5}$ cm^{3}STP/g/atm, which is slightly lower than that reported by Javoy and Pineau[9] ($\sim 3.6 \times 10^{-9}$ mol/g/atm). The solubility of argon is obtained to be $(2.58-4.47) \times 10^{-9}$ mol/g/atm or $(5.78-10.0) \times 10^{-5}$ cm^{3}STP/g/atm, which is consistent with the previous data[1,2,3,4,5]. There are still considerable variations (about a factor of two) in the observed solubilities. At present, the cause for these variations is unclear. Heterogeneity in the samples (i.e., some extent of disequilibrium between the samples and the gas phase) may be one of the explanations. However, such a heterogeneity, if present, might not be large judging from the consistency of the present argon solubility and the previous data (see Fig.1). Another possibility is contribution of CO, which we ignored in the present calculation. Although this effect may not be large for most fractions judging from the peak height of mass 31, this possibility cannot be ruled out at present. If this is the case, solubility of nitrogen must be slightly lower than that obtained here. The apparent disagreement between the obtained nitrogen solubilities for SLB-2-1 and SLB-2-2 is rather strange. This may be due to contribution of CO. However, we also suspect that a change in the relative sensitivity of nitrogen and argon for QMS2 just after the accidental electricity shut down (between the first 1000°C and the second 600°C fractions of SLB-2-2) could be a cause for the disagreement. Although we checked the sensitivity change and made its correction, the sensitivity might not be stable enough during the analysis of SLB2-2. Because of the ambiguities as to the variations in the obtained solubilities, the present data should be considered preliminary. However, if we take the ratios of K(N$_2$)/K(Ar) (0.64, 0.38 and 0.50, respectively), they consistently show lower solubility for nitrogen. Therefore, it is reasonable to conclude from the present results that the solubility of nitrogen in basalt melt is lower than (about a half of) that of argon under the present (highly oxidizing) conditions.

The nitrogen solubility in basaltic melt obtained in the present study ($(1.30-2.86) \times 10^{-9}$ mol/g/atm) is slightly lower than that reported by Javoy and Pineau[9] ($\sim 3.6 \times 10^{-9}$

Table 3. Analytical data of nitrogen and argon with the calculated solubilities.

Sample (weight)	Temp. (°C) [1]	QMS Method [2]	Nitrogen [mol/g] 28N2 (E-9)	29N2 (E-12)	30N2 (E-12)	Excess 15N [mol/g] [3] mass 29 (E-12)	mass 30 (E-12)	(%)	total (E-12)	40Ar [mol/g] (E-12)	K(N2) [mol/g/atm] (E-9) [4]	K(Ar) (E-9)
SLB-1	600	2 C	0.770	5.82	0.055	(0.17)[5]	(0.09)[5]		(0.26)[5]			
(0.1550 g)	800	2 C	0.052	0.40	0.012	(0.01)	(0.02)		(0.03)			
	1000	2 C	0.182	1.51	0.904	0.17	1.80	91.2	1.98			
	1200	2 C	0.578	5.28	3.240	1.04	6.46	86.2	7.50			
	1200	2 C	0.218	2.10	0.975	0.50	1.94	79.5	2.44			
	1200	2 C	0.070	0.88	0.635	0.36	1.27	77.8	1.63			
	1200	2 C	0.165	1.81	0.879	0.60	1.75	74.5	2.35			
	1600	1 L	1.410	12.90	0.631	2.55	1.22	32.4	3.77			
	1600	1 L	0.725	7.15	0.402	1.83	0.78	30.0	2.61			
	1600	1 L	0.096	1.01	0.063	0.30	0.12	29.0	0.43			
	1800	1 L	0.939	8.19	0.240	1.30	0.45	25.9	1.75			
	total			5.206	47.04	8.036	8.65	15.82	64.6	24.47	n.m.	2.01	n.m.
SLB-2-1	600	2 C	0.786	5.88	0.031	(0.06)[5]	(0.02)[5]		(0.08)[5]	(0.05)[5]		
(0.1062 g)	800	2 C	0.067	0.54	0.213	(0.02)	(0.21)		(0.23)	(0.13)		
	1000	2 C	0.197	1.91	4.590	0.46	9.17	95.2	9.64	2.21		
	1200	2 C	0.532	7.04	16.200	3.14	32.39	91.2	35.52	7.05		
	1200	2 C	0.127	2.59	5.880	1.66	11.76	87.6	13.41	2.70		
	1200	2 C	0.000	0.41	1.990	0.41	3.98	90.7	4.39	0.67		
	1200	2 C	0.199	2.16	2.190	0.70	4.37	86.2	5.07	2.29		
	1600	1 L	9.550	75.90	2.240	5.80	4.22	42.1	10.03	4.05		
	1600	1 L	0.980	8.40	0.361	1.21	0.70	36.6	1.90	0.62		
	1700	1 L	0.431	6.06	0.800	2.90	1.59	35.4	4.48	1.83		
	1800	1 L	0.227	2.73	0.203	1.06	0.40	27.3	1.46	0.91		
	total			13.096	113.62	34.698	17.34	68.58	79.8	85.91	22.33	2.86	4.47
SLB-2-2	600	2 C	0.745	5.40	0.000	(0.00)[5]	(0.00)[5]		(0.00)[5]	(0.00)[5]		
(0.1417 g)	800	2 C	0.272	1.94	0.029	(0.00)	(0.04)		(0.04)	(0.47)		
	1000	2 C	0.176	1.42	2.520	0.13	5.04	97.5	5.16	1.80		
	600	2 C	0.052	0.42	0.000	(0.00)	(0.00)		(0.00)	(0.00)		
	800	2 C	0.009	0.08	0.011	(0.00)	(0.00)		(0.00)	(0.00)		
	1000	2 C	0.040	1.34	0.756	1.05	1.51	59.0	2.56	0.93		
	1200	2 C	0.283	3.85	7.820	1.77	15.63	89.8	17.41	5.86		
	1200	2 C	0.062	0.96	2.180	0.51	4.36	89.6	4.87	1.51		
	1200	2 C	0.039	0.65	1.590	0.37	3.18	89.7	3.54	1.09		
	1200	2 C	0.096	1.30	2.030	0.59	4.06	87.2	4.65	2.57		
	1100	1 L	0.521	3.77	0.000	0.00	0.00		0.00	0.05		
	1700	1 L	0.578	6.03	1.029	1.79	2.04	53.3	3.83	3.25		
	1700	1 L	0.170	2.43	0.601	1.18	1.20	50.3	2.38	2.04		
	1700	1 L	0.068	0.91	0.179	0.41	0.36	46.2	0.77	0.73		
	total			3.111	30.50	18.745	7.80	37.37	82.7	45.17	19.84	1.51	3.97
SLB-3	600	1 C	8.030	72.80	0.213	(13.86)[6]	(0.21)[6]		(14.07)[6]	(0.00)[6]		
(0.1510 g)	800	1 C	1.260	11.30	0.518	(2.05)	(1.00)		(3.05)	(0.11)		
	1000	1 C	0.731	7.43	0.269	2.06	0.52	20.1	2.58	1.83		
	1200	1 C	1.040	11.30	0.734	3.67	1.44	28.2	5.11	6.09		
	1200	1 C	0.904	10.60	0.875	3.96	1.73	30.3	5.69	8.13		
	1100	1 L	0.180	1.30	0.003	0.00	0.00		0.00	0.00		
	1400	1 L	2.970	22.20	0.225	0.40	0.37	48.0	0.77	1.69		
	1700	1 L	1.500	14.40	0.576	3.39	1.11	24.7	4.50	7.26		
	1700	1 L	0.147	1.16	0.017	0.08	0.03	27.3	0.11	0.29		
	total			16.762	152.49	3.430	13.57	5.20	27.7	18.76	25.29	1.30	2.58

(1) QMS: 1=QMS1, 2=QMS2.

(2) Method: C=stepwise combustion in a quartz furnace, L=laser extraction.

(3) Excess 15N observed in mass 29 and mass 30; fractions of excess 15N in mass 30 are also shown in %.

(4) Solubilities of nitrogen and argon; K= concentration in melt (mol/g)/ partial pressure in the gas phase (atm).

(5) Gases in 600°C and 800°C fractions are omitted from the calculation as atmospheric contamination.

(6) High nitrogen release at the 600°C fraction is most likely due to organic contamination. Apparent high excess 15N is due to mass fractionation caused by partial loss of the sample gas prior to the analysis.

mol/g/atm), though experimental details are not given for the latter. The solubilities of nitrogen and argon obtained in the present study are illustrated in Fig.1, along with the previously reported data. Solubilities (physical solubilities) of gases may be interpreted in terms of the molecular sizes of the gases. As has been pointed out by several authors[5,12], (physical) solubility of noble gases in melts show negative correlation with their atomic sizes. The apparent molecular "diameters" of nitrogen and other gases can be estimated from various methods: from viscosity, thermal conductivity, the second virial coefficient of the gas, and so on. For example, the molecular "diameters" estimated from the viscosity of gases are 2.86Å for Ar, 3.16Å for N_2, and 3.18Å for Kr [13]. Hence, we may expect that solubility of nitrogen is also between those of Ar and Kr, which is quite consistent with the present results (Fig. 1).

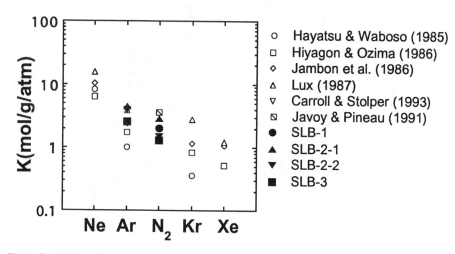

Fig.1 Solubilities (Henry's constants) of noble gases and nitrogen in basalt melts. Gases are in the order of their molecular sizes.

An important observation in the present results is that the gas recovered from the samples also showed isotopic disequilibrium like that in the ambient gas. In Table 3, the fractions (%) of excess ^{15}N observed in mass 30 are listed. It is remarkable that excess ^{15}N was released mostly as $^{15}N^{15}N$ for 1000-1200°C fractions (75-98%) for the samples analyzed with QMS2 (SLB-1, SLB-2-1 and SLB-2-2). This suggests that isotopic exchange has not occurred during dissolution of the gas in the melt. In other words, nitrogen physically dissolved in the basalt melt as molecules. For the sample SLB-3 and for 1100-1800°C fractions by laser extraction, both of which were analyzed with QMS1, the effect of such isotopic disequilibrium is smaller. For example, the fraction of excess ^{15}N in mass 30 is generally less than 50% for these cases. This suggests that significant degree of isotopic exchange among nitrogen molecules occurred in QMS1 between the time 'zero' and the time of the first cycle of the analysis (~70 sec). (It should be noted that such an isotopic exchange will not affect the amount of *total* excess ^{15}N calculated in the present study as long as we sum up the amount of excess ^{15}N in mass 30 and in mass 29.)

The present study proved that the use of a $^{15}N^{15}N$-labeled gas gave a great advantage in obtaining nitrogen solubility in basalt melt. If we calculate nitrogen

. solubility using $^{28}N_2$ instead of excess ^{15}N (excluding 600-800°C fractions), we would obtain 5.5, 29.4, 4.9 and 9.1 x10^{-9} mol/g/atm for SLB-1, SLB-2-1, SLB-2-2 and SLB-3, respectively, which are significantly higher than the values obtained above. Even we exclude data for the first fraction of laser extraction to further avoid air contamination, the obtained solubility for $^{28}N_2$ would be 3.7, 6.5, 3.6 and 8.9 x10^{-9} mol/g/atm, respectively, which are still about 2 to 8 times higher than the values obtained for excess ^{15}N. This suggests that atmospheric contamination could not be removed completely by omitting the data for low temperature fractions (600-800°C). Another advantage using the $^{15}N^{15}N$-labeled gas having isotopic disequilibrium is, as shown above, to give important information about the dissolution mechanisms of nitrogen in the melt.

In conclusion, under highly oxidizing conditions, nitrogen dissolves in basalt melt as molecules and its solubility is slightly lower than that of argon.

ACKNOWLEDGMENT

We thank Lin Huang and Des Patterson for their careful reviews, which improved the manuscript. This work was financially supported in part by a Grant in Aid for Scientific Research from the Ministry of Education, Science and Culture.

REFERENCES

1. A. Hayatsu and C.E. Waboso, *Chem. Geol. (Isotope Geosci. Sec.)* **52**, 97 (1985).
2. H. Hiyagon and M. Ozima, *Geochim. Cosmochim. Acta*, **50**, 2045 (1986).
3. A. Jambon, H. Weber and O. Braun, *Geochim. Cosmochim. Acta*, **50**, 401 (1986).
4. G. Lux, *Geochim. Cosmochim. Acta*, **51**, 1549 (19867.
5. M. R. Carrol and E. M. Stolper, *Geochim. Cosmochim. Acta*, **57**, 5039 (1993).
6. H.-O. Mulfinger, *J. Amer. Ceramic Soc.*, **49**, 462 (1966).
7. S. Shilobreeva, A. Kadic, S. Matveev and B. Chapyzhniv, *Yamada Conference XXXVII (abstract)*, 21 (1994).
8. R. A. Fogel, *LPSC XXV (abstract)*, 384 (1994).
9. M. Javoy and F. Pineau, *Earth Planet. Sci. Lett.* **107**, 598 (1991).
10. K. Hashizume and N. Sugiura, *Mass Spectroscopy*, **38**, 269 (1990).
11. G.H. Frischat, O. Buschmann and H. Meyer, *Glastechn. Ber.* **51**, 321 (1962).
12. M. Blander, W. R. Grims, N. V. Smith and G. N. Watson, *J. Phys.Chem.*, **63**, 1164 (1959).
13. S. Dushman and J. M. Lafferty, *Scientific Foundation of Vacuum Technique*, 2nd.ed., (Willey, New York), 32, (1962).

LOSS OF SOLAR HE AND NE FROM IDPS IN SUBDUCTING SEDIMENTS: DIFFUSION AND THE EFFECT OF PHASE CHANGES

Hajime Hiyagon
Department of Earth and Planetary Physics University of Tokyo, Tokyo 113, Japan.

ABSTRACT

Three samples of a magnetic fraction of Pacific Ocean sediment were heated to 500°C, 800°C and 950°C, respectively, for two hours in a vacuum furnace in the same condition as that for the diffusion experiment[1]. The run products as well as an unheated sample were examined with an X-ray diffraction method. The results show that (1) the major magnetic mineral was magnetite for all the samples, (2) no peaks of hematite, wustite, nor metallic iron were observed, (3) however, magnetite in the unheated sample had been partly oxidized to form maghemite, which has essentially the same crystal structure as that of magnetite. These observations suggest that the effect of phase changes, if present, would be negligibly small in the diffusion experiment. The present results further support the reliability of the diffusion data obtained in the previous study[1] and support the conclusion that solar He and Ne would be lost from subducting slabs at shallow depths.

INTRODUCTION

Hiyagon[1] conducted a diffusion experiment for solar He and Ne in IDPs (Interplanetary Dust Particles) in a magnetic separate from Pacific Ocean sediment. The results suggest that solar He and Ne would easily be released from IDPs and lost from subducting slabs at shallow depths and cannot be transported to the mantle[1,2]. However, since the diffusion experiment was conducted under high vacuum, there is a possibility that magnetite grains, which are supposed to be the main carrier of the solar noble gases[3], would have been partly reduced to form metallic iron due to low oxygen fugacity in the experimental condition as suggested by Craig[4]. If this is the case, such a phase change might affect the gas release and hence the results of the calculated diffusion coefficients. In order to see whether or not such a phase change really occurred in the samples, I conducted again a heating experiment on a magnetic separate of deep sea sediment exactly in the same condition as that in the diffusion experiment. The run products as well as an unheated sample were examined with an X-ray diffraction method[5].

EXPERIMENTAL

A magnetic fraction was separated from the Pacific Ocean sediment by using hand magnets exactly in the same way as in the diffusion experiment[1]. Three samples (about 60mg each) were prepared from the magnetic separate, wrapped with platinum-foil and put in a noble gas extraction system. Each sample was dropped in a Mo-crucible, which was placed at the bottom of a Ta-crucible, and heated to 500°C, 800°C and 950°C, respectively, for two hours under static vacuum. Note that 500°C is the temperature where decomposition of magnetite into metallic iron may take place as pointed out by Craig[4], and 800°C and 950°C are those where good linearity was obtained for the diffusion data plotted in an Arrhenius diagram[1,2]. A titanium furnace connected to the vacuum line was heated to 800°C during heating the sample just in the same way as in the diffusion experiment. The experimental condition is illustrated in Fig.1. After cooled down to room temperature, the sample was taken out from the vacuum furnace. The color of the sample changed from brown (the unheated sample) to dark gray (the sample heated at 950°C). The run products as well as an unheated sample were analyzed with an X-ray diffraction method at the Radio-Isotope Centre of the University of Tokyo. For comparison, magnetite, hematite, and metallic iron were also analyzed with the X-ray diffractometer.

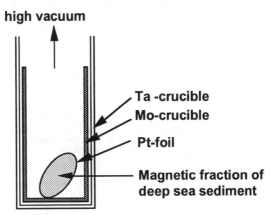

Fig.1 Experimental condition for noble gas extraction under high vacuum.

RESULTS AND DISCUSSION

Figure 2 shows X-ray diffraction spectra of the samples heated at 950°C (a), at 800°C (b), at 500°C (c), respectively, and that of an unheated sample (d). For comparison, the positions of peaks for maghemite (e), magnetite (f), hematite (g), and metallic iron (h) are also shown. Apparently the magnetic separate contained various silicate minerals as well as magnetic minerals. This can be seen in rather complicated X-ray diffraction spectra. Among possible candidates for the magnetic minerals, magnetite seems to be the major one in the samples. Peaks corresponding to magnetite are shown by the shadowed peaks in Fig.2. No recognizable peaks of hematite, wustite, nor metallic iron were observed in any of the heated and unheated samples. This suggests that magnetite stayed magnetite as the major magnetic mineral in the samples throughout the heating experiment and that no phase changes of magnetite seems to have occurred in the present experimental condition.

For the unheated sample, however, the peaks of the spectrum corresponding to magnetite seem to be shifted slightly to higher diffraction angles (Fig.2 (d)). This may be explained by the existence of maghemite (γFe_2O_3) instead of magnetite (Fe_3O_4) in the unheated sample (see Fig.2 (e) and (f)). This may also explain at least partly the brownish color of the unheated sample compared with the dark gray color of the heated samples. Maghemite is considered to have formed at the ocean floor due to low temperature oxidation of magnetite. It has a crystal structure essentially the same as that of magnetite except that 1/9 of the Fe sites are occupied by vacancies[6]. This can also be seen in the X-ray diffraction spectrum of maghemite, which is very similar to that of magnetite. Since a volume change due to conversion of maghemite into magnetite is

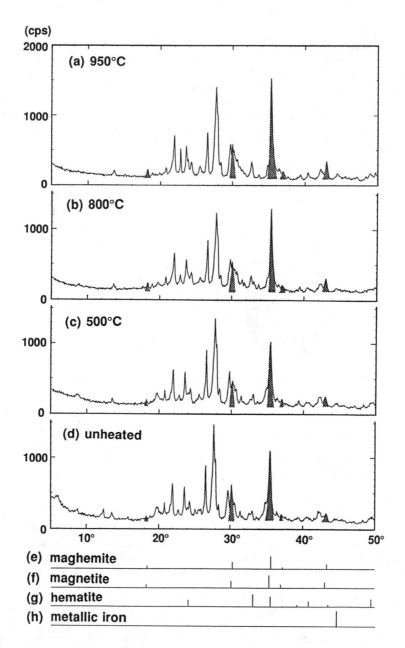

Fig.2   X-ray diffraction patterns (obtained with Cu-K$_\alpha$ line) for an unheated sample (d), and for samples heated at 950°C (a), at 800°C (b), and at 500°C (c); horizontal axis indicates X-ray diffraction angle and vertical axis intensity (cps). For comparison, positions of peaks for maghemite (e), magnetite (f), hematite (g), and metallic iron (h) are also shown. Shadowed peaks correspond to magnetite. For the unheated sample, the shadowed peaks are more likely to be maghemite rather than magnetite.

very small, such a phase change would not have caused significant degassing of solar noble gases from magnetite (maghemite) grains during the heating experiment.

The X-ray diffraction patterns (Fig.2) show that there are also many silicate minerals in the magnetic fraction of the deep sea sediment. Among them, peaks of quartz and albite and/or anorthite (not shown here) are recognizable in the obtained spectra. Phase changes are also noticeable in some silicate peaks during the heating experiment. For example, the intensity of the peak at ~23° increased significantly at 950°C. If solar noble gases are also trapped in these silicate minerals, they may affect the results of the diffusion experiment. However, contribution of such solar noble gases would be very minor judging from the results of the previous works[3,7,8], in which even a factor of 100 enrichment of solar He relative to bulk sediment was observed in a magnetic fraction. Also a phase change at 800-950°C suggested by a significant increase of the peak at ~23°, possibly caused by decomposition of some hydrous minerals, may not contribute much to the degassing of solar noble gases, because good linearity was obtained in the Arrhenius diagram in this temperature range[1]. Because of the above reasons, contribution of solar noble gases in silicate minerals may be ignored. However, a possibility of a minor contribution from silicates cannot be excluded.

From the present results, I conclude that no significant phase changes occurred in the magnetic phase of the deep sea sediment in the present experimental condition and hence in the previous diffusion experiment. Only a minor phase change from maghemite to magnetite seems to have occurred at low temperatures, which, however, would have negligible contribution to the gas release. Hence, the release of solar noble gases from IDPs in the stepwise-heating experiment might be controlled by diffusion. It should be noted, however, that an X-ray diffraction analysis provides only qualitative information and hence a small amount of phase changes cannot be ruled out completely. However, this would not change the conclusion, anyway.

The present results further support the reliability of the previous diffusion experiment and hence support the conclusion that solar He and Ne would be easily lost from subducting slabs at shallow depths. If any phase changes of magnetite occur, in turn, in the real subduction system, it would even enhance the gas release from IDPs and, as a result, strengthen the above conclusion.

ACKNOWLEDGMENT

I thank Shun'ichi Azuma in the University of Tokyo for his help in the X-ray diffraction analysis. The sediment sample was supplied by Kazuo Yamakoshi in Cosmic Ray Research Institute, the University of Tokyo. A brief summary of the present results is also presented in *Science* [5].

REFERENCES

1. H. Hiyagon, *Science* **263**, 1257 (1994)
2. H. Hiyagon, *Noble gas geochemistry and cosmochemistry* (ed. J.Matsuda, Terra Sci. Publ. Co., Tokyo, 1994) p.67.
3. S. Amari and M. Ozima, *Geochim. Cosmochim. Acta* **52**, 1087 (1985).
4. H. Craig, (Technical Comment) *Science* **265**, 1892 (1994).
5. H. Hiyagon, (Response) *Science* **265**, 1893 (1994).
6. T. Nagata, *Rock magnetism.* (Maruzen Co. Ltd., Tokyo, 1961).
7. H. Fukumoto et al., *Geochim. Cosmochim. Acta* **50**, 2245 (1986).
8. J. Matsuda and M. Murota, *J. Geophys. Res.* **95**, 7111 (1990).

Author Index

A

Azuma, S., 270

B

Bar-Nun, A., 123
Bose, K., 221
Bounama, Ch., 45
Burgess, R., 91

C

Cruikshank, D. P., 143

D

Dai, W., 117
Donahue, T. M., 154
Duffy, T. S., 211, 250

E

Ebrahim, F., 117
Eggert, J. H., 250

F

Farley, K. A., 81
Fegley, B., Jr., 99
Franck, S., 45
Fukuoka, K., 270
Futagami, T., 270

G

Goncharov, A. F., 250

H

Hanfland, M., 250
Hemley, R. J., 211, 250
Hervig, R. L., 229
Hiyagon, H., 270, 276, 284
Holloway, J. R., 229
Hunten, D. M., 200

I

Igarashi, G., 70
Ita, J. J., 33

J

Jeanloz, R., 240

K

Kadik, A. A., 106
Kaula, W. M., 139
Kawamoto, T., 229
Kerridge, J. F., 167
King, S. D., 33

L

Leinenweber, K., 229
Li, M., 250
Li, X., 240
Lodders, K., 99
Lunine, J. I., 117

M

Mao, H. K., 211, 250
McInnes, B., 81
Miyazaki, A., 276

289

AIP Conference Proceedings